Lecture Notes in Artificial Intelligence   9321

Subseries of Lecture Notes in Computer Science

More information about this series at http://www.springer.com/series/1244

Vicenç Torra · Yasuo Narukawa (Eds.)

Modeling Decisions for Artificial Intelligence

12th International Conference, MDAI 2015
Skövde, Sweden, September 21–23, 2015
Proceedings

 Springer

Editors
Vicenç Torra
University of Skövde
Skövde
Sweden

Yasuo Narukawa
Toho Gakuen
Tokyo
Japan

ISSN 0302-9743 ISSN 1611-3349 (electronic)
Lecture Notes in Artificial Intelligence
ISBN 978-3-319-23239-3 ISBN 978-3-319-23240-9 (eBook)
DOI 10.1007/978-3-319-23240-9

Library of Congress Control Number: 2015947408

LNCS Sublibrary: SL7 – Artificial Intelligence

Springer Cham Heidelberg New York Dordrecht London
© Springer International Publishing Switzerland 2015

Printed on acid-free paper

Springer International Publishing AG Switzerland is part of Springer Science+Business Media
(www.springer.com)

Preface

This volume contains papers presented at the 12th International Conference on Modeling Decisions for Artificial Intelligence (MDAI 2015), held in Skövde, Sweden, September 21–23. This conference followed MDAI 2004 (Barcelona, Catalonia), MDAI 2005 (Tsukuba, Japan), MDAI 2006 (Tarragona, Catalonia), MDAI 2007 (Kitakyushu, Japan), MDAI 2008 (Sabadell, Catalonia), MDAI 2009 (Awaji Island, Japan), MDAI 2010 (Perpinyà, France), MDAI 2011 (Changsha, China), MDAI 2012 (Girona, Catalonia), MDAI 2013 (Barcelona, Catalonia) and MDAI 2014 (Tokyo, Japan) with proceedings also published in the LNAI series (Vols. 3131, 3558, 3885, 4617, 5285, 5861, 6408, 6820, 7647, 8234, 8825).

The aim of this conference is to provide a forum for researchers to discuss theory and tools for modeling decisions as well as applications that encompass decision-making processes and information-fusion techniques.

The organizers received 38 papers from 14 different countries, 18 of which are published in this volume. Each submission received at least two reviews from the Program Committee and a few external reviewers. We would like to express our gratitude to them for their work. The plenary talks presented at the conference are also included in this volume.

The conference was supported by the University of Skövde, the School of Informatics, the Japan Society for Fuzzy Theory and Intelligent Informatics (SOFT), the Catalan Association for Artificial Intelligence (ACIA), the European Society for Fuzzy Logic and Technology (EUSFLAT), and the UNESCO Chair in Data Privacy.

July 2015

Vicenç Torra
Yasuo Narukawa

Organization

General Chair

Vicenç Torra   University of Skövde, Skövde, Sweden

Program Chairs

Vicenç Torra   University of Skövde, Skövde, Sweden
Yasuo Narukawa   Toho Gakuen, Tokyo, Japan

Advisory Board

Didier Dubois	Institut de Recherche en Informatique de Toulouse, CNRS, France
Lluis Godo	IIIA-CSIC, Catalonia, Spain
Kaoru Hirota	Tokyo Institute of Technology, Japan
Janusz Kacprzyk	Systems Research Institute, Polish Academy of Sciences, Poland
Sadaaki Miyamoto	University of Tsukuba, Japan
Michio Sugeno	European Centre for Soft Computing, Spain
Ronald R. Yager	Machine Intelligence Institute, Iona Collegue, NY, USA

Program Committee

Eva Armengol	IIIA-CSIC, Catalonia, Spain
Edurne Barrenechea	Universidad Pública de Navarra, Spain
Gleb Beliakov	Deakin University, Australia
Gloria Bordogna	Consiglio Nazionale delle Ricerche, Italia
Humberto Bustince	Universidad Pública de Navarra, Spain
Francisco Chiclana	De Montfort University, UK
Anders Dahlbom	University of Skövde, Sweden
Susana Díaz	Universidad de Oviedo, Spain
Josep Domingo-Ferrer	Universitat Rovira i Virgili, Catalonia
Jozo Dujmovic	San Francisco State University, California
Yasunori Endo	University of Tsukuba, Japan
Zoe Falomir	Universität Bremen, Germany
Katsushige Fujimoto	Fukushima University, Japan
Michel Grabisch	Université Paris I Panthéon-Sorbonne, France
Enrique Herrera-Viedma	Universidad de Granada, Spain
Aoi Honda	Kyushu Institute of Technology, Japan

Masahiro Inuiguchi	Osaka University, Japan
Marie-Jeanne Lesot	Université Pierre et Marie Curie (Paris VI), France
Xinwang Liu	Southeast University, China
Jun Long	National University of Defense Technology, China
Jean-Luc Marichal	University of Luxembourg, Luxembourg
Radko Mesiar	Slovak University of Technology, Slovakia
Andrea Mesiarová-Zemánková	Slovak Academy of Sciences, Slovakia
Tetsuya Murai	Hokkaido University, Japan
Toshiaki Murofushi	Tokyo Institute of Technology, Japan
Guillermo Navarro-Arribas	Universitat Autònoma de Barcelona, Catalonia, Spain
Jordi Nin	Universitat Politècnica de Catalunya, Spain
Gabriella Pasi	Università di Milano Bicocca, Italia
Susanne Saminger-Platz	Jihannes Kepler University, Austria
Sandra Sandri	Instituto Nacional de Pesquisas Espaciais, Brazil
Roman Słowiński	Poznan University of Technology, Poland
László Szilágyi	Sapientia-Hungarian Science University of Transylvania, Hungary
Aida Valls	Universitat Rovira i Virgili, Catalonia, Spain
Vilem Vychodil	Palacky University, Czech Republic
Zeshui Xu	Southeast University, China
Yuji Yoshida	University of Kitakyushu, Japan

Local Organizing Committee Chairs

Göran Falkman	University of Skövde, Skövde, Sweden
Anders Dahlbom	University of Skövde, Skövde, Sweden

Local Organizing Committee

Tina Erlandsson
Tove Helldin
Alexander Karlsson
Gunnar Mathiason
Maria Riveiro
Joe Steinhauer
Klara Stokes

Additional Reviewers

Montserrat Batet
Marco Viviani
Luis Gonzalez-Abril

Supporting Institutions

University of Skövde
The Catalan Association for Artificial Intelligence (ACIA)
The European Society for Fuzzy Logic and Technology (EUSFLAT)
The Japan Society for Fuzzy Theory and Intelligent Informatics (SOFT)
The UNESCO Chair in Data Privacy

Abstracts of Invited Talks

Modeling the Complex Search Space of Data Privacy Problems

Bradley A. Malin

Department of Biomedical Informatics, School of Medicine,
Vanderbilt University, Nashville, TN, USA

Abstract. Data privacy protection models often dictate towards worst-case adversarial assumptions, assuming that a recipient of data will always attack and exploit it. Yet, adversaries are agents who must make a wide range of decisions before running an attack, and accounting for the outcomes of such decisions indicate data sharing frameworks, can, at times, be more safe than expected under traditional beliefs. In this presentation, various models of data protection and attacks are reviewed, followed by illustrations of how decision making can be integrated as two-party games, leading to unexpected "no attack" scenarios of data. This presentation concludes with a review of challenges and opportunities for the AI community in data privacy, including modeling the search over large data protection spaces for defenders and decision making under uncertainty for attackers.

Extended Abstract

The past several decades have led to dramatic advances in our ability to collect, store, and analyze personal data. At the same time, there is a push to make data available beyond the confines of the individuals who generate it, as well as the organizations that initial collect it, for some predefined *primary* purpose (e.g., biomedical research, service optimization of mobile applications, and creditworthiness reviews). The motivation behind such *secondary* data sharing is based on many factors, including open data and transparency initiatives, the need to support validation of scientific findings, and increasing recognition that data holds remarkable value and can be commercialized in a market-like setting. Despite the opportunities such data provides, there are concerns that making information widely available substantially increases the chances that privacy rights will be infringed upon. In recognition of such rights, various laws and policies suggest organizations should "de-identify", "depersonalize", or "anonymise" data prior to disclosing it.

Efforts to do so have led to the establishment of an alphabet soup of data manipulation schemes (e.g., redaction of explicit identifiers, k-anonymity, l-diversity, and t-closeness) [1], but also detective-like investigations that illustrate where holes reside in the current arsenal of protection methods [2]. Based on these presumed failures of data protection, there are cries for changes in the manner by which data sharing is accomplished, ranging from foundational legal revisions to new computational definitions of privacy. In many respects, such calls may be in the right and society may need to revamp our perspectives on privacy. However, it must be recognized that such calls are often predicated on the belief that there is some adversary who will exploit the holes in the system to begin with - and this is certainly not a guarantee. Rather than dictate towards worst-case scenarios and aim to protect accordingly (or suggest the futility of protection), we need to be more pragmatic. Specifically, we must model tradeoffs between the utility in sharing data and the anticipated threats towards exploiting it in a manner that erodes privacy. In this regard, the databases and privacy communities have an opportunity to adapt risk management frameworks to enable data custodians to quantify the privacy risks in a much broader context. There are several specific aspects of this problem that are notable for the artificial intelligence community to be aware of.

First, the space of potential data protection models is massive which implies decision making requires efficient exploration strategies. As an example, let us consider a traditional view on the *re-identification* problem. In this problem, the data holder suspects that certain attributes about an individual may be exploited by some adversary. As such, the data holder strips away all explicitly identifying attributes, such as Social Security Number, phone number, and personal names, but then has to decide which remaining quasi-identifying attributes (i.e., the values of which can uniquely represent an individual and be linked to resources that contain an individual's identity) can be shared and at what level of fidelity. If the data holder is permitted to use generalization (e.g., by making values more coarse) and a generic information loss measure (e.g., KL-divergence), then the search space can be ordered and we can search for solutions that optimize the balance between privacy and utility because of a monotonicity in the system [3, 4]. However, if we allow for suppression (e.g., removal of a specific value for an individual), then monotonicity may no longer hold and we may need to use approximation strategies to guide the search.

Second, and perhaps more importantly, the decision making aspects of the system need to be modeled, which often entails accounting for uncertainty. Recently, we showed that the re-identification problem can be modeled as a basic Stackelberg (i.e., leader-follower) game, where the data holder attempts to publish data in a manner that limits the chance that an adversary will commit a re-identification attempt [5]. Notably, we illustrated that, the defender can discover policies in which substantial quantities of personal data are shared and, despite the potential for re-identification, the adversary is not sufficiently incentivized to mount an attack. This is notable because it also implies that the space of possible solutions is not just data-based, but also policy-based, such that we could make data more general, but also raise the threat of fines (or some other penalties) to disincentive malevolent behavior in a system. Yet to deploy such an approach in practice, we need allow for uncertainty in the risks and utility functions. It is unlikely that such functions yield exact values, such that we will need to reason over distributions. At the same time, to date, such modeling has assumed a two-player game, but it is clear that the data about an individual is distributed across different organizations, each of which may have their own utility function and have limited ability to coordinate before making data disclosure decisions [6]. Similarly, there may be more than one adversary in the system and coalitions may be formed in a manner that lead to more effective attack strategies [7].

Despite the challenges associated with modeling large search spaces and adversarial behavior, it is critical to ensure claims about data privacy and policy decisions based on such claims are grounded. Now that a foundation for privacy concerns has been established and there is a general idea of the types of protections that can be invoked, it is time to start mapping these protections into optimization problems that account for active decisions on data holders and adversaries alike.

Acknowledgments. The author would like to thank the various individuals who have informed this idea, including his collaborators at the University of Texas (Murat Kantarcioglu) and Vanderbilt University (Ellen Wright Clayton, Raymond Heatherly, Yevgeniy Vorobeychik, Zhiyu Wan, and Weiyi Xia).

References

1. Fung, B., Wang, K., Chen, R., Yu, P.: Privacy-preserving data publishing: a survey of recent developments. ACM Comput. Surv. **42**, 14 (2010)
2. Narayanan, A., Shmatikov, V.: Myths and fallacies of "personally identifiable information". Commun. ACM **53**, 24–26 (2010)
3. Xia, W., Heatherly, R., Ding, X., Li, J., Malin. B.: R-U policy frontiers for health data de-identification. J. Am. Med. Inform. Assoc. (2015, in press). doi:10.1093/jamia/ocv004
4. Xia, W., Heatherly, R., Ding, X., Li, J., Malin, B.: Efficient discovery of de-identification policies through a risk-utility frontier. In: Proceedings of the 3rd ACM Conference on Data and Applications Security, pp. 59–70 (2013)
5. Wan, Z., Vorobeychik, Y., Xia, W., Clayton, E.W., Kantarcioglu, M., Ganta, R., Heatherly, R., Malin, B.: A game theoretic framework for analyzing re-identification risks. PLoS One **10**, e0120592 (2015)
6. Sattar, S., Li, J., Liu, J., Heatherly, R., Malin, B.: A probabilistic approach to mitigate composition attacks on privacy in non-coordinated environments. Knowl.-Based Syst. **67**, 361–372 (2014)
7. Duong, Q., LeFevre, K., Wellman, M.P.: Strategic modeling of information sharing among data privacy attackers. Informatica **34**, 151–158 (2010)

Classifying Large Graphs
with Differential Privacy

Fredrik D. Johansson, Otto Frost, Carl Retzner, and Devdatt Dubhashi

Chalmers University of Technology, 412 58, Göteborg, Sweden
{frejohk,dubhashi}@chalmers.se
{ottfro,krettan}@gmail.com

Abstract. We consider classification of graphs using graph kernels under differential privacy. We develop differentially private mechanisms for two well-known graph kernels, the random walk kernel and the graphlet kernel. We use the Laplace mechanism with restricted sensitivity to release private versions of the feature vector representations of these kernels. Further, we develop a new sampling algorithm for approximate computation of the graphlet kernel on large graphs with guarantees on sample complexity, and show that the method improves both privacy and computation speed. We also observe that the number of samples needed to obtain good accuracy in practice is much lower than the bound. Finally, we perform an extensive empirical evaluation examining the trade-off between privacy and accuracy and show that our private method is able to retain good accuracy in several classification tasks.

Statistical Forecasting Using Belief Functions

Thierry Denœux

Sorbonne Universités, Université de Technologie de Compiègne, CNRS, France
tdenoeux@utc.fr

Forecasting may be defined as the task of making statements about events that have not yet been observed. When the events can be considered as generated by some random process, we need a statistical model that has to be fitted to the data and used to predict the occurrence of future events. In this process, the quantification of uncertainty is of the utmost importance. In particular, to be useful for making decisions, forecasts need to be accompanied with some confidence measure.

In this talk, the limits of classical approaches to the quantification of statistical forecasting are discussed, and we advocate a new approach based on the Dempster-Shafer theory of belief functions [6]. A belief function can be viewed both as a non additive measure, and as a random set. Dempster-Shafer reasoning thus extends both Bayesian reasoning and set-membership computation. More specifically, the method presented in this talk consists in modeling estimation uncertainty using a consonant belief function constructed from the likelihood [2, 3], and combining it with random uncertainty represented by an error probability distribution [4, 5]. A predictive belief function can be approximated to any desired accuracy using a combination of Monte Carlo simulation and interval computation. When prior probabilistic information is available, the method boils down to Bayesian prediction. It is, however, more widely applicable, as it does not require such precise prior information.

One of the advantages of the proposed methodology is that it allows us to combine statistical data and expert judgements within a single general framework. We illustrate this process using several examples, including prediction using a regression model with partially known cofactors [5], and the combination of expert opinions and statistical data for extreme sea level prediction, taking into account climate change [1].

References

1. Ben Abdallah, N., Mouhous Voyneau, N., Denoeux, T.: Combining statistical and expert evidence using belief functions: application to centennial sea level estimation taking into account climate change. Int. J. Approx. Reas. **55**, 341–354 (2014)
2. Denoeux, T.: Likelihood-based belief function: justification and some extensions to low-quality data. Int. J. Approx. Reas. **55**(7), 1535–1547 (2014)
3. Denoeux, T.: Rejoinder on "likelihood-based belief function: justification and some extensions to low-quality data". Int. J. Approx. Reas. **55**(7), 1614–1617 (2014)
4. Kanjanatarakul, O., Sriboonchitta, S., Denoeux, T.: Forecasting using belief functions: an application to marketing econometrics. Int. J. Approx. Reas. **55**(5), 1113–1128 (2014)

5. Kanjanatarakul, S.S.O., Denoeux, T.: Statistical estimation and prediction using belief functions: principles and application to some econometric models. Int. J. Approx. Reas. (2015, under review)
6. Shafer, G.: A Mathematical Theory of Evidence. Princeton University Press, Princeton (1976)

Preference Learning: Machine Learning Meets Preference Modeling

Eyke Hüllermeier

Department of Computer Science, University of Paderborn, Paderborn, Germany
eyke@upb.de

The notion of "preferences" has a long tradition in operational research, economics and the social sciences, where it has been formalised in various ways and studied extensively from different points of view [2, 5, 7]. It is a topic of key importance in fields such as game theory, social choice and the decision sciences. In these fields, much emphasis is put on properly *modeling* a decision maker's preferences, and on deriving and (axiomatically) characterizing rational decision rules.

In machine learning, like in artificial intelligence and computer science in general, the interest in the topic of preferences arose much more recently [1]. The emerging field of *preference learning* [3] is concerned with methods for learning preference models from explicit or implicit preference information, which are typically used for predicting the preferences of an individual or a group of individuals in new decision contexts. While research on preference learning has been specifically triggered by applications such as recommender systems and "learning to rank" for information retrieval [6], the methods developed in this field are useful in many other domains as well.

Obviously, preference modeling and preference learning can ideally complement and mutually benefit from each other. In particular, the suitable specification of an underlying model class is a key prerequisite for successful machine learning, that is to say, successful learning presumes appropriate modeling. Likewise, data-driven approaches for preference elicitation are becoming more and more important in preference modeling and decision analysis nowadays, mainly due to large scale applications and the increasing availability of preference data.

The goal of this talk is to provide a brief introduction to the field of preference learning and, moreover, to elaborate on its connection to preference modeling. In this regard, specific emphasis will be put on preference learning based on aggregation operators.

First, learning with the Choquet integral will be discussed [9, 10]. The (discrete) Choquet integral is an established aggregation function that has been used in various fields of application, including multi-criteria decision problems and information fusion [11]. It can be seen as a generalization of the weighted arithmetic mean that is not only able to capture the importance of individual criteria but also information about the interaction (e.g., redundancy or complementarity) between different criteria. Moreover, it obeys monotonicity properties in a rather natural way. Due to these properties, the Choquet integral appears to be a very appealing modeling tool in the context of preference learning.

Second, the (data-driven) modeling of utility functions by means of fuzzy pattern trees will discussed [4, 8]. A fuzzy pattern tree is a hierarchical, tree-like structure, whose inner nodes are marked with generalized (fuzzy) logical and arithmetic aggregation operators, and whose leaf nodes are associated with fuzzy predicates evaluating individual criteria. Thus, a pattern tree recursively decomposes the evaluation of a criterion into sub-criteria. To evaluate an alternative, it propagates information from the bottom to the top of the hierarchy: A node takes the values of its descendants as input, combines them using the respective operator, and submits the output to its predecessor. This way, a pattern tree implements a monotone, nonlinear utility function that flexibly combines evaluations on individual criteria into an overall evaluation assuming values in the unit interval.

References

1. Domshlak, C., Hüllermeier, E., Kaci, S., Prade, H.: Preferences in AI: an overview. Artif. Intell. **175**(7–8), 1037–1052 (2011)
2. Fishburn, P.C.: Utility-Theory for Decision Making. Wiley (1969)
3. Fürnkranz, J., Hüllermeier, E. (eds.) Preference Learning. Springer (2011)
4. Hüllermeier, E., Rifqi, M., Henzgen, S., Senge, R.: Comparing fuzzy partitions: a generalization of the Rand index and related measures. IEEE Trans. Fuzzy Syst. **20**(3), 546–556 (2012)
5. Keeney, R.L., Raiffa, H.: Decisions with Multiple Objectives-Preferences and Value Tradeoffs. Wiley (1976)
6. Liu, T.Y.: Learning to Rank for Information Retrieval. Springer (2011)
7. Von Neumann, J., Morgenstern, O.: Theory of Games and Economic Behavior. John Wiley and Sons (1953)
8. Senge, R., Hüllermeier, E.: Top-down induction of fuzzy pattern trees. IEEE Trans. Fuzzy Syst. **19**(2), 241–252 (2011)
9. Fallah Tehrani, A., Cheng, W., Dembczynski, K., Hüllermeier, E.: Learning monotone nonlinear models using the Choquet integral. Mach. Learn. **89**(1), 183–211 (2012)
10. Fallah Tehrani, A., Cheng, W., Hüllermeier, E.: Preference learning using the Choquet integral: the case of multipartite ranking. IEEE Trans. Fuzzy Syst. **20**(6), 1102–1113 (2012)
11. Torra, V., Narukawa, Y.: Modeling Decisions: Information Fusion and Aggregation Operators. Springer (2007)

Game-Theoretic Approaches to Decision Making in Cyber-Physical Systems Security (Extended Abstract)

Weiru Liu

School of Electronics, Electrical Engineering and Computer Science
Queen's University Belfast, Belfast, UK

Recent years have seen a significant increase in research activities in game-theoretic approaches to security, covering a variety of areas, such as cyber-security (e.g., network intrusion detection), information security (e.g., fraudulent transactions), physical security (the protection of citizens and critical infrastructure/assets, e.g., smart grids, airports, government buildings) [1, 2, 16, 22]. One of the main focuses in game-theoretic approaches to physical security is to strategically allocate security resources to protect assets. Most of the existing work so far tackles this issue with the Bayesian Stackelberg game framework (also referred to as *security games*). A Bayesian Stackelberg game has two players: a defender (e.g., a security force, a leader) and an attacker (or a follower). Uncertainty over the type of attacker that a defender may face is modelled by a probability distribution over a set of possible attacker types. A typical *solution concept*, specifying assumptions under which a defender can determine a mixed strategy[1] maximizing their expected utility, is the *Strong Stackelberg Equilibrium* (SSE).

There are three assumptions underpinning the SSE solution concept:

- An attacker can observe a defender's mixed strategy and responds optimally. This assumption states that an attacker has the full observability and knowledge of a defender's strategy, and will act rationally.
- A defender can precisely execute its optimal mixed strategy. This assumption requires that there will be no failures or exceptions during the execution of a chosen strategy.
- Each player has a point payoff (precise payoff) value for each pure strategy profile. This assumption rules out imprecise or absent payoff values. It also rules out the situation where two or more pure strategy profiles cannot be separated when

This talk presents a summary of joint research work with Wenjun Ma and Kevin McAreavey, as part of the CSIT project (Centre for Secure Information Technologies) funded by the UK EPSRC, TSB and Industry, and PACES project funded by the UK EPSRC.

[1] A pure strategy for a player is the complete set of actions assigning an action to each possible state. A pure strategy profile is a pair of pure strategies, one from each player. A mixed strategy is a probability distribution over the set of all pure strategies of a given player.

assigning payoff values (e.g., the payoff values for these pure strategy profiles cannot be individually identified).

Clearly, these assumptions have severe limitations. For the first assumption, it is unrealistic to assume that an attacker always has full knowledge of a defender's mixed strategy (e.g., how to allocate security resources to protect assets). It has been argued that human attackers usually act on partial information of a defender's mixed strategy and that an attacker's rationality is bounded [5, 20, 22, 23]. As for the second assumption, recent research has been conducted in analyzing stochastic effects when a defender executes its mixed strategy [6, 25]. The third assumption has attracted much research effort due to various factors affecting the assignment of point payoff values to individual pure strategy profiles. These factors, for example, include reliability of expert assessment, accuracy of analyzing historical data or qualitative judgements on security risks. In addition, understanding the motivations of possible attackers and the levels of threat of different types of attackers also have an effect on the payoff value assignment. Therefore, such assignments are often pervaded with ambiguity and the requirement of precise point payoff values for pure strategy profiles is unrealistic in many real-world applications.

Consequently, there is a growing emphasis on the need to address uncertainty over payoffs [4, 7, 8, 19]. In fact, the Bayesian Stackelberg game already provides some handling of a specific type of uncertainty related to payoffs, i.e., the uncertainty over the possible types of attacker (as different types of attackers have different motivations and hence project different levels of threat). That is, Bayesian Stackelberg games assume the existence of a probability distribution over attacker types, where a probability value is interpreted as risk, associated with that particular type of attacker.

However, another important type of uncertainty related to this assumption (point payoff value assignment) which cannot be handled by the Bayesian Stackelberg game is ambiguity. Ambiguity refers to situations where there is a lack of information to justify such probabilities. Two main classes of ambiguity exist in security games: (i) ambiguous payoff values where payoffs cannot be represented by point values or a probability distribution over point payoff values; (ii) ambiguous payoff assignments where two or more pure strategy profiles cannot be distinguished when assigning payoff values.

In this talk, I will first present our general decision-theoretic framework for ambiguous security games based on Dempster-Shafer (D-S) theory [18]. An instantiation of this framework based on the ambiguity-aversion principle of minimax regret [11] will be introduced and its properties discussed. We prove that when all the payoff values are point values on individual pure strategy profiles, our framework is reduced to a standard security game. When there are no absent payoff values and no payoff values being assigned to subsets of pure strategy profiles, we prove the equivalence between the outcome of our entire framework and the outcome of a simplified version of the framework that is specific to handling all types of ambiguity except these two types. In addition, six existing decision methods (including Γ-maximin, Maximax, Maximin regret, Hurwitz criterion, Transferable Belief Model, and Ordered Weighted Average) [3, 17, 21, 24] will be used to instantiate our framework, and extensive experiments on evaluating the performance of all these different decision methods (including ours) will be presented.

I will then present our ongoing work on game-theoretic approaches to cyber-physical systems security utilizing real-time intelligent surveillance information, to dynamically determine the most plausible type of attacker, and to differentiate (or rank) potential multiple simultaneous (and independent) attacks based on the assessment of levels of threat [12–14]. Current Bayesian Stackelberg games assume that a prior probability distribution over attacker types is known. We argue that the degrees of plausibility of possible types of attacker (as well as the consequence of such an attack) shall be based on the analytical results and fusion of real-time surveillance information [9, 10] rather than a pre-defined probability distribution. Finally, if we do not assume that an attacker has the knowledge of a defender's payoff values assigned to its strategy profiles, nor a defender's knowledge about the plausibilities of their types, then our approaches developed under the notion of Surveillance Driven Security Resource Allocation (SDSRA) framework in [15] shall be adopted.

Acknowledgement. This work has been supported by the CSIT project with references EP/G034303/1; EP/H049606/1; and EPSRC PACES project with reference EP/D070864/1.

References

1. Abrahamyan, A.: Application of stackelberg security games in information security. In: Proceedings of the 9th Annual Symposium on Information Assurance (ASIA 2014), pp. 62–55 (2014)
2. Brown, G., Carlyle, M., Salmerón, J., Wood, K.: Defending critical infrastructure. Interfaces **36**(6), 530–544 (2006)
3. Jaffray, J.Y., Jeleva, M.: Information processing under imprecise risk with the Hurwicz criterion. In: Proceedings of the Fifth International Symposium on Imprecise Probability: Theories and Applications, pp. 233–242 (2007)
4. Jain, M., Kiekintveld, C., Tambe, M.: Quality-bounded solutions for finite bayesian stackelberg games: scaling up. In: AAMAS, pp. 997–1004 (2011)
5. Jiang, A.X., Nguyen, T.H., Tambe, M., Procaccia, A.D.: Monotonic maximin: a robust stackelberg solution against boundedly rational followers. In: Das, S.K., Nita-Rotaru, C., Kantarcioglu, M. (eds.) GameSec 2013. LNCS, vol. 8252, pp. 119–139. Springer, Heildelberg (2013)
6. Jiang, A.X., Yin, Z., Zhang, C., Tambe, M., Kraus, S.: Game-theoretic randomization for security patrolling with dynamic execution uncertainty. In: AAMAS, pp. 207–214 (2013)
7. Kiekintveld, C., Islam, T., Kreinovich, V.: Security games with interval uncertainty. In: AAMAS, pp. 231–238 (2013)
8. Kiekintveld, C. Marecki, J., Tambe, M.: Approximation methods for infinite Bayesian Stackelberg games: modeling distributional payoff uncertainty. In: AAMAS, pp. 1005–1012 (2011)
9. Ma, J., Liu, W., Miller, P.C., Yan, W.: Event composition with imperfect information for bus surveillance. In: AVSS, pp. 382–387 (2009)

10. Ma, J., Liu, W., Miller, P.C.: Event modelling and reasoning with uncertain information for distributed sensor networks. In: Deshpande, A., Hunter, A. (eds.) SUM 2010. LNCS, vol. 6379, pp. 236–249. Springer, Heildelberg (2010)
11. Ma, W., Luo, X., Liu, W.: An ambiguity aversion framework of security games under ambiguities. In: IJCAI, pp. 271–278 (2013)
12. Ma, W., Liu, W., Miller, P.C., Luo, X.: A game-theoretic approach for threats detection and intervention in surveillance. In: AAMAS, pp. 1565–1566 (2014)
13. Ma, W., Liu, W.: An intelligent threat prevention framework with heterogeneous information. In: ECAI, pp. 1061–1062 (2014)
14. Ma, W., Liu, W., Ma, J., Miller, P.C.: An extended event reasoning framework for decision support under uncertainty. In: Laurent, A., Strauss, O., Bouchon-Meunier, B., Yager, R.R (eds.) IPMU 2014, Part III. CCIS, vol. 444, pp. 335–344. Springer, Heildelberg (2014)
15. Ma, W., Liu, W. McAreavey, K.: Game-theoretic resource allocation with real-time probabilistic surveillance information. In: ECSQARU (2015, to appear)
16. Manshaei, M., Zhu, Q., Alpcan, T., Basar, T., Hubaux, J.P.: Game theory meets network security and privacy. ACM Comput. Surv. **45**(3), 25 (2013)
17. Smets, P., Kennes, R.: The transferable belief model. Artif. Intell. **66**, 191–234 (1994)
18. Shafer, G.: A Mathematical Theory of Evidence. Princeton University Press (1976)
19. Paruchuri, P., Pearce, J.P., Marecki, J., Tambe, M., Ordóñez, F., Kraus, S.: Playing games for security: an efficient exact algorithm for solving bayesian stackelberg games. In: AAMAS, pp. 895–902 (2008)
20. Pita, J., Jain, M., Ordóñez, F., Tambe, M., Kraus, S., Magori-Cohen, R.: Effective solutions for real-world stackelberg games: when agents must deal with human uncertainties. In: AAMAS, pp. 369–376 (2009)
21. Seidenfeld, T.: A contrast between two decision rules for use with (convex) sets of probabilities: Γ-maximin versus e-admissibilty. Synthese **140**, 69–88 (2004)
22. Tambe, M.: Security and Game Theory: Algorithms, Deployed Systems, Lessons Learned. Cambridge University Press, Cambridge (2011)
23. Yang, R., Ordóñez, F., Tambe, M.: Computing optimal strategy against quantal response in security games. In: AAMAS, pp. 847–854 (2012)
24. Yager, R.R.: Families of owa operators. Fuzzy Sets Syst. **59**, 125–148 (1993)
25. Yin, Z., Jain, M., Tambe, M., Ordóñez, F.: Risk-averse strategies for security games with execution and observational uncertainty. In: AAAI (2011)

Contents

Data Mining and Data Privacy

Logics

Invited Paper

Classifying Large Graphs
with Differential Privacy

Fredrik D. Johansson[✉], Otto Frost, Carl Retzner, and Devdatt Dubhashi

Chalmers University of Technology, SE-412 58 Göteborg, Sweden
{frejohk,dubhashi}@chalmers.se,
{ottfro,krettan}@gmail.com

Abstract. We consider classification of graphs using graph kernels under differential privacy. We develop differentially private mechanisms for two well-known graph kernels, the random walk kernel and the graphlet kernel. We use the Laplace mechanism with restricted sensitivity to release private versions of the feature vector representations of these kernels. Further, we develop a new sampling algorithm for approximate computation of the graphlet kernel on large graphs with guarantees on sample complexity, and show that the method improves both privacy and computation speed. We also observe that the number of samples needed to obtain good accuracy in practice is much lower than the bound. Finally, we perform an extensive empirical evaluation examining the trade-off between privacy and accuracy and show that our private method is able to retain good accuracy in several classification tasks.

1 Introduction

Data containing personal information about individuals are increasingly often analyzed using machine learning methods. In many applications, such data are well represented by large graphs, notably in the cases of social networks, email correspondance and telephone traffic. Sources like these have created vast collections of sensitive graph data, with growing concerns about privacy.

For example, analyzing how information spreads in a social network can provide great value to social scientists and advertisers alike, but may also leak sensitive information, such as the sexual orientation, of its users [11]. Simply removing names and identifiers from the network does often not provide sufficient anonymity [19]. This problem has received considerable attention in various research communities. *Differential privacy* [5] offers a quantitative way of trading off analytical accuracy for privacy of individuals in a database. Intuitively, the knowledge about an individual, gained from a differentially private query to a database, should be approximately insensitive to whether the individual is present in the database or not. Recently, there have been attempts to apply the framework of differential privacy to graphs. In social networks, this corresponds to not disclosing the identity of a user, or the relationship between two people. While methods have been developed for releasing several different graph statistics, practical applications of these are largely unexplored.

© Springer International Publishing Switzerland 2015
V. Torra and Y. Narakawa (Eds.): MDAI 2015, LNAI 9321, pp. 3–17, 2015.
DOI: 10.1007/978-3-319-23240-9_1

Graph classification is a common analysis problem in diverse application areas such as social sciences, bioinformatics and information retrieval [27]. When data are represented by graphs, this has traditionally been solved by extracting features of graph structure and attributes, and applying standard classification tools such as support vector machines. More recently, kernel methods have been used for this purpose, giving rise to a large family of *graph kernels* – positive semi-definite similarity measures on graphs [27]. As of yet, the privacy aspects of these methods have not been examined.

In many applications, graphs are large, consisting of thousands or millions of nodes [16]. Most social networks, for example, are orders of magnitude larger than graphs traditionally used to evaluate graph classification mehods [24]. Working with large graphs is beneficial from a privacy perspective, as the influence of a single individual or link is less significant, but challenging from a classification perspective as rich features become expensive to compute.

Main Contributions. We bring classification of large graphs and differential privacy together, showing how sampling improves both privacy and computational complexity. We examine the privacy aspects of graph kernels, state-of-the-art tools for graph classification. We design mechanisms to release two graph kernels under differential privacy. We conclude that a good graph kernel is not necessarily a good private kernel. To the best of our knowledge, this is the first application of graph differential privacy to a learning problem. In contrast to general results on learning with differential privacy we release features used for classification themselves, not the output of a learning algorithm. This has the advantage of not limiting the analysis to a single learning method.

We develop efficient methods for computing approximations of graphlet kernels on large graphs using edge sampling, with guarantees on sample complexity. We make an extensive empirical evaluation comparing our private kernels to their original counterparts, showing that our methods retain good accuracy in several tasks. For the evaluation, we compile three new datasets for graph classification. Our results show that both privacy and computational complexity is improved when sampling and that only a small number samples is sufficient for high accuracy.

2   Background

We review results on classification with graph kernels, and differential privacy for graphs.

2.1   Graph Kernels

Graph kernels have been introduced as efficient and accurate tools for classifying graphs [27]. Graph kernels are positive definite similarity measures on graphs, $K(G, G') = \langle \phi(G), \phi(G') \rangle$, for some embedding $\phi(G)$ into a Hilbert space. As such, they can be used with general kernel methods such as support vector machines or kernel principal component analysis [22]. For some kernels, there exist an explicit, *finite dimensional* mapping $\phi(G)$. We show that some of these kernels can be made private using standard mechanisms for differential privacy.

A common pattern in the design of graph kernels is to extract, aggregate and compare features of subgraphs [24]. For example, shortest-path kernels [3] compare features of the shortest paths between all pairs of nodes in two graphs. A recent trend is to focus on leveraging attributes on nodes and edges of graphs [15,24,29]. Privacy mechanisms for kernels on attributed graphs is beyond the scope of this paper however, and is left for future work.

Graphlet Kernels. Graphlet kernels compare counts or distributions of small subgraph patterns [25]. The size of the patterns is usually limited to $k \in \{3,4,5\}$ nodes to maintain efficiency. Let $\mathcal{H}^k = \{H_1^k, \ldots, H_a^k\}$ denote the set of patterns of k nodes that are unique modulo isomorphism. Given a pair of graphs G, G', the k-graphlet kernel is defined by

$$K_{gl}^k(G, G') = \mathbf{f}_G^{k\top} \mathbf{f}_{G'}^k \tag{1}$$

$$f_G^k(i) = |\{U \subset V : |U| = k, G[U] \cong H_i^k\}| \tag{2}$$

where $G[U]$ denotes the subgraph of G induced by U and $G \cong G'$ denotes that G and G' are isomorphic. $f_G^k(i)$ is the number of subgraphs in G that are isomorphic to graphlet H_i^k. A common variant is the normalized kernel $\tilde{K}_{gl}(G, G') = \mathbf{f}_G^\top \mathbf{f}_G / (\|\mathbf{f}_G\| \|\mathbf{f}_G\|)$ where counts are replaced by distributions over graphlets. The exact kernel can be computed in $O(nd^{k-1})$ time for $k \in \{3,4,5\}$ [23].

Graphlet kernels have several other variations, including sample approximations and ones that only consider *connected* subgraph patterns. Shervashidze et al. [25] showed that the features of the standard graphlet kernel can be approximated by repeatedly sampling k nodes uniformly at random and identifying the subgraph pattern induced by the sampled nodes. Specifically, the distribution of all graphlet types of size k present in G, including disconnected ones, can be approximated with L_1-error less than γ, with probability $p \geq 1 - \rho$ using

$$m = \lceil \frac{2(\log 2 \cdot a + \log(\frac{1}{\rho}))}{\gamma^2} \rceil \tag{3}$$

samples, where $a = |\mathcal{H}^k|$ is the number of unique k-node subgraph patterns.

Random Walk Kernels. Random walk kernels compare graphs based on features of node walks [7,12]. The p-random walk kernel compares numbers of walks of length up to p and is defined by

$$K_{prw}(G, G') = \sum_{i,j=1}^{|V||V'|} \left[\sum_{l=0}^{p} \lambda^l (A \otimes A')^l \right]_{ij}, \tag{4}$$

where $A \otimes A'$ is the Kronecker product of A and A', and λ a constant. The full random walk kernel of Gärtner et al. [7] is recovered when $p = \infty$. An equivalent formulation for undirected, unweighted graphs, is

$$K_{prw}(G, G') = \sum_{l=0}^{p} (\lambda^{\frac{l}{2}} w_G(l))(\lambda^{\frac{l}{2}} w_{G'}(l)),$$

where $\mathbf{w}_G = [w_G(1), \ldots, w_G(p)]^\top$ is the vector with elements $w_G(l)$, the number of walks of length l in G. When $p = \infty$, the kernel can be computed exactly in $O(n^3)$ time by solving a Sylvester equation [28]. While polynomial, this is still too expensive for many applications. For finite p, the complexity can be reduced to $O(pm)$, or $O(pDn)$ in graphs of maximum degree, D [8].

2.2 Differential Privacy and Graphs

Differential privacy (DP) offers a quantitative means of managing the trade-off between privacy and utility when releasing sanitized statistics of a database [5]. Intuitively, an adversary should not learn much more from a query, about an individual present in the database, than she/he would if the individual was not present. Equivalently, an adversary should not detect, with high probability, that an individual is removed or included in the database. In DP, this is achieved by making sure small changes to an individuals data does not have large effects on the output of queries to the database. In tabular data, individuals are represented by records and changes to an individual correspond to additions or removal of rows in the table. For graph data, there are several interpretations of what constitutes an individuals data [14]. In *node* differential privacy, an algorithm is private if it does not reveal the inclusion or removal of a single node and all its edges in the graph. In *edge* differential privacy, an algorithm is differentially private if it does not reveal the inclusion or removal of a single edge.

Differential privacy requires a distance function between datasets to make precise what is a small change to a database. In node differential privacy, this is usually the *rewiring* distance, $d_{node}(G, G')$, equal to the number of nodes in G that need to be rewired to obtain G'. In edge differential privacy, it is the *edge* distance, $d_{edge}(G, G')$, equal to the number of edges that need to be added to or deleted from G to obtain G'. Two graphs are said to be *neighbors* if their distance is 1. An algorithm is differentially private if the output distribution of the algorithm applied to two neighboring graphs does not differ much.

Definition 21 (Edge/Node-differential privacy [20]). *A randomized algorithm $\mathcal{A}$ is (ϵ, δ)-edge(node)-differentially private if for all events S in the output space of $\mathcal{A}$ and for all neighbor graphs G, G',*

$$Pr(\mathcal{A}(G) \in S) \leq \exp(\epsilon) \times Pr(A(G') \in S) + \delta.$$

A common means of achieving differential privacy is to perturb the output of a function f, by adding random noise. In the frequently used *Laplace mechanism* [5], the magnitude of the noise is proportional to the *global sensitivity*

of the function, $S_f = \max_{G,G':d(G,G')=1} |f(G) - f(G')|$. Many graph proper-
ties, including graphlet counts, have high global sensitivity [14]. This is partly
because the maximization in S_f includes graphs of unbounded degree. To this
end, Kasiviswanathan et al. [14] and Blocki et al. [2] independently proposed
schemes to provide differential privacy by bounding or truncating the degree
of graphs. In this work we adopt the *restricted sensitivity* of Blocki et al. [2].
There, a query on G is guaranteed to have a private response for any G, but
high accuracy only if G is part of an hypothesis set $\mathcal{G}^H \subset \mathcal{G}$ (a subset of the
space of graphs).

Definition 22 (Restricted sensitivity. Blocki et al. (2013) [2]). *For a
query f over $\mathcal{G}^H \subset \mathcal{G}$, with distance metric $d(G, G')$, the restricted sensitivity is*

$$RS_f(\mathcal{G}) = \max_{G,G' \in \mathcal{G}^H} \left(\frac{|f(G) - f(G')|}{d(G, G')} \right).$$

Let $\mathcal{G}^H = \mathcal{G}_D$ be the set of graphs of maximum degree D. Blocki et al. [2]
showed that differential privacy can be achieved for $f(G)$ by first applying a
projection $\mu_D : \mathcal{G} \to \mathcal{G}_D$ to G and then adding Laplace noise to $f(\mu_D(G))$,
proportional to the restricted sensitivity of f. In the projection $\mu_D(G)$, edges
are removed from G, one by one, in a particular order until the maximum degree
is $d \leq D$. We use the following result.

Theorem 21 (Laplace mechanism with restricted sensitivity [2]). *The
mechanism $\mathcal{A}(f, G) = f(\mu_D(G)) + Lap\left(3 \cdot RS_f(\mathcal{G}_D)/\epsilon\right)$ for the projection function $\mu_D(G)$ as described in Blocki et al. [2] (Claim 13) preserves $(\epsilon, 0)$-edge
differential privacy for any graph G.*

In the sequel we make use of the following result on releasing multiple statistics in sequence (often called composition).

Theorem 22 (Composition. McSherry and Mironov (2009) [18]). *If $\mathcal{A}_1$
and $\mathcal{A}_2$ satisfy (ϵ_1, δ_1) and (ϵ_2, δ_2)-differential privacy respectively, their sequential composition satisfies $(\epsilon_1 + \epsilon_2, \delta_1 + \delta_2)$.*

3   Private Kernels for Large Graphs

We introduce a framework for releasing differentially private graph kernels in
the edge model. We adopt the Laplace mechanism combined with restricted sensitivity, see Theorem 21. Specifically, we consider the case where the hypothesis
space is the set of graphs $\mathcal{G}_D$ with degree bounded by D and use the simple
projection scheme μ_D proposed by [2] (Claim 13). Projecting graphs to lower
maximum degree is advantageuous not just for function sensitivity but for the
complexity of computing the released features themselves, as seen in Sect. 2.1.
Further, as we will see in Sect. 5.3, classification accuracy is fairly insensitive to
the projection itself. A description of the framework can be seen in Algorithm 1.
We can state the following result.

Algorithm 1. EDGE-PRIVATE GRAPH FEATURES

Input: $G = (V, E)$, truncation level D
Input: Privacy level ϵ
Input: Queries $f_G(i) : \mathcal{G} \to \mathbb{R}$, $i = 1, ..., a$

$G_D := \mu_D(G)$ (see Sect. 3)
$\mathbf{f}_{G_D} := [f_{G_D}(1), \ldots, f_{G_D}(a)]^\top$
$\tilde{\mathbf{f}}_G := \mathbf{f}_{G_D} + \mathbf{e}$, $e(l) \sim \mathrm{Lap}(3a \cdot RS_{f(i)}(\mathcal{G}_{G_D})/\epsilon)$

Output: Private counts $\tilde{\mathbf{f}}_G$

Corollary 31. *The output of Algorithm 1 is $(\epsilon, 0)$-differentially private.*

Proof. Apply Theorems 21 and 22.   □

We note in Algorithm 1 that the magnitude of noise added to each component i of the released feature vector is proportional both to the restricted sensitivity $RS_{f(i)}(\mathcal{G}_D)$ of the feature $f(i)$ and the number of features, a. This highlights an interesting trade-off between the richness of representations and the privacy loss of releasing them. Graph kernels that are well suited for classification may not be well suited for private release.

The shortest-path kernel [3] is a popular graph kernel. It compares graphs by counts of shortest-paths of length k for $k = 1, 2, \ldots, n$. Let $f_G(k)$ be the number of unordered pairs of nodes such that the shortest path between them have length k. Now consider a path graph, G, of $n = 2n'$ nodes and let $k = \frac{n}{2} + 1$. It is fairly easy to see that $f_G(k) = \frac{n}{2} - 1$. Now, construct G' by adding an edge between the two end nodes of the path, forming a cycle. Then, $f_{G'}(k) = 0$. Clearly $d_{edge}(G, G') = 1$, but $|f_G(k) - f_{G'}(k)| \geq \frac{n}{2} - 2$. Hence, $RS_{f_G(k)}(\mathcal{G}_D)$ is in $\Omega(n)$, for $D \geq 2$, and thus unbounded in general.

We seek to release features of graphs that both have bounded restricted sensitivity and work well for classification. We proceed to show that the graphlet kernel and p-random walk kernel fulfill these requirements.

3.1 Private Graphlet Kernels

We define a differentially private version of the graphlet kernel, using the general framework of Algorithm 1. We address the unnormalized version of the kernel, see (1). This is because we can easily estimate the restricted sensitivity of the counts, but not of the distribution. Note however that we can normalize the private counts released by our algorithm.

We begin our treatment by noting that counts of disconnected graphlets typically have very large or unbounded sensitivity. Consider a graph $G = (V, E)$ and the count $f_G^3(1)$ of empty 3-node subgraphs of G. Removing a single edge $e \in E$ from the graph may add as much as $n - 2$ to $f_G^3(1)$ in the worst case. Further, consider the distribution of subgraph patterns in the graphlet kernel. For large, sparse graphs, the components corresponding to empty subgraph patterns will vastly dominate the mass of the distribution. Hence, when comparing graphs by such distributions, the actual structure of the graph has little influence on the comparison. For these reasons, we consider only connected graphlets in this work. Similar to previous research [25], we limit ourselves to graphlets of k nodes with $k \in \{3, 4, 5\}$.

As shown by Blocki et al. [2] (Claim 22), the restricted sensitivity of a counting query f^k of graphlets of size k under $\mathcal{G}_D$ is

$$RS_{f^k}(\mathcal{G}_D) = kD^{k-1}.$$

This result is in contrast to the shortest-path kernel, see the previous section, as the restricted sensitivity of graphlet counts is bounded by a constant.

Let $f_G^k(\cdot)$ be the graphlet counting function of (2). Then, with $\tilde{\mathbf{f}}_G^k$ and $\tilde{\mathbf{f}}_{G'}^k$, the private counts produced by Algorithm 1 with f^k for G and G' respectively, Then, $\tilde{K}_{gl} = \tilde{\mathbf{f}}_G^{k\top} \tilde{\mathbf{f}}_G^k$ is an $(\epsilon, 0)$-differentially private version of the graphlet kernel in (1).

3.2 Private p-random Walk Kernels

We consider a differentially private version of the p-random walk kernel. The full random walk kernel, comparing walks of infinite length, cannot be represented explicitly as finite-dimensional feature vectors. Clearly, this kernel is not a good candidate for our framework. Further, the $O(n^3)$ time complexity of the full kernel is prohibitively expensive for large graphs. Instead, we consider the p-random walk kernel, comparing walks of length only up to p.

We define a differentially private version of the p-random walk kernel by applying Algorithm 1 to the counting function $w_G(l)$ of (4). We note in passing that λ is a kernel parameter, and therefore public. We can thus release the unnormalized counts as well as the actual kernel.

We show below that the restricted sensitivity of walk counts is similar to that of counts of connected subgraphs. The sensitivity is a factor 2 higher because of walks being ordered.

Proposition 31. *Under the bounded-degree hypothesis $\mathcal{G}_D$, the restricted sensitivity of counts of walks of length l, in the edge adjacency model, is*

$$RS_{w(l)}(\mathcal{G}_D) \leq 2lD^{l-1}.$$

Proof. Let $e = (i, j)$ be any edge. Then, consider any walk of length l passing through e. In such a walk, either the sequence (i, j) or (j, i) must be present at least once. There are $2l$ choices for e to be placed at. Then, at each of the remaining $l - 1$ steps of the walk, there are at most D nodes to choose from. Thus, there are at most $2lD^{l-1}$ walks passing through e. □

Applying Theorem 21 and Proposition 31 to the walk counts of Algorithm 1 results in an $(\epsilon, 0)$-differentially private version of the p-random walk kernel.

Algorithm 2. APPROXIMATE GRAPHLET COUNT, $\hat{\mathbf{f}}$

Input: $G = (V, E)$, sample size s

$\hat{\mathbf{f}} = \mathbf{0}$, Approximate count for each graphlet type

for $j = 1 \ldots, s$ **do**

    Sample $e_j \in E$ uniformly at random

    $\hat{\mathbf{f}} \leftarrow \hat{\mathbf{f}} +$ counts of graphlets containing e

end for

$\hat{f}(i) \leftarrow \frac{\hat{f}(i)}{s} \frac{m}{|m_{H_i}|}$ for all i

---

4 Scaling Private Graphlet Kernels

As remarked in Sect. 2.1, computing the exact counts of graphlets is often prohibitively expensive for large real-world graphs, even ones with bounded degree. Instead, we rely on approximate counting using sampling. In this section, we define a new sampling scheme for counting connected graphlets. We then derive privacy guarantees for the approximate counts. We show that sampling improves both privacy and computational complexity.

We exploit the fact that our graphs have bounded degree due to the projection μ_D applied to guarantee differential privacy with restricted sensitivity, see Sect. 2.2. Exact counting of graphlets in graphs of bounded degree can be done in $O(nD^{k-1})$ time [23]. Although linear in n, this is typically a slow process. Several schemes to compute approximate counts using sampling have been proposed [21,25], including MCMC methods [9].

Shervashidze et al. [25] sample k nodes uniformly at random, without replacement, and determine to which pattern the subgraph induced by those nodes belong. This method is not suitable for sampling connected graphlets of large graphs of low bounded degree, as most induced subgraphs will be empty. While the error of the sampled distribution of graphlets decreases quickly, see (3), the actual values contains little information. Rejection sampling using the same sampling scheme, but counting only connected subgraphs, is prohibitively slow for large sparse graphs.

The existing methods referred to above give only approximate *distributions* of connected graphlets, not approximate *counts*. Distributions cannot trivially be released using the Laplace mechanism as the restricted sensitivity of the distribution is inversely proportional to the total number of connected graphlets, which is unknown in general. We address the issues with existing methods mentioned above, devising a new sampling scheme for connected graphlets in order to guarantee both privacy and utility for large graphs. Let m_{H_i} denote the number of edges in graphlet H_i. Our scheme counts graphlets containing edges sampled uniformly at random with replacement, see Algorithm 2.

Theorem 41 (Approximation error of edge graphlet sampling). *Consider $G = (V, E)$ with degree bounded by D, and let $Z_e(i)$ be the number of graphlets in G of type i that contains $e \in E$. For any $\gamma > 0, 0 < \rho < 1,$*

the counts, $\hat{\mathbf{f}} = [\hat{f}(1), \ldots, \hat{f}(a)]^{\top}$ produced by Algorithm 2, have the following property.

$$Pr\left(\left|\hat{f}(i) - \mathbb{E}[\hat{f}(i)]\right| \geq \gamma \mathbb{E}[\hat{f}(i)]\right) \leq \rho, \quad i \in [a] \tag{5}$$

using $s_i = 3\alpha_i \frac{\log \frac{2}{\rho}}{\gamma^2}$ samples with $\alpha_i = \frac{\max_e Z_e(i)}{\mathbb{E}[Z_e(i)]}$. Due to space constraints, the proof is deferred to the full version of this paper.

Observation. It is easily shown that $\alpha_i \leq m/m_{H_i}$. Given an edge e, it is possible to compute the number of k-node graphlets containing e exactly in $O(D^{k-2})$ time [23]. This leads to an overall complexity of $O(sD^{k-2})$ for fixed γ and ρ. The worst-case bound of $\alpha \in O(m)$ results in a complexity of $O(mD^{k-2})$ or $O(nD^{k-1})$ which is equivalent to that of exact counting. *However, in practice, α is usually much smaller than the bound,* see Table 1. For example, on the ROADS dataset, the maximum number of edges is 13284 while $\max_i \alpha_i^3 = 14.9$.

As we will see next, counting graphlets by edge sampling, instead of node sampling, also gives a natural means of analysing the effects of sampling on privacy.

4.1   Differential Privacy with Sampling

Intuitively, sampling graphlets rather than enumerating all of them, should increase the privacy of a particular edge, as the edge may be left out of sample. Results on differential privacy for statistics computed on samples from graphs are largely unexplored [1]. For tabular data, there are several results of this nature [6], none of which are easily applied to graph data. The problem is that in graphs, there is no straight-forward analogue to partitioning the rows of a database table. Privacy may be achieved simply by partitioning the node set, but then utility becomes difficult to analyze.

We record the following general results about releasing noisy statistics under (ϵ, δ)-differential privacy.

Theorem 42 (Private release of noisy data). *Consider a function $\hat{f}(x)$ such that $\mathbb{E}[\hat{f}(x)] = f(x)$, and $Pr(|\hat{f}(x) - f(x)| \leq \gamma) \geq 1 - \rho$. Further, let S be the global sensitivity of $f(x)$. Then, $\mathcal{A}(x) := \hat{f}(x) + Lap(\frac{S+2\gamma}{\epsilon})$ is (ϵ, δ)-private mechanism for releasing $\hat{f}(x)$, for $\delta = \rho(2 - \rho)$. The proof is deferred to the full version of this paper.*

Unfortunately, applying Theorem 42 to statistics with multiplicative error, like the approximate graphlet count, the noise added for privacy is too large to be practical. Instead of using a result for general graph statistics, we use the results of Li et al. [17], specifically for graphlet counting queries. In their work, they show that if any record is sampled with probability β, the increase in privacy (decrease in ϵ) is proportional to β. For our sampling mechanism, Algorithm 2, an edge can affect the statistics of the sample in two ways, either from being sampled itself, or by being part of a graphlet containing a sampled edge. We show below that by bounding the probability of this event, we can achieve increased privacy via sampling, in line with the intuition expressed above.

Proposition 41 (Increasing privacy of sampled graphlet counts). *Let* $G = (V, E)$ *and* $m = |E|$. *The mechanism* $\mathcal{A}(\hat{\mathbf{f}})$ *applying Algorithm 1 with* ϵ_1 *to the output of Algorithm 2 is an* (ϵ_2, δ_2)-*private algorithm for graphlet counts, for* $\epsilon_2 \geq \log(1 + \beta_u e^{\epsilon_1} - \beta_l)$ *and* $\delta_2 \geq \beta_u \delta_1$, *with* $\beta_u = 1 - \left(\frac{(m-1)(m-2(D-1)^{t-1})}{m^2} \right)^s$ *and* $\beta_l = 1 - (1 - \frac{1}{m})^s$. *The proof is deferred to the full version of this paper.*

We note that applying Proposition 41 does not always result in $\epsilon_2 < \epsilon_1$. In particular, ϵ_2/ϵ_1 grows exponentially with the number of samples s. However, as seen in Fig. 1, a small number of samples is usually sufficient. We also note that the privacy of a single edge increases with the total number of edges m.

5   Experiments

We evaluate the performance of our private kernels comparing to state-of-the-art methods. Particularly, we study the effects of degree bounding and sampling.

5.1   Datasets

We have selected four datasets for evaluation, denoted PROTO, ROADS, SOCIAL and D & D. Most existing benchmark datasets used for classification consists of graphs that are too small [24] for the private mechanism to provide useful output. To this end, PROTO, ROADS and SOCIAL were compiled for the purpose of this paper[1]. All of the sets are divided into two classes.

PROTO is a set of connected induced subgraphs sampled using random vertex expansion from synthesized population interaction networks[2] [26] of Portland and Montgomery County. The class of each graph corresponds to the city representing the network from which it has been sampled. ROADS is a set of connected induced subgraphs sampled using an MCMC scheme [9], from road network graphs of Texas and California available from SNAP[3]. The class of each graph corresponds to the city representing the network from which it has been sampled. SOCIAL is a collection of Twitter and Google+ graphs available from SNAP (see footnote 3). We divide the set into two classes of equal size: Google+ graphs in one and Twitter graphs in the other. We use only the largest (by number of nodes) Twitter networks to obtain equally large classes, we also remove the *ego-nodes* of the networks and direction of edges. Last, D & D [4] is a dataset commonly used as a benchmark for graph classification. D & D consists of graphs representing proteins classified according to whether they are enzymes or not. A set of statistics for the datasets can be found in Table 1.

---

[1] http://www.cse.chalmers.se/~frejohk/data/graphdata_aistats2015.zip.
[2] http://www.vbi.vt.edu/ndssl.
[3] http://snap.stanford.edu/.

Table 1. Statistics of datasets. Number of graphs N, number of nodes n, number of edges m and with α_k^* the maximum α_i over k-node graphlets, with α_i as in Theorem 41. ‡Computation did not finish within 2 days.

Dataset	N	Pos./Neg.	α_3^*	α_4^*	m_{max}	m_{avg}	n_{max}	n_{avg}	d_{max}	d_{avg}
D&D	1178	691/487	4.3	50.6	14267	715.7	5748	284.3	19	5.5
PROTO	200	100/100	14.9	1689	10308	4321.4	1000	1000	176	9.5
ROADS	200	100/100	33.9	2291	13973	13283.7	10000	10000	12	2.7
SOCIAL	262	132/132	‡	‡	1473709	104026.1	4938	1072.2	2971	80.6

5.2 Experimental Setup

We use a binary C-SVM with 10-fold cross validation. For each fold, we optimize C w.r.t classification accuracy. We repeat this experiment 10 times and report the average accuracy and standard error. As baselines, we use the p-random walk (PRW) kernel [7], the Weisfeiler-Lehman (WL) kernel [24] and the graphlet (GK-Ck or GK-Ak, $k \in \{3,4,5\}$) kernels [25]. We select the number of samples for GK-Ak, based on (3) with $\gamma = \rho = 0.1$.

We denote our private kernels introduced in Sect. 3 by (DPPRW) and (DPGK-Ck, $k \in \{3,4\}$), for the private p-random walk kernel the private graphlet kernels respectively. The sampled version of (DPGK-Ck) is denoted (DPGK-Sk). Motivated by the results in Fig. 1, the sampled kernels used 50 samples for each count. We use the unnormalized feature vector representation of the graphlet kernels as we observe that it yields better classification accuracy in general.

To make use of the restricted sensitivity as in Definition 21 we employ the projection μ_D as described in Sect. 2.2. We choose D w.r.t. maximum classification accuracy. We use the following values of D, in the order of D & D, PROTO and SOCIAL; PRW: $(9, 9, 50)$, GK-C3: $(6, 9, 100)$ and GK-C4: $(9, 7, 100)$. For the private classification we give the results at the optimal degree truncation level and with a fixed value of $\epsilon = 0.5$. We compare the baselines to their private counterparts at the same truncation level for the GK-Ck and the PRW kernel. As a reference, we give results for all kernels without truncation.

For the computation of the PRW-kernel we selected the parameter λ from the set $\{1, 0.1, 0.01, 0.001\}$ w.r.t optimal accuracy. For D & D and SOCIAL, $\lambda = 0.1$ and for PROTO, $\lambda = 0.01$ gave the best results. To highlight the effects of the private mechanism, we exclude walks of length 0 in the computation of the PRW-kernel as they can be released independent of ϵ.

5.3 Classification Results

The results of the classification experiments are found in Table 2. We observe for all datasets that using a low bounding degree in comparison to the maximum degree of the dataset does not significantly affect accuracy. None of the results worsen by more than 1.5 %. In fact, for PRW on PROTO, the classification accuracy significantly increases as the data is truncated from 83.8 % to 89.0 % when

Table 2. Classification accuracy. The four groups are (1) baselines, (2) results on truncated graphs, (3) (0.5,0)-private kernels and (4) private and sampled kernels. [‡]Computation time exceeded two days.

Kernel	D & D	PROTO	SOCIAL
PRW	75.4 ± 0.6	83.8 ± 1.2	83.0 ± 0.4
WL	74.9 ± 0.6	93.7 ± 5.1	79.8 ± 1.8
GK-C3	74.4 ± 1.0	98.4 ± 1.1	89.0 ± 0.7
GK-C4	73.3 ± 1.0	99.9 ± 0.2	‡
GK-C5	74.1 ± 0.7	‡	‡
GK-A3	74.4 ± 0.4	74.0 ± 1.1	71.8 ± 1.7
GK-A4	74.7 ± 0.5	83.8 ± 0.9	76.2 ± 2.0
GK-A5	74.6 ± 0.5	85.0 ± 1.9	81.5 ± 1.8
PRW	75.2 ± 0.9	89.0 ± 1.1	82.7 ± 0.9
GK-C3	74.2 ± 1.3	97.8 ± 2.9	88.7 ± 0.7
GK-C4	73.8 ± 0.7	99.5 ± 0.0	‡
GK-C5	73.6 ± 1.2	‡	‡
DPPRW	68.4 ± 1.1	86.9 ± 3.0	68.9 ± 1.9
DPGK-C3	59.3 ± 0.5	73.3 ± 1.6	77.0 ± 0.5
DPGK-C4	58.6 ± 0.1	51.0 ± 3.1	‡
DPGK-S3	58.8 ± 0.1	74.0 ± 2.6	77.2 ± 0.7
DPGK-S4	58.7 ± 0.1	53.0 ± 2.6	52.7 ± 2.6

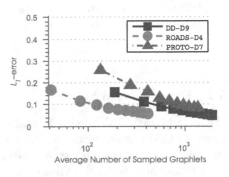

Fig. 1. Average L_1-error in estimated 4-graphlet distributions of three datasets, for varying numbers of sampled graphlets. Each marker corresponds to an addition of 10 sampled edges.

bounding the maximum degree from 159 to 9. As expected, the addition of noise to the feature vectors of the graphs have a negative effect on the classification accuracy.

For all datasets, the DPPRW kernel achieves non-trivial results with accuracies higher than the label distributions. On PROTO, the result is actually better than for PRW. This suggests that the DPPRW kernel is a good candidate for private graph classification. The DPGK-C3 kernel performs better than the label distribution for PROTO and SOCIAL, using only 2 private features. For social, this is noteworthy as it is competititve to the WL kernel. We also observe, that as the sizes of the graphs increase, the DPGK-C3 perform better, approaching the baseline accuracy. For D & D we can not classify better than the label distribution with either of the private graphlet kernels. This is likely caused by the small size of the D & D graphs. For each dataset, DPGK-C4-kernel fails to achieve accuracy significantly higher than the label distribution. This is likely due to the large amounts of noise added due to both graphlet size and composition. ROADS is not included in Table 2 as none of the private kernels were able to classify better than the label distribution. In fact, even for the baselines, only WL and GK-C5 were able to achieve higher accuracy than the label distribution.

5.4 Sampling the Private Graphlet Kernel

In Fig. 1 we show the average L_1-error, the L_1-norm of the difference between the true and estimated 4-graphlet distributions over all graphs of each dataset for which we sampled $s_i = 10, 20, 30, ..., 100$ edges. We note that the number of samples needed to obtain a low L_1-error is significantly lower than the sample complexity of Theorem 41. From Table 2 we can see that the loss of accuracy when using approximate counts instead of true counts is negligible. Notable is also that these results correspond to sampling far less edges than the worst case sample complexity.

6   Related Work

The combination of privacy and learning has been approached in several contexts. Kasiviswanathan 2011 et al. [13] addressed the general problem of learning a concept class under differential privacy. Jain & Thakurta [10] studies the problem of releasing a differentially private predictor in the framework of kernel methods. The setting of these works is different from ours as we do not release the output of a learning algorithm. Rather, we output private statistics used as input to learning algorithms.

Rahman et al. [21] compute approximate graphlet counts through an edge sampling scheme, but do not provide any guarantees on the approximation error. Also, their method differs from ours in that they sample without replacement and do not count every graphlet containing an edge.

7   Conclusions

We have proposed new mechanisms for releasing two well-known graph kernels under differential privacy, the graphlet kernel and the p-random walk kernel. We have given a new algorithm for approximating the graphlet kernel by edge sampling and show that this improves both privacy and computational complexity. Further, we perform an extensive empirical evaluation of our private kernels showing that they provide privacy while retaining good accuracy on several classification tasks.

We note both theoretically and empirically that a good graph kernel does not always make a good private graph kernel. This is because expressive features are often strongly connected with highly sensitive functions, that are not suitable for private release. Future work includes applying these ideas to other graph kernels, and perhaps more importantly, to kernels on attributed graphs.

References

1. Blocki, J., Blum, A., Datta, A., Sheffet, O.: The johnson-lindenstrauss transform itself preserves differential privacy. In: 2013 IEEE 54th Annual Symposium on Foundations of Computer Science, pp. 410–419 (2012)
2. Blocki, J., Blum, A., Datta, A., Sheffet, O.: Differentially private data analysis of social networks via restricted sensitivity. In: ITCS, pp. 87–96 (2013)
3. Borgwardt, K.M., Kriegel, H.-P.: Shortest-path kernels on graphs. In: Proceedings of ICDM, pp. 74–81 (2005)

4. Dobson, P.D., Doig, A.J.: Distinguishing enzyme structures from non-enzymes without alignments. J. Mol. Biol. **330**(4), 771–783 (2003)
5. Dwork, C., McSherry, F., Nissim, K., Smith, A.: Calibrating noise to sensitivity in private data analysis. In: Halevi, S., Rabin, T. (eds.) TCC 2006. LNCS, vol. 3876, pp. 265–284. Springer, Heidelberg (2006)
6. Dwork, C., Roth, A.: The algorithmic foundations of differential privacy. Theoret. Comput. Sci. **9**(3–4), 211–407 (2013)
7. Gärtner, T., Flach, P.A., Wrobel, S.: On graph kernels: hardness results and efficient alternatives. In: Schölkopf, B., Warmuth, M.K. (eds.) COLT/Kernel 2003. LNCS (LNAI), vol. 2777, pp. 129–143. Springer, Heidelberg (2003)
8. Hermansson, L., Kerola, T., Johansson, F., Jethava, V., Dubhashi, D.: Entity disambiguation in anonymized graphs using graph kernels. In: CIKM 2013, pp. 1037–1046. ACM (2013)
9. Hubler, C., Kriegel, H.-P., Borgwardt, K., Ghahramani, Z.: Metropolis algorithms for representative subgraph sampling. In: Eighth IEEE International Conference on Data Mining, ICDM 2008, pp. 283–292. IEEE (2008)
10. Jain, P., Thakurta, A.: Differentially private learning with kernels. In: JMLR Proceedings of ICML 2013, vol. 28, pp. 118–126 (2013). JMLR.org
11. Jernigan, C., Mistree, B.F.: Gaydar: facebook friendships expose sexual orientation. First Monday 14(10) (2009)
12. Kashima, H., Tsuda, K., Inokuchi, A.: Marginalized kernels between labeled graphs. In: Proceedings of the 20th International Conference on Machine Learning (2003)
13. Kasiviswanathan, S.P., Lee, H.K., Nissim, K., Raskhodnikova, S., Smith, A.: What can we learn privately? In: CoRR (2008)
14. Kasiviswanathan, S.P., Nissim, K., Raskhodnikova, S., Smith, A.: Analyzing graphs with node differential privacy. In: Sahai, A. (ed.) TCC 2013. LNCS, vol. 7785, pp. 457–476. Springer, Heidelberg (2013)
15. Kriege, N., Mutzel, P.: Subgraph matching kernels for attributed graphs. In: ICML. icml.cc / Omnipress (2012)
16. Leskovec, J., Lang, K.J., Dasgupta, A., Mahoney, M.W.: Community structure in large networks: natural cluster sizes and the absence of large well-defined clusters. Internet Math. **6**(1), 29–123 (2009)
17. Li, N., Qardaji, W., Su, D.: On sampling, anonymization, and differential privacy or, k-anonymization meets differential privacy. In: ASIACCS 2012, pp. 32–33. ACM, New York (2012)
18. McSherry, F., Mironov, I.: Differentially private recommender systems: building privacy into the net. In: IV Elder, J.F., Fogelman-Souli, F., Flach, P.A., Zaki, M. (eds.) KDD, pp. 627–636. ACM (2009)
19. Narayanan, A., Shmatikov, V.: Robust de-anonymization of large sparse datasets. In: 2013 IEEE Symposium on Security and Privacy, pp. 111–125 (2008)
20. Nissim, K., Raskhodnikova, S., Smith, A.: Smooth sensitivity and sampling in private data analysis. In: STOC 2007, pp. 75–84. ACM (2007)
21. Rahman, M., Bhuiyan, M., Hasan, M.A.: Graft: an approximate graphlet counting algorithm for large graph analysis. In: CIKM 2012, pp. 1467–1471. ACM (2012)
22. Schölkopf, B., Smola, A.J.: Learning With Kernels: Support Vector Machines, Regularization, Optimization, And Beyond. MIT Press, Cambridge (2001)
23. Shervashidze, N.: Scalable graph kernels. Ph.D. thesis, Eberhard Karls Universitt Tbingen (2012)
24. Shervashidze, N., Schweitzer, P., van Leeuwen, E.J., Mehlhorn, K., Borgwardt, K.M.: Weisfeiler-lehman graph kernels. J. Mach. Learn. Res. **12**, 2539–2561 (2011)

25. Shervashidze, N., Vishwanathan, S., Petri, T., Mehlhorn, K., Borgwardt, K.M.: Efficient graphlet kernels for large graph comparison. In: Proceedings of AISTATS (2009)
26. Swarup, S., Eubank, S.G., Marathe, M.V.: Computational epidemiology as a challenge domain for multiagent systems. In: AAMAS 2014, pp. 1173–1176, Richland, SC (2014)
27. Vishwanathan, S., Schraudolph, N.N., Kondor, R., Borgwardt, K.M.: Graph kernels. J. Mach. Learn. Res. **11**, 1201–1242 (2010)
28. Vishwanathan, S.V.N., Borgwardt, K.M., Schraudolph, N.N.: Fast computation of graph kernels. In: Schkopf, B., Platt, J., Hoffman, T. (eds.) NIPS, pp. 1449–1456. MIT Press (2006)
29. Woznica, A., Kalousis, A., Hilario, M.: Matching based kernels for labeled graphs. In: ECML/PKDD Workshop on Mining and Learning with Graphs (2006)

Aggregation Operators
and Decision Making

Extremal Completions of Triangular Norms Known on a Subregion of the Unit Interval

Andrea Mesiarová-Zemánková[⊠]

Mathematical Institute, Slovak Academy of Sciences,
Bratislava, Slovakia
zemankova@mat.savba.sk

Abstract. The strongest and the weakest t-norms that coincide with the given t-norm on a subregion of the unit interval are discussed. The question whether such a t-norm can be obtained as a limit of the sequence of continuous t-norms that coincide with the original t-norm on the given subregion is investigated.

Keywords: t-norm · Ordinal sum · Additive generator

1 Introduction

The (left-continuous) t-norms and their dual t-conorms are special aggregation functions and they have an indispensable role in many domains [2,4,5,10,11]. In real-world applications it can happen that only a part of the aggregation function is known, either it is observed from the input-output relationships in the training data or implied by the requirements of the modelled problem. With additional requirement that the aggregation function has to be a t-norm we are looking for the weakest and the strongest t-norms which coincide with the original t-norm which is known only on the interval $[a,b]^2 \subsetneq [0,1]^2$. In the case when a and b are idempotent elements this problem was studied in [8] (see also [4]), in a broader context of aggregation functions. We would like to extend these results also for the case when a or b (or both) is not an idempotent element. In [9] we have studied continuous t-norms which coincide with the original t-norm known on $[a,b]^2$. In this contribution we will focus on extremal t-norms and we will be also interested whether these t-norms can be approximated, i.e., obtained as a limit of a sequence of continuous t-norms T_i that coincide with the t-norm T_1 on $[a,b]^2$. Therefore we will focus on such t-norms T_1 which are continuous on $[a,b]^2$.

After recalling several basic notions and results in Sect. 2, we will focus on the case when there is no non-trivial idempotent point of T_1 in $[a,b]^2$ (Sect. 3). We will study several special cases: when $a = b$, when $a = 0$, when $b = 1$ and a general case when $0 < a < b < 1$. In Sect. 4 we will discuss the case when there is a non-trivial idempotent point of T_1 in $[a,b]^2$. We give our conclusions in Sect. 5.

© Springer International Publishing Switzerland 2015
V. Torra and Y. Narakawa (Eds.): MDAI 2015, LNAI 9321, pp. 21–32, 2015.
DOI: 10.1007/978-3-319-23240-9_2

2   Basic Notions and Results

Let us recall several useful definitions and results on t-norms (see [1,7]).

Definition 1. *(i) A binary function $T\colon [0,1]^2 \longrightarrow [0,1]$ is a t-norm if it is commutative, associative, non-decreasing in both variables and 1 is its neutral element.*
(ii) A binary function $C\colon [0,1]^2 \longrightarrow [0,1]$ is a t-conorm if it is commutative, associative, non-decreasing in both variables and 0 is its neutral element.
(iii) A binary function $S\colon [0,1]^2 \longrightarrow [0,1]$ is a t-subnorm if it is commutative, associative, non-decreasing in both variables and $S(x,y) \leq \min(x,y)$ for all $(x,y) \in [0,1]^2$.

Thus every t-norm is also a t-subnorm. Due to the associativity, n-ary form of any t-norm (t-conorm) is uniquely given and thus it can be extended to an aggregation function working on $\bigcup_{n\in\mathbb{N}}[0,1]^n$. The duality between t-norms and t-conorms is expressed by the fact that from any t-norm T we can obtain its dual t-conorm C by the equation

$$C(x,y) = 1 - T(1 - x, 1 - y)$$

and vice-versa. Therefore all results that we obtain for t-norms can be immediately obtained also for t-conorms.

A t-norm T is called *Archimedean* if for all $x,y \in \,]0,1[$ there exists an $n \in \mathbb{N}$ such that $x_T^{(n)} < y$, where $x_T^{(n)} = T(\underbrace{x, T(x, \ldots)}_{n\text{-times}})$.

A continuous t-norm is Archimedean if and only if it has only trivial idempotent points 0 and 1. A continuous Archimedean t-norm T is either strict, i.e., strictly increasing on $]0,1]^2$, or nilpotent, i.e., there exists $(x,y) \in \,]0,1]^2$ such that $T(x,y) = 0$.

For every t-norm it holds

$$T_{\mathbf{D}}(x,y) \leq T(x,y) \leq T_{\mathbf{M}},$$

where $T_{\mathbf{D}}$ is the drastic product t-norm given by $T_{\mathbf{D}}(x,y) = 0$ if $\max(x,y) < 1$ and $T_{\mathbf{D}}(x,y) = \min(x,y)$ otherwise, and $T_{\mathbf{M}}$ is the minimum t-norm given by $T_{\mathbf{M}}(x,y) = \min(x,y)$ for all $(x,y) \in [0,1]^2$.

Proposition 1. *Let $t\colon [0,1] \longrightarrow [0,\infty]$ be a continuous strictly decreasing function such that $t(1) = 0$. Then the binary operation $T\colon [0,1]^2 \longrightarrow [0,1]$ given by*

$$T(x,y) = t^{-1}(\min(t(0), t(x) + t(y)))$$

is a continuous t-norm. The function t is called an additive generator *of T.*

Note that every continuous Archimedean t-norm possesses a continuous additive generator. Non-continuous t-norms can be additively generated by non-continuous additive generators.

Definition 2. *(i) Let $t: [0,1] \longrightarrow [0,\infty]$ be a non-increasing function. Then the function $t^{(-1)}: [0,\infty] \longrightarrow [0,1]$ given by*

$$t^{(-1)}(x) = \sup\{y \in [0,1] \mid t(y) > x\}$$

is called the pseudo-inverse *of t.*

(ii) A strictly decreasing function $t: [0,1] \longrightarrow [0,\infty]$, $t(1) = 0$, is called an additive generator of a t-norm $T: [0,1]^2 \longrightarrow [0,1]$ if

$$T(x,y) = t^{(-1)}(t(x) + t(y)))$$

for all $(x,y) \in [0,1]^2$.

Further we recall a construction of t-norms via the ordinal sum. The basic stones for construction of t-norms via the ordinal sum (see [3]) are t-subnorms (see [6]).

Proposition 2. *Let $(]a_k, b_k[)_{k \in K}$ be a disjoint system of open subintervals of $[0,1]$, where K is a finite or countably infinite index set. Let $(S_k)_{k \in K}$ be a system of left-continuous t-subnorms such that if $b_{k_0} = 1$ for some $k_0 \in K$ then S_{k_0} is a t-norm, and if $b_{k_1} = a_{k_2}$ for some $k_1, k_2 \in K$ then either S_{k_2} has no zero divisors or S_{k_1} is a t-norm. Then the ordinal sum $T = (\langle a_k, b_k, S_k \rangle \mid k \in K)$ given by*

$$T(x,y) = \begin{cases} a_k + (b_k - a_k)T_k\left(\frac{x-a_k}{b_k-a_k}, \frac{y-a_k}{b_k-a_k}\right) & \text{if } (x,y) \in \,]a_k, b_k]^2, \\ \min(x,y) & \text{else} \end{cases}$$

is a left-continuous t-norm.

Recall that each continuous t-norm can be expressed as an ordinal sum of continuous Archimedean t-norms. In the following sections $\{\varepsilon_i\}_{i \in \mathbb{N}}$ will always be a sequence of small enough $\varepsilon_i > 0$ which converges to 0.

3   Extremal Extensions of t-norms Without a Non-trivial Idempotent Element in $[a,b]$

Let the t-norm T_1 be known only on $[a,b]^2$ and let it be continuous on $[a,b]^2$. In this section we will suppose that there is no non-trivial idempotent point of T_1 in $[a,b]$, i.e., that $T_1(x,x) < x$ for all $x \in [a,b] \setminus \{0,1\}$. We will suppose several subcases: when $a = b$, when $a = 0$, when $b = 1$ and a general case when $0 < a < b < 1$.

3.1   Case When $a = b$

In this case T_1 is known only in one point (a,a). Since all t-norms coincide on the boundary of the unit square if $a \in \{0,1\}$ then the strongest t-norm that coincides with T_1 in (a,a) is the minimum t-norm and the weakest is the drastic product t-norm. Note that the drastic product can be obtained as a limit of the sequence of continuous t-norms (see [7]). Similarly, since the minimum t-norm is continuous it can be obtained as a limit of the sequence of continuous t-norms.

Suppose $a \in\ ]0,1[$. Then since a is not an idempotent element we have $T_1(a,a) = q$ for some $q \in [0,a[$. Due to the monotonicity the strongest t-norm T_2 that coincides with T_1 in (a,a) is given by

$$T_2(x,y) = \begin{cases} q & \text{if } (x,y) \in [q,a]^2, \\ \min(x,y) & \text{otherwise.} \end{cases}$$

We see that T_2 is an ordinal sum on the zero t-subnorm Z on $[q,a]$, where Z is given by $Z(x,y) = 0$ for all $(x,y) \in [0,1]^2$. The t-norm T_2 can be obtained as a limit of the sequence of continuous t-norms T_i that coincide with T_1 in (a,a). Let $\{V_i\}_{i\in\mathbb{N}}$ be the sequence of t-norms that converges to the drastic product t-norm (see [7]). Then we define a sequence of ordinal sum t-norms $T_i = (\langle q, a + \varepsilon_i, V_i \rangle)$. Then $\{T_i\}_{i\in\mathbb{N}}$ converges to T_2.

The weakest t-norm T_3 that coincides with T_1 in (a,a) is given by

$$T_3(x,y) = \begin{cases} 0 & \text{if } \min(x,y) < a, \max(x,y) < 1, \\ \min(x,y) & \text{if } \max(x,y) = 1, \\ q & \text{otherwise.} \end{cases}$$

Next we would like to know whether T_3 can be obtained as a limit of the sequence of continuous t-norms that coincide with T_1 in (a,a). This sequence can be obtained as a sequence of continuous Archimedean t-norms T_i with additive generators t_i with $t_i(q) = 2 \cdot t_i(a)$ which converges to a function $t \colon [0,1] \longrightarrow [0,\infty]$ which is linear on $[0,q^-]$, on $[q^+, a^-]$ and on $[a^+, 1^-]$ with $t(0) = 2.5$, $t(q^-) = 2.1$, $t(q) = 2$, $t(q^+) = 1.7$, $t(a^-) = 1.6$, $t(a) = t(a^+) = 1$, $t(1^-) = 0.9$ and $t(1) = 0$. Then $\{T_i\}_{i\in\mathbb{N}}$ converges to T_3.

3.2   Case when $a = 0$

Now we will focus on the interval $[0,b]$. Let the t-norm T_1 be known and continuous on $[0,b]^2$. Then we will use the following result.

Lemma 1. *Let $S \colon [0,b]^2 \longrightarrow [0,b]$ be a t-subnorm for some $b \in [0,1]$. We define the binary operation $T^* \colon [0,1]^2 \longrightarrow [0,1]$ by*

$$T^*(x,y) = \begin{cases} S(x,y) & \text{if } \max(x,y) \leq b, \\ \min(x,y) & \text{otherwise.} \end{cases}$$

Then T^* is a t-norm.

The t-norm T^* from the previous result is an ordinal sum of a t-subnorm S_1 on $[0,b]$, where S_1 on $[0,1]^2$ is linearly isomorphic with S on $[0,b]^2$. Then we get the following.

Proposition 3. *Let* $T_1 \colon [0,1]^2 \longrightarrow [0,1]$ *be a t-norm. Then the strongest t-norm* T_2 *which coincides with* T_1 *on* $[0,b]^2$ *is given by*

$$T_2(x,y) = \begin{cases} T_1(x,y) & \text{if } \max(x,y) \leq b, \\ \min(x,y) & \text{otherwise.} \end{cases}$$

Now the question is whether T_2 can be obtained as a limit of a sequence of continuous t-norms T_i which coincide with T_1 on $[0,b]^2$. This is possible only if T_1 fulfills the conditions necessary for existence of a continuous t-norm which coincides with T_1 on $[0,b]^2$ (for more details see [9]). In such a case every continuous t-norm that coincides with T_1 on $[0,b]^2$ is Archimedean on $[0,b]^2$ and thus it possess an additive generator s on $[0,b]$. Then we construct a sequence $\{T_i\}_{i\in\mathbb{N}}$ of t-norms where $T_i = (\langle 0, b + \varepsilon_i, V_i \rangle)$, where V_i is a continuous, Archimedean t-norm and additive generator s_i of T_i on $[0, b + \varepsilon_i]$ satisfies $s_i(x) = s(x)$ for all $x \in [0,b]$. Then the sequence $\{T_i\}_{i\in\mathbb{N}}$ converges to T_2.

For the weakest extension we will use the following result.

Lemma 2. *Let* $S \colon [0,b]^2 \longrightarrow [0,b]$ *be a t-subnorm for some* $b \in [0,1]$. *We define the binary operation* $T_* \colon [0,1]^2 \longrightarrow [0,1]$ *by*

$$T_*(x,y) = \begin{cases} \min(x,y) & \text{if } \max(x,y) = 1, \\ S(\min(x,b), \min(y,b)) & \text{otherwise.} \end{cases}$$

Then T_* *is a t-norm.*

We get the following.

Proposition 4. *Let* $T_1 \colon [0,1]^2 \longrightarrow [0,1]$ *be a t-norm. Then the weakest t-norm* T_3 *which coincides with* T_1 *on* $[0,b]^2$ *is given by*

$$T_3(x,y) = \begin{cases} \min(x,y) & \text{if } \max(x,y) = 1, \\ T_1(\min(x,b), \min(y,b)) & \text{otherwise.} \end{cases}$$

Similarly as above, if T_1 fulfills the conditions necessary for existence of a continuous t-norm which coincides with T_1 on $[0,b]^2$ then T_1 has an additive generator s on $[0,b]$. We will construct a sequence of continuous, Archimedean t-norms $\{T_i\}_{i\in\mathbb{N}}$, with additive generators t_i such that $t_i(x) = s(x)$ for all $x \in [0,b]$, t_i is linear on $[b, 1 - \varepsilon_i]$ and on $[1 - \varepsilon_i, 1]$ and $t_i(1 - \varepsilon_i) = s(b) - \varepsilon_i$. Then the sequence $\{T_i\}_{i\in\mathbb{N}}$ converges to T_3.

3.3   Case When $b = 1$

Now we will focus on the interval $[a,1]$. Here we recall a result from [9].

Lemma 3. *Let* $T_1 \colon [0,1]^2 \longrightarrow [0,1]$ *be a t-norm continuous on* $[a,1]^2$, $0 < a < 1$. *Then for any t-norm* T_2 *such that* T_1 *coincides with* T_2 *on* $[a,1]^2$, *we have* $T_2(x,y) = T_2(y,x) = T_1(x,y) = T_1(y,x)$ *for all* $(x,y) \in A$, *where*

$$A = \{(x,y),(y,x) \in [0,1]^2 \mid \text{ there exists a } z \in [0,1], T_1(z,a) = x, T_1(z,y) \geq a\}.$$

From the previous lemma we see that T_2 is on A uniquely given by values of T_1 on $[a, 1]^2$. Moreover, since T_1 is continuous on $[a, 1]^2$ for every $x \in [T_1(a, a), a]$ there exists a $z \in [a, 1]$ such that $x = T_1(a, z) = T_2(a, z)$. By continuity again, considering the fact that $T_1(a, z) \leq a$, $T_1(z, 1) = z$, we see that there exists a $p \in [a, 1]$ such that $T_1(z, p) = a$. Then $T_1(z, y) \geq a$ for all $y \geq p$. Thus the set A is a symmetric connected set which contains all points from $[0, 1]^2$ greater than the points from the lower border of A, where the lower border of A is the set $B = \{(x, y), (y, x) \in [0, 1]^2 \mid \text{there exists a } z \in [0, 1], T_1(z, a) = x, T_1(z, y) = a\}$. Note that if $(x, y) \in B$ then $T_2(x, y) = T_1(a, a)$ (see [9]). Therefore we get the following.

Proposition 5. *Let* $T_1 \colon [0, 1]^2 \longrightarrow [0, 1]$ *be a t-norm. Then the strongest t-norm* T_2 *which coincides with* T_1 *on* $[a, 1]^2$ *is given by*

$$T_2(x, y) = \begin{cases} T_1(a, T_1(z, y)) & \text{if } (x, y) \in A, x = T_1(a, z), \\ T_1(a, T_1(z, x)) & \text{if } (x, y) \in A, y = T_1(a, z), \\ T_1(a, a) & \text{if } (x, y) \in [T_1(a, a), 1]^2 \setminus A, \\ \min(x, y) & \text{otherwise.} \end{cases}$$

Here $A = \{(x, y), (y, x) \in [0, 1]^2 \mid \text{there exists a } z \in [0, 1], T_1(a, z) = x, T_1(z, y) \geq a\}$.

Since T_2 is continuous it is easy to see that T_2 can be obtained as a limit of a sequence of continuous t-norms that coincide with T_1 on $[a, 1]^2$.

Example 1. Assume that T is the product t-norm and $a = \frac{1}{2}$. Then the strongest t-norm that coincide with T on $[a, 1]^2$ is given by

$$T^*(x, y) = \begin{cases} x \cdot y & \text{if } x \cdot y \geq \frac{1}{4}, \\ \frac{1}{4} & \text{if } (x, y) \in \left[\frac{1}{4}, 1\right]^2, x \cdot y < \frac{1}{4} \\ \min(x, y) & \text{otherwise.} \end{cases}$$

Proposition 6. *Let* $T_1 \colon [0, 1]^2 \longrightarrow [0, 1]$ *be a t-norm. Then the weakest t-norm* T_3 *which coincides with* T_1 *on* $[a, 1]^2$ *is given by*

$$T_3(x, y) = \begin{cases} T_1(a, T_1(z, y)) & \text{if } (x, y) \in A, x = T_1(a, z), \\ T_1(a, T_1(z, x)) & \text{if } (x, y) \in A, y = T_1(a, z), \\ \min(x, y) & \text{if } \max(x, y) = 1, \\ 0 & \text{otherwise.} \end{cases}$$

Here again $A = \{(x, y), (y, x) \in [0, 1]^2 \mid \text{there exists a } z \in [0, 1], T_1(a, z) = x, T_1(z, y) \geq a\}$.

Since the strongest t-norm T_2 which coincides with T_1 on $[a, 1]^2$ is continuous and Archimedean on $[T_1(a, a), 1]^2$ it has an additive generator s on $[T_1(a, a), 1]$.

We will define a sequence $\{T_i\}_{i \in \mathbb{N}}$ of continuous, Archimedean t-norms which are generated by respective additive generators t_i such that $t_i(x) = s(x)$ for all $x \in [T_1(a, a), 1]$ and t_i is linear on $[0, T_1(a, a)]$ and $t_i(0) = s(T_1(a, a)) + \varepsilon_i$. Then the sequence $\{T_i\}_{i \in \mathbb{N}}$ converges to T_3.

Example 2. Let us again assume that T is the product t-norm and $a = \frac{1}{2}$. Then the weakest t-norm that coincide with T on $[a, 1]^2$ is given by

$$T_*(x, y) = \begin{cases} x \cdot y & \text{if } x \cdot y \geq \frac{1}{4}, \\ \min(x, y) & \text{if } \max(x, y) = 1, \min(x, y) < \frac{1}{4}, \\ 0 & \text{otherwise.} \end{cases}$$

3.4   Case When $0 < a < b < 1$

Let T_1 be known and continuous on $[a, b]^2$. First we will focus on the strongest and the weakest extensions of T_1 to $[0, b]^2$. Since each t-norm on $[0, b]^2$ is linearly isomorphic with some t-subnorm S on $[0, 1]^2$ we will use the following result.

Lemma 4. *Let $S_1 \colon [0, 1]^2 \longrightarrow [0, 1]$ be a t-subnorm. Then for any t-subnorm S_2 such that S_1 coincides with S_2 on $[a, 1]^2$, we have $S_2(x, y) = S_2(y, x) = S_1(x, y) = S_1(y, x)$ for all $(x, y) \in \overline{A}$, where*

$$\overline{A} = \{(x, y), (y, x) \in [0, 1]^2 \mid \text{ there exists } z, q \in [a, 1], S_1(z, q) = x, S_1(z, y) \geq a\}.$$

Suppose that $T_1(a, b) = a$. Then $a = T_1(a, \underbrace{T_1(b, \ldots, b)}_{n\text{-times}})$ and since T_1 is continuous on $[a, b]^2$ and has no idempotents in $[a, b]$ there exists an $n \in \mathbb{N}$ such that $T_1(\underbrace{b, \ldots, b}_{n\text{-times}}) = u < a$. Then, however, $a = T_1(a, u) \leq u < a$ what is a contradiction. Thus we have always $T_1(a, b) < a$. We have now two possibilities: either $T_1(b, b) \geq a$, or $T_1(b, b) < a$.

First suppose $T_1(b, b) \geq a$. Then there exists an $r \in [a, b]$ such that $T_1(r, r) = a$. Since $T_1(a, b) < a$ also $T_1(a, r) < a$. We then have the following.

Lemma 5. *Let $T \colon [0, 1]^2 \longrightarrow [0, 1]$ be a t-norm such that T is continuous on $[a, b]^2$ for $0 < a < b < 1$, and $T(a, b) < a$, $T(r, r) = a$ for some $r \in [a, b]$. Then the values of T on $A_1 \cup A_2$ are determined by the values of T on $[a, b]^2$, where $A_1 = \{(x, y), (y, x) \in [0, b]^2 \mid \text{ there exists } z \in [a, r], T_1(r, z) = x, T_1(z, y) \geq T_1(a, r)\}$, $A_2 = \{(x, y), (y, x) \in [0, b]^2 \mid \text{ there exists } z \in [a, r], T_1(a, z) = x, T_1(z, y) \geq a\}.$*

If we combine this with Lemma 1 we see that the strongest t-norm T_2 which coincides with T_1 on $[a, b]^2$, if $T(a, b) < a$ and $T(b, b) \geq a$ is given by

$$
T_2(x,y) = \begin{cases}
\min(x,y) & \text{if } (x,y) \in [0,1]^2 \setminus [T_1(a,a),b]^2, \\
T_1(x,y) & \text{if } (x,y) \in [a,b]^2, \\
T_1(a,w) & \text{if } T_1(r,z) = x, T_1(z,y) = T_1(r,w), z,w \in [a,r], \\
T_1(a,w) & \text{if } T_1(r,z) = y, T_1(z,x) = T_1(r,w), z,w \in [a,r], \\
T_1(r,T_1(z,y)) & \text{if } T_1(r,z) = x, z \in [a,b], T_1(z,y) > a, \\
T_1(r,T_1(z,x)) & \text{if } T_1(r,z) = y, z \in [a,b], T_1(z,x) > a, \\
T_1(a,T_1(z,y)) & \text{if } T_1(a,z) = x, z \in [a,r], T_1(z,y) \geq a, \\
T_1(a,T_1(z,x)) & \text{if } T_1(a,z) = y, z \in [a,r], T_1(z,x) \geq a, \\
T_1(a,a) & \text{otherwise.}
\end{cases}
$$

The t-norm T_2 is an ordinal sum t-norm and $T_1(a,a)$ is its idempotent point. If there exists a continuous t-norm that coincides with T_1 on $[a,b]^2$ then such a t-norm has an additive generator s on $[T_1(a,a),b]$. We construct a sequence of t-norms T_i which are ordinal sums of one continuous, Archimedean summand on $[T_1(a,a),b+\varepsilon_i]$ and T_i is generated on $[T_1(a,a),b+\varepsilon_i]$ by an additive generator t_i such that $t_i(x) = s(x)$ for all $x \in [T_1(a,a),b]$ and t_i is linear on $[b,b+\varepsilon_i]$ with $t_i(b+\varepsilon_i) = 0$. Then the sequence $\{T_i\}_{i\in\mathbb{N}}$ converges to T_2.

Similarly, using Lemma 2 we see that the weakest t-norm T_3 which coincides with T_1 on $[a,b]^2$, if $T(a,b) < a$ and $T(b,b) \geq a$ is given by

$$
T_3(x,y) = \begin{cases}
\min(x,y) & \text{if } \max(x,y) = 1, \\
T_2(\min(x,b),\min(y,b)) & \text{if } (x,y) \in [T_1(a,p),1]^2 \setminus [T_1(a,p),b]^2, \\
T_2(x,y) & \text{if } (x,y) \in A_1 \cup A_2 \cup [a,b]^2, \\
0 & \text{otherwise,}
\end{cases}
$$

where p is the smallest point from $[a,b]$ such that $T_1(b,p) = a$ and $A_1 = \{(x,y),(y,x) \in [0,b]^2 \mid \text{there exists } z \in [a,r], T_1(r,z) = x, T_1(z,y) \geq T_1(a,r)\}$, and $A_2 = \{(x,y),(y,x) \in [0,b]^2 \mid \text{there exists } z \in [a,r], T_1(a,z) = x, T_1(z,y) \geq a\}$.

If there exists a continuous t-norm that coincides with T_1 on $[a,b]^2$ then such a t-norm has an additive generator s on $[T_1(a,a),b]$. We construct a sequence of t-norms T_i with respective additive generators t_i such that $t_i(x) = s(x)$ for all $x \in [T_1(a,a),b]$ and t_i is linear on $[0,T_1(a,a)]$, $[b,1-\varepsilon_i]$ and $[1-\varepsilon_i,1]$ with $t_i(0) = s(T_1(a,a)) + \varepsilon_i$, $t_i(1-\varepsilon_i) = s(b) - \varepsilon_i$, $t_i(1) = 0$. Then the sequence $\{T_i\}_{i\in\mathbb{N}}$ converges to T_3.

Finally, we will suppose that $T_1(b,b) < a$.

Then using Lemma 2 we see that the weakest t-norm T_3 which coincides with T_1 on $[a,b]^2$, if $T(a,b) < a$ and $T(b,b) < a$ is given by

$$
T_3(x,y) = \begin{cases}
\min(x,y) & \text{if } \max(x,y) = 1, \\
T_1(\min(x,b),\min(y,b)) & \text{if } (x,y) \in [a,1]^2, \\
0 & \text{otherwise.}
\end{cases}
$$

If there exists a continuous t-norm that coincides with T_1 on $[a,b]^2$ then this t-norm has an additive generator s on $[T_1(a,a), b]$. Then its values on $[T_1(a,a), T_1(b,b)] \cup [a,b]$ determines T_1 on $[a,b]^2$. However, in [9] we have shown that values of such a generator on $[T_1(a,a), T_1(b,b)] \cup [a,b]$ are not uniquely determined by T_1 on $[a,b]^2$. More precisely, we can obtain a whole class of such additive generators with $s(a) = 1$ which are dependent on a parameter $s(b) = w$, where $1 > w > \frac{1}{2}$. We will select such an additive generator for which $3s(b) > 2s(a)$. Then we will define a sequence of t-norms T_i generated by respective additive generators t_i, where $t_i(x) = s(x)$ for $x \in [T_1(a,a), T_1(b,b)] \cup [a,b]$ and t_i is linear on $[0, T_1(a,a)]$, on $[T_1(b,b), a - \varepsilon_i]$, on $[a - \varepsilon_i, a]$, on $[b, 1 - \varepsilon_i]$, and on $[1 - \varepsilon_i, 1]$, with $t_i(0) = s(T_1(a,a)) + \varepsilon_i$, $t_i(a - \varepsilon_i) = s(T_1(b,b)) - \varepsilon_i$, $t_i(1 - \varepsilon_i) = s(b) - \varepsilon_i$, $t_i(1) = 0$. Then the sequence $\{T_i\}_{i \in \mathbb{N}}$ converges to T_3.

In the case of the strongest extension the situation is more complicated. Let $T \colon [0,1]^2 \longrightarrow [0,1]$ be a t-norm that coincides with T_1 on $[a,b]^2$. Then it is clear that for all $(x,y) \in [T_1(a,a), a]^2$ we have $T(x,y) \leq T_1(a,a)$. Further, $T(x,y) \leq T_1(a,y)$ for all $x \leq a \leq y$. If $T_1(a,b) = T_1(a,a)$ then the strongest t-norm that coincides with T_1 on $[a,b]^2$ is given by

$$T_2(x,y) = \begin{cases} \min(x,y) & \text{if } (x,y) \in [0,1]^2 \setminus [T_1(a,a), b]^2, \\ T_1(\max(a,x), \max(a,y)) & \text{otherwise.} \end{cases}$$

If T_1 on $[a,b]^2$ can be extended to a continuous t-norm then this t-norm has an additive generator s on $[T_1(a,a), b]$ and values of s on $[T_1(a,a), T_1(b,b)] \cup [a,b]$ determine the values of T_1 on $[a,b]^2$. Since $T_1(a,b) = T_1(a,a)$ we see that $T_1(a,a)$ is an idempotent element of such a t-norm and $s(a) + s(b) \geq 2 \cdot s(a)$. Thus if we define a sequence of t-norms T_i which are ordinal sums of one continuous, Archimedean summand on $[T_1(a,a), b + \varepsilon_i]$, where T_i is on $[T_1(a,a), b + \varepsilon_i]$ generated by respective additive generators t_i, where $t_i(x) = s(x)$ for $x \in [T_1(a,a), T_1(b,b)] \cup [a,b]$ and t_i is linear on $[T_1(b,b), a]$ and on $[b, b + \varepsilon_i]$, with $t_i(b + \varepsilon_i) = 0$, then $\{T_i\}_{i \in \mathbb{N}}$ converges to T_2.

Further, let $c = T_1(a,b) > T_1(a,a)$. Then $T(c,b) = T(a, T(b,b)) \leq T_1(a,a)$, and we get the following:

Lemma 6. *Let* $T_1 \colon [0,1]^2 \longrightarrow [0,1]$ *be a t-norm which is continuous on* $[a,b]^2$, $T_1(b,b) < a$, $T_1(a,b) > T_1(a,a)$. *Then each t-norm* T *which coincides with* T_1 *on* $[a,b]^2$ *is smaller than or equal to the binary function* $T^* \colon [0,1]^2 \longrightarrow [0,1]$ *given by*

$$T^*(x,y) = \begin{cases} \min(x,y) & \text{if } (x,y) \in [0,1]^2 \setminus [T_1(a,a), b]^2, \\ T_1(x,y) & \text{if } (x,y) \in [a,b]^2, \\ T_1(a,y) & \text{if } y \in [a,b], x \in \,]T_1(b,b), a[, \\ T_1(a,x) & \text{if } x \in [a,b], y \in \,]T_1(b,b), a[, \\ T_1(a,w) & \text{if } x = T_1(b,q), T_1(y,q) = T_1(b,w), q, w \in [a,b], \\ T_1(a,w) & \text{if } y = T_1(b,q), T_1(x,q) = T_1(b,w), q, w \in [a,b], \\ T_1(a,a) & \text{otherwise.} \end{cases}$$

The binary function T^* is commutative, non-decreasing and has 1 as a neutral element. However, it need not be associative.

The binary function T^* is associative only if the inequalities $T_1(x,y) = T_1(b,u)$, $T_1(u,z) = T_1(b,v)$, $T_1(x,z) = T_1(b,w)$, $T_1(w,y) = T_1(b,c)$, imply $v = c$ for all $x,y,z,u,v,w,c \in [a,b]$. This is satisfied, for example, if there exists a continuous t-norm which coincides with T_1 on $[a,b]^2$, which follows from the existence of an additive generator of such a t-norm on $[T_1(a,a),b]$. Thus in this case T^* is the strongest t-norm which coincides with T_1 on $[a,b]^2$.

Suppose that there exists a continuous t-norm that coincides with T_1 on $[a,b]^2$. Then similarly as above the values of the corresponding additive generator s on $[T_1(a,a),T_1(b,b)] \cup [a,b]$ are not determined uniquely by values of T_1 on $[a,b]^2$, but for $s(a) = 1$ we have $s(b) = w$ for $1 > w > \frac{1}{2}$. We select a sequence of such additive generators s_i, where for the respective parameter $s_i(b) = w_i$ we have $w_i = \frac{1+\varepsilon_i}{2}$. We now define a sequence of t-norms T_i which are ordinal sums of one continuous, Archimedean summand on $[T_1(a,a),b+\varepsilon_i]$, and T_i is on $[T_1(a,a),b+\varepsilon_i]$ generated by respective additive generators t_i such that $t_i(x) = s_i(x)$ for $x \in [T_1(a,a),T_1(b,b)] \cup [a,b]$, and t_i is linear on $[T_1(b,b),a]$ and on $[b,b+\varepsilon_i]$, with $t_i(b+\varepsilon_i) = 0$. Then the sequence $\{T_i\}_{i\in\mathbb{N}}$ converges to T_2.

4   Extremal Extensions of t-norms with a Non-trivial Idempotent Element in $[a,b]$

Assume that the t-norm T_1 is known only on $[a,b]^2$ and is continuous on $[a,b]^2$. In this section we will suppose that there is a non-trivial idempotent in $[a,b]$. First suppose that $a = b$, i.e., T_1 is known only in one point (a,a) and $T_1(a,a) = a$. Then it is evident that the strongest t-norm which satisfies $T(a,a) = a$ is the minimum t-norm. The weakest t-norm T_3 with $T_3(a,a) = a$ is given by

$$T_3(x,y) = \begin{cases} \min(x,y) & \text{if } \max(x,y) = 1, \\ a & \text{if } (x,y) \in [a,1]^2, \max(x,y) < 1, \\ 0 & \text{otherwise.} \end{cases}$$

However, T_3 cannot be obtained as a limit of a sequence of continuous t-norms T_i with $T_i(a,a) = a$ as for each such a continuous t-norm T_i there is $T_i(x,y) = x$ for all $x \le a \le y$.

From now on we will suppose that $a < b$. Let $q \in [a,b]$ be an idempotent point of T_1, i.e., $T_1(q,q) = q$. Then we have the following.

Lemma 7. *Let $T\colon [0,1]^2 \longrightarrow [0,1]$ be a t-norm which is continuous on $[a,b]^2$ and let $q \in [a,b]$ be an idempotent point of T_1. Then $T_1(x,y) = x$ for all $x,y \in [a,b]$, $x \le q \le y$.*

From the previous result we see that if any t-norm T coincides with T_1 on $[a,b]^2$ then for all $x \in [a,q]$ we have $T(x,z) = x$ for all $z \in [q,1]$. Recall that for every t-norm T we have $T(x,y) \le \min(x,y)$. Thus we get the following result.

Proposition 7. *Let* $T: [0,1]^2 \longrightarrow [0,1]$ *be a t-norm which is continuous on* $[a,b]^2$ *and let* $q_1, q_2 \in [a,b]$ *be respectively the smallest and the biggest idempotent point of* T_1 *in* $[a,b]$. *Then the strongest t-norm* T_2 *which coincides with* T_1 *on* $[a,b]^2$ *is equal to an ordinal sum* $T_2 = (\langle 0, q_1, T_2^1\rangle, \langle q_1, q_2, T_1^2\rangle, \langle q_2, 1, T_2^3\rangle)$. *Here* T_2^1 *is the strongest t-norm that coincides with the t-norm* T_1^1 *on* $\left[\frac{a}{q_1}, 1\right]$, *where* T_1^1 *is linearly isomorphic with* T_1 *on* $[0, q_1]^2$, *and* T_2^3 *is the strongest t-norm that coincides with the t-norm* T_1^3 *on* $\left[0, \frac{b-q_2}{1-q_2}\right]$, *where* T_1^3 *is linearly isomorphic with* T_1 *on* $[q_2, 1]^2$. *The t-norm* T_1^2 *is linearly isomorphic with* T_1 *on* $[q_1, q_2]^2$.

Since in the previous result q_1 and q_2 are the smallest and the biggest idempotent point in $[a,b]$ thus in $[a, q_1[$ and in $]q_2, b]$ there is no non-trivial idempotent point of T_1. Therefore T_2^1 and T_2^3 can be determined from the previous section. We get

$$T_2^3(x,y) = \begin{cases} T_1^3(x,y) & \text{if } (x,y) \in \left[0, \frac{b-q_2}{1-q_2}\right]^2, \\ \min(x,y) & \text{otherwise,} \end{cases}$$

and

$$T_2^1(x,y) = \begin{cases} T_1^1(\frac{a}{q_1}, T_1^1(y,z)) & \text{if } (x,y) \in A, x = T_1^1(z, \frac{a}{q_1}), \\ T_1^1(\frac{a}{q_1}, T_1^1(x,z)) & \text{if } (x,y) \in A, y = T_1^1(z, \frac{a}{q_1}), \\ \min(x,y) & \text{if } \min(x,y) \le \frac{T_1(a,a)}{q_1}, \\ \frac{T_1(a,a)}{q_1} & \text{otherwise,} \end{cases}$$

with $A = \{(x,y), (y,x) \in [0,1]^2 \mid \text{there exists } z \in [0,1], x = T_1^1(z, \frac{a}{q_1}), T_1^1(y,z) \ge \frac{a}{q_1}\}$.

As T_2^1 and T_2^2 are continuous and T_2^3 can be obtained as a limit of continuous t-norms that coincide on the corresponding interval, also T_2 can be obtained as a limit of a sequence of continuous t-norms that coincide with T_1 on $[a,b]^2$.

The monotonicity and the results obtained in previous section gives us for the weakest extension the following.

Proposition 8. *Let* $T_1: [0,1]^2 \longrightarrow [0,1]$ *be a t-norm which is continuous on* $[a,b]^2$ *and let* $q_1, q_2 \in [a,b]$ *be respectively the smallest and the biggest idempotent point of* T_1 *in* $[a,b]$. *Then the weakest t-norm* T_3 *which coincides with* T_1 *on* $[a,b]^2$ *is given by*

$$T_3(x,y) = \begin{cases} \min(x,y) & \text{if } \max(x,y) = 1, \\ \min(x,y) & \text{if } (x,y) \in [q_1, 1] \times [T_1(a,a), q_2] \setminus [q_1, q_2]^2, \\ \min(x,y) & \text{if } (y,x) \in [q_1, 1] \times [T_1(a,a), q_2] \setminus [q_1, q_2]^2, \\ T_1(x,y) & \text{if } (y,x) \in [q_1, q_2]^2, \\ T_1(\min(x,b), \min(y,b)) & \text{if } (x,y) \in [q_2, 1]^2 \\ T_1(a, T_1(z,y)) & \text{if } (x,y) \in A, x = T_1(a,z), \\ T_1(a, T_1(z,x)) & \text{if } (x,y) \in A, y = T_1(a,z), \\ 0 & \text{otherwise,} \end{cases}$$

with $A = \{(x,y),(y,x) \in [0,q_1]^2 \mid there\ exists\ z \in [0,q_1], x = T_1(a,z), T_1(z,y) \geq a\}$.

If $T_1(a,a) > 0$ then T_3 cannot be obtained as a limit of continuous t-norms that coincide with T_1 on $[a,b]^2$. This is fact that if q_1 is an idempotent element of a continuous t-norm T then $T(x,y) = x$ for all $x \leq q_1 \leq y$. However, we have $T_3(x,y) = 0$ for $x \in [0,T_1(a,a)]$ and $y \in [q_1,1]$. If $T_1(a,a) = 0$ then T_3 is similarly as T_2 an ordinal sum of three t-norms and the result can be composed from results of the previous section.

5 Conclusion

We have described the strongest and the weakest t-norms that coincide with the given t-norm T_1 on $[a,b]^2$. These results can be applied everywhere when we search for an extremal t-norm that coincides with the given values on some subinterval of the unit interval.

Acknowledgement. This work was supported by grant VEGA 2/0049/14 and Program Fellowship of SAS.

References

1. Alsina, C., Frank, M.J., Schweizer, B.: Associative Functions: Triangular Norms and Copulas. World Scientific, Singapore (2006)
2. Beliakov, G., Pradera, A., Calvo, T.: Aggregation Functions: A Guide for Practitioners. Springer-Verlag, New York (2007)
3. Clifford, A.H.: Naturally totally ordered commutative semigroups. Am. J. Math. **76**, 631–646 (1954)
4. Grabisch, M., Marichal, J.-L., Mesiar, R., Pap, E.: Aggregation Functions. Cambridge University Press, Cambridge (2009)
5. Hájek, P.: Metamathematics of fuzzy logic. Kluwer Academic Publishers, Dordrecht (1998)
6. Jenei, S.: A note on the ordinal sum theorem and its consequence for the construction of triangular norms. Fuzzy Sets Syst. **126**, 199–205 (2002)
7. Klement, E.P., Mesiar, R., Pap, E.: Triangular norms. Kluwer Academic Publishers, Dordrecht (2000)
8. Mesiar, R., Baets, B. D.: New construction methods for aggregation operators. In: IPMU 2000, pp. 701–706. Madrid (2000)
9. Mesiarová-Zemánková A.: Continuous completions of triangular norms known on a subregion of the unit interval, Fuzzy Sets and Systems. http://www.mat.savba.sk/zemankova/unpublished.htm
10. Schweizer, B., Sklar, A.: Probabilistic Metric Spaces. North-Holland, New York (1983)
11. Sugeno, M.: Industrial Applications of Fuzzy Control. Elsevier, New York (1985)

The Notion of Pre-aggregation Function

Giancarlo Lucca[1], José Antonio Sanz[2], Graçaliz Pereira Dimuro[1],
Benjamín Bedregal[3], Radko Mesiar[4,5], Anna Kolesárová[6],
and Humberto Bustince[2]([✉])

[1] Centro de Ciencias Computacionais, Universidade Federal do Rio Grande,
Rio Grande, Brazil
[2] Departamento of Automática y Computación and with the Institute
of Smart Cities, Universidad Publica de Navarra, 31006 Navarra, Spain
bustince@unavarra.es
[3] Departamento de Informática e Matemática Aplicada, Universidade Federal do Rio
Grande do Norte, Natal, Brazil
[4] Slovak University of Technology, Radlinskeho 11, Bratislava, Slovakia
[5] Institute of Information Theory and Automation, Academy of Sciences
of the Czech Republic, 18208 Prague, Czech Republic
[6] Institute of Information Engineering, Automation and Mathematics,
Slovak University of Technology, 810 05 Bratislava, Slovakia

Abstract. In this work we consider directional monotone functions and
use this idea to introduce the notion of pre-aggregation function. In particular, we propose an example of such functions inspired on Choquet
integrals.

Keywords: Pre-aggregation functions · Aggregation functions · Directional monotonicity · Fuzzy measures · Choquet integral

1 Introduction

Aggregation functions [1,2] are crucial tools nowadays to deal with many computation problems [3–9]. However, and since some very relevant operators such as
the mode are not monotone, in recent years there exists an increasing interest on
the relaxation of the monotonicity property in order to recover operators which
may be useful for applications. For instance, in [10], Wilkin and Beliakov proposed the notion of weak monotonicity, which amounts to consider monotonicity
along the fixed ray defined by the first quadrant diagonal. The consideration of
monotonicity along any other fixed ray led Bustince et al. [11] to introduce the
notion of directional monotonicity. Note that monotone functions in the usual
sense are both weakly monotone and directionally monotone.

In this work we propose to define pre-aggregation function, which are functions which satisfy the same boundary conditions as aggregation functions but
for which only directional monotonicity is considered. In particular, and for the
present work, we discuss a particular instance of these pre-aggregations, which

© Springer International Publishing Switzerland 2015
V. Torra and Y. Narakawa (Eds.): MDAI 2015, LNAI 9321, pp. 33–41, 2015.
DOI: 10.1007/978-3-319-23240-9_3

is obtained by replacing the product by an appropriate aggregation in the definition of the Choquet integral [12–15]. The usefulness of this approach has been made clear in recent works [16,17] for classification problems.

This paper is organized as follows. In Sect. 2, we present some related preliminary concepts that are necessary to understand the paper. In Sect. 3 we introduce the notion of pre-aggregation function, discussing some properties. A specific method of construction of pre-aggregation functions is described in Sect. 4. We finish with some conclusions and references.

2   Preliminaries

We recall here some relevant notions and definitions which are useful for subsequent developments.

Definition 1. *A function $A : [0,1]^n \to [0,1]$ is said to be an n-ary aggregation function if the following conditions hold:*

(A1) *A is increasing[1] in each argument: for each $i \in \{1, \ldots, n\}$, if $x_i \leq y$, then*
$A(x_1, \ldots, x_n) \leq A(x_1, \ldots, x_{i-1}, y, x_{i+1}, \ldots, x_n);$
(A2) *A satisfies the boundary conditions: $A(0, \ldots, 0) = 0$ and $A(1, \ldots, 1) = 1$.*

Definition 2. *A bivariate aggregation function $T : [0,1]^2 \to [0,1]$ is a t-norm if, for all $x, y, z \in [0,1]$, it satisfies the following properties:*

(T1) *Commutativity: $T(x,y) = T(y,x)$;*
(T2) *Associativity: $T(x, T(y, z)) = T(T(x,y), z)$;*
(T3) *Boundary condition: $T(x, 1) = x$.*

If T satisfies (T3) (and also (T(1,x) = x) only, then it is called a semi-copula.

Some important examples of t-norms are the following

1. Minimum $T_M(x,y) = \min\{x, y\}$
2. Algebraic Product $T_P(x,y) = xy$
3. Łukasiewickz $T_Ł(x,y) = \max\{0, x + y - 1\}$
4. Drastic Product $T_{DP}(x,y) = \begin{cases} x & \text{if } y = 1 \\ y & \text{if } x = 1 \\ 0 & \text{otherwise} \end{cases}$
5. Nilpotent Minimum $T_{NM}(x,y) = \begin{cases} \min\{x, y\} & \text{if } x + y > 1 \\ 0 & \text{otherwise} \end{cases}$
6. Hamacher Product $T_{HP}(x,y) = \begin{cases} 0 & \text{if } x = y = 0 \\ \frac{xy}{x+y-xy} & \text{otherwise} \end{cases}$

Let's consider now the set $N = \{1, \ldots, n\}$ for an arbitrary positive integer n.

---

[1] In this paper, an increasing (decreasing) function does not need to be strictly increasing (decreasing).

Definition 3. *A function* $\mathfrak{m} : 2^N \to [0,1]$ *is a fuzzy measure if, for all* $X, Y \subseteq N$, *it satisfies the following properties:*

(m1) *Increasing: if* $X \subseteq Y$, *then* $\mathfrak{m}(X) \leq \mathfrak{m}(Y)$;
(m2) *Boundary conditions:* $\mathfrak{m}(\emptyset) = 0$ *and* $\mathfrak{m}(N) = 1$.

Some examples of measures are the following

Uniform measure:

$$\mathfrak{m}_U(A) = \frac{|A|}{n}. \tag{1}$$

Dirac's measure: For a previously fixed $i \in N$, chosen according to a certain methodology,

$$\mathfrak{m}_D^i(A) = \begin{cases} 1 \text{ if } i \in A \\ 0 \text{ if } i \notin A. \end{cases} \tag{2}$$

Additive measure (Wmean):   Take $W = (w_1, \ldots, w_n) \in [0,1]^n$ such that $\sum_{i=1}^n w_i = 1$. Consider

$$\mathfrak{m}_W(\{i\}) = w_i$$

Then, for $|A| > 1$, define:

$$\mathfrak{m}_W(A) = \sum_{i \in A} w_i. \tag{3}$$

Symmetric measure (OWA):   Take $W = (w_1, \ldots, w_n) \in [0,1]^n$ such that $\sum_{i=1}^n w_i = 1$. Then, for any non-empty subset A, define:

$$\mathfrak{m}_{sW}(A) = \sum_{i=1}^{|A|} w_i. \tag{4}$$

Power measure:

$$\mathfrak{m}_{PM}(A) = \left(\frac{|A|}{n}\right)^q, \text{ with } q > 0. \tag{5}$$

Finally, we recall the definition of Choquet integral.

Definition 4. *[1, Definition 1.74] Let* $\mathfrak{m} : 2^N \to [0,1]$ *be a fuzzy measure. The discrete Choquet integral of* $x = (x_1, \ldots, x_n) \in [0,1]^n$ *with respect to* $\mathfrak{m}$ *is defined as a function* $C_\mathfrak{m} : [0,1]^n \to [0,1]$, *given by*

$$C_\mathfrak{m}(x) = \sum_{i=1}^n \left(x_{(i)} - x_{(i-1)}\right) \cdot \mathfrak{m}\left(A_{(i)}\right), \tag{6}$$

where $\left(x_{(1)}, \ldots, x_{(n)}\right)$ *is an increasing permutation on the input* x, *that is,* $0 \leq x_{(1)} \leq \ldots \leq x_{(n)}$, *with the convention that* $x_{(0)} = 0$, *and* $A_{(i)} = \{(i), \ldots, (n)\}$ *is the subset of indices of* $n - i + 1$ *largest components of* x.

Now we recall the notion of directional monotonicity [11].

Definition 5. *Let $r = (r_1, \ldots, r_n)$ be a real n-dimensional vector, $r \neq 0$. A function $F : [0,1]^n \to [0,1]$ is r-increasing if for all points $(x_1, \ldots, x_n) \in [0,1]^n$ and for all $c > 0$ such that $(x_1 + cr_1, \ldots, x_n + cr_n) \in [0,1]^n$ it holds*

$$F(x_1 + cr_1, \ldots, x_n + cr_n) \geq F(x_1, \ldots, x_n).$$

Example 1. – Fuzzy implication functions (see [18]) are $(-1,1)$-increasing functions.
  – Weakly increasing functions are a particular case of directionally increasing functions, with $r = (1, \ldots, 1)$.

3   Pre-aggregation Functions

In this section we introduce the notion of pre-aggregation function and discuss some properties and construction methods.

Definition 6. *A function $F : [0,1]^n \to [0,1]$ is said to be an n-ary pre-aggregation function if the following conditions hold:*

  (PA1) *There exists a real vector $r \in [0,1]^n$ $(r \neq 0)$ such that F is r-increasing.*
  (PA2) *F satisfies the boundary conditions: $F(0, \ldots, 0) = 0$ and $F(1, \ldots, 1) = 1$.*

Example 2. Some examples of pre-aggregation functions are the following.

  (i) Consider the mode, $Mod(x_1, \ldots, x_n)$, defined as the function that gives back the value which appears most times in the considered n-tuple, or the smallest of the values that appears most times, in case there is more than one. Then, the mode is $(1, \ldots, 1)$-increasing, and it is a particular case of pre-aggregation function.
  (ii) $F(x, y) = x - (\max\{0, x - y\})^2$ is, for instance, $(0,1)$-increasing, and then it is an example of pre-aggregation function.

If F is a pre-aggregation function with respect to a vector r we will just say that F is an r-pre-aggregation function.

Remark 1. Note that if $A : [0,1]^n \to [0,1]$ is an aggregation function, then A is also a pre-aggregation function.

We can use aggregation functions to obtain directionally increasing functions as follows.

Proposition 1. *Let $A : [0,1]^m \to [0,1]$ be an aggregation function. Let $F_i : [0,1]^n \to [0,1]$ $(i \in \{1, \ldots, m\})$ be a family of m r-pre-aggregation functions for the same vector $r \in [0,1]^n$. Then, the function $A(F_1, \ldots, F_m) : [0,1]^n \to [0,1]$, defined as*

$$A(F_1, \ldots, F_m)(x_1, \ldots, x_n) = A(F_1(x_1, \ldots, x_n), \ldots, F_m(x_1, \ldots, x_n))$$

is also an r-pre-aggregation function.

Proof. Assume that $r = (r_1, \ldots, r_n)$. Then, from the r-increasingness, it follows that, for every $(x_1, \ldots, x_n) \in [0, 1]^n$ and $c > 0$ such that $(x_1 + cr_1, \ldots, x_n + cr_n) \in [0, 1]^n$,

$$F_i(x_1 + cr_1, \ldots, x_n + cr_n) \geq F(x_1, \ldots, x_n)$$

for every $i \in \{1, \ldots, m\}$. Since any aggregation function is increasing, the result follows.

The following result is straightforward.

Proposition 2. *Let $F_1, F_2 : [0, 1]^n \to [0, 1]$ be two r-pre-aggregation functions for the same vector $r \in [0, 1]^n$. Then:*

(i) $\frac{F_1 + F_2}{2}$ is also an r-pre-aggregation function.
(ii) $F_1 F_2$ is also an r-pre-aggregation function.

Regarding duality, we can state the following.

Proposition 3. *Let $F : [0, 1]^n \to [0, 1]$ be an r-pre-aggregation function for $r \in [0, 1]^n$. Then, the function*

$$F^d(x_1, \ldots, x_n) = 1 - F(1 - x_1, \ldots, 1 - x_n)$$

is also an r-pre-aggregation function.

Proof. Let $r = (r_1, \ldots, r_n)$. Take $(x_1, \ldots, x_n) \in [0, 1]^n$ and $c > 0$ such that $(x_1 + cr_1, \ldots, x_n + cr_n) \in [0, 1]^n$. Then it holds that

$$F^d(x_1 + cr_1, \ldots, x_n + cr_n) = 1 - F(1 - x_1 - cr_1, \ldots, 1 - x_n - cr_n)$$
$$\geq 1 - F(1 - x_1, \ldots, 1 - x_n) = F^d(x_1, \ldots, x_n)$$

so the result holds.

4   A Way to Build Pre-aggregation Functions

In this section, we introduce a method to build pre-aggregation functions. This method is inspired in the way the Choquet integral is built, replacing the product operation in Eq. (6) by other aggregation functions.

Let $\mathfrak{m} : 2^N \to [0, 1]$ be a fuzzy measure and $M : [0, 1]^2 \to [0, 1]$ be a function. Taking as basis the Choquet integral, we define the function $C_\mathfrak{m}^M : [0, 1]^n \to [0, n]$ by

$$C_\mathfrak{m}^M(x) = \sum_{i=1}^n M\left(x_{(i)} - x_{(i-1)}, \mathfrak{m}\left(A_{(i)}\right)\right), \tag{7}$$

where $N = \{1, \ldots, n\}$, $(x_{(1)}, \ldots, x_{(n)})$ is an increasing permutation on the input x, that is, $0 \leq x_{(1)} \leq \cdots \leq x_{(n)}$, with the convention that $x_{(0)} = 0$, and $A_{(i)} = \{(i), \ldots, (n)\}$ is the subset of indices of $n - i + 1$ largest components of x.

First of all, we have the following results.

Proposition 4. *Let* $M : [0,1]^2 \to [0,1]$ *be a function such that* $M(x,y) \le x$ *for every* $x, y \in [0,1]$. *Then*

$$C_{\mathfrak{m}}^M(x_1, \ldots, x_n) \le \max(x_1, \ldots, x_n)$$

for every $x_1, \ldots, x_n \in [0,1]$.

Proof. Note that

$$C_{\mathfrak{m}}^M(x_1, \ldots, x_n) = \sum_{i=1}^n M\left(x_{(i)} - x_{(i-1)}, \mathfrak{m}\left(A_{(i)}\right)\right)$$

$$\le \sum_{i=1}^n (x_{(i)} - x_{(i-1)}) = x_{(n)} = \max(x_1, \ldots, x_n)$$

so the result follows.

Proposition 5. *Let* $M : [0,1]^2 \to [0,1]$ *be a function such that* $M(x,1) = x$ *for every* $x \in [0,1]$. *Then*

$$C_{\mathfrak{m}}^M(x_1, \ldots, x_n) \ge \min(x_1, \ldots, x_n)$$

for every $x_1, \ldots, x_n \in [0,1]$.

Proof. Note that

$$\min(x_1, \ldots, x_n) = x_{(1)} = M\left(x_{(1)} - x_{(0)}, \mathfrak{m}\left(A_{(1)}\right)\right) \le C_{\mathfrak{m}}^M(x_1, \ldots, x_n)$$

so the result follows.

Proposition 6. *Assume that* $M : [0,1]^2 \to [0,1]$ *is a function such that* $M(x,1) = x$ *and* $M(0,y) = 0$ *for every* $x, y \in [0,1]$. *Then, the function* $C_{\mathfrak{m}}^M$ *is idempotent. That is,* $C_{\mathfrak{m}}^M(x, \ldots, x) = x$ *for every* $x \in [0,1]$.

Proof. For every $(x, \ldots, x) \in [0,1]^n$ it holds that

$$C_{\mathfrak{m}}^M(x, \ldots, x) = M(x,1) + \sum_{i=2}^n M(0, m(A_{(i)})) = x .$$

Proposition 7. *For any function* $M : [0,1]^2 \to [0,1]$ *such that* $M(0,1) = 0$, $M(1,1) = 1$ *and which is* $(1,0)$-*increasing,* $C_{\mathfrak{m}}^M$ *is* $(1, \ldots, 1)$-*increasing.*

Proof. Take as $r = (1, \ldots, 1)$. Note that in Eq. (7), for $i \ge 2$, it follows that, for any $c > 0$

$$M\left(x_{(i)} + c - (x_{(i-1)} + c), \mathfrak{m}\left(A_{(i)}\right)\right) = M\left(x_{(i)} - x_{(i-1)}, \mathfrak{m}\left(A_{(i)}\right)\right)$$

whereas, for $i = 1$

$$M\left(x_{(1)} + c - x_{(0)}, \mathfrak{m}\left(A_{(1)}\right)\right) = M\left(x_{(1)} + c, \mathfrak{m}\left(A_{(1)}\right)\right) \ge M\left(x_{(1)}, \mathfrak{m}\left(A_{(1)}\right)\right)$$

so $C_{\mathfrak{m}}^M$ is r-increasing.

Now the following result is straight.

Theorem 1. *Let $M : [0,1]^2 \to [0,1]$ be a function such that for all $x, y \in [0,1]$ it satisfies $M(x,y) \leq x$, $M(x,1) = x$, $M(0,y) = 0$ and M is (1,0)-increasing. Then, for any fuzzy measure $\mathfrak{m}$, $C_\mathfrak{m}^M$ is a pre-aggregation function which is idempotent and averaging, i.e., $\min(x_1, \ldots, x_n) \leq C_\mathfrak{m}^M(x_1, \ldots, x_n) \leq \max(x_1, \ldots, x_n)$.*

Proof. It follows from Propositions 4, 5 and 7.

Taking into account that a semi-copula is an aggregation function M such that $M(1,x) = M(x,1) = x$ for every $x \in [0,1]$, we have the following corollary.

Corollary 1. *Let $M : [0,1]^2 \to [0,1]$ be a function. If M is a semi-copula, then, for any measure $\mathfrak{m}$, $C_\mathfrak{m}^M$ is a pre-aggregation function which is idempotent and averaging.*

Remark 2. Under the constraints of Theorem 1, we cannot ensure the monotonicity of $C_\mathfrak{m}^M$, i.e., $C_\mathfrak{m}^M$ is, in general, a proper pre-aggregation function. To see it, observe the following:

(i) Take $M(x,y) = T_M(x,y)$. Consider $N = \{1,2,3,4\}$ and the uniform measure $\mathfrak{m} = \mathfrak{m}_U$ given in Eq. (1). Then, we have that

$$C_\mathfrak{m}^{T_M}(0.05, 0.1, 0.7, 0.9) = 0.8\,,$$

whereas

$$C_\mathfrak{m}^{T_M}(0.05, 0.1, 0.8, 0.9) = 0.7\,,$$

so $C_\mathfrak{m}^{T_M}$ is not an increasing function and hence it is not an aggregation function.

(ii) Consider the Łukasiewicz t-norm $T_L(x,y) = \max\{0, x+y-1\}$. Again, for $N = \{1,2,3,4\}$ and the uniform measure $\mathfrak{m} = \mathfrak{m}_U$ we have that

$$C_\mathfrak{m}^{T_L}(0.05, 0.1, 0.7, 0.9) = 0.15\,,$$

whereas

$$C_\mathfrak{m}^{T_L}(0.05, 0.2, 0.7, 0.9) = 0.05\,,$$

so $C_\mathfrak{m}^{T_L}$ is not an increasing function and hence it is not an aggregation function.

(iii) Regarding the drastic product $T_{DP}(x,y) = \min\{x,y\}$ if $\max\{x,y\} = 1$ and 0 in other case, consider $N = \{1,2,3\}$ and the uniform measure $\mathfrak{m} = \mathfrak{m}_U$. Then

$$C_\mathfrak{m}^{T_{DP}}(0, 0, 1) = 0.33\,,$$

whereas

$$C_\mathfrak{m}^{T_{NM}}(0, 0.5, 1) = 0\,,$$

so $C_\mathfrak{m}^{T_{NM}}$ is not an increasing function and hence it is not an aggregation function.

(iv) Consider the nilpotent minimum t-norm $T_{NM}(x,y) = \min\{x,y\}$ if $x+y > 1$ and 0 in other case. Again, for $N = \{1,2,3,4\}$ and the uniform measure $\mathfrak{m} = \mathfrak{m}_U$ we have that

$$C_{\mathfrak{m}}^{T_{NM}}(0.05, 0.1, 0.7, 0.9) = 0.55 \,,$$

whereas

$$C_{\mathfrak{m}}^{T_{NM}}(0.05, 0.2, 0.7, 0.9) = 0.5 \,,$$

so $C_{\mathfrak{m}}^{T_{NM}}$ is not an increasing function and hence it is not an aggregation function.

(v) Consider the Hamacher product $T_{HP}(x,y) = \frac{xy}{x+y-xy}$ if $(x,y) \neq (0,0)$ and $T(0,0) = 0$. If we consider again $N = \{1,2,3,4\}$ and the uniform measure $\mathfrak{m} = \mathfrak{m}_U$ given in Eq. (1), we see that

$$C_{\mathfrak{m}}^{T_{HP}}(0.05, 0.1, 0.7, 0.9) = 0.5991 \,,$$

whereas

$$C_{\mathfrak{m}}^{T_{HP}}(0.05, 0.1, 0.8, 0.9) = 0.5877 \,,$$

so $C_{\mathfrak{m}}^{T_{HP}}$ is not an increasing function and hence it is not an aggregation function.

5   Conclusion

In this paper, based on the notion of an aggregation function, we have introduced the concept of a pre-aggregation function. We have described a construction method for such functions from the Choquet integral by using other t-norms in the place of the product t-norm considered in the standard definition of the Choquet integral.

Although due to the lack of space it has not been possible to include a detailed experimental study, it has been shown in [16] that the use of Choquet-like pre-aggregation functions allows to improve in some situations the behaviour of methods such as FARC-HD. In this sense, we intend in the future to study in a more general way the construction of pre-aggregation functions with an eye kept on its use for specific problems.

Acknowledgment. This work is supported by CNPq (Proc. 305131/10-9, 481283/ 2013-7, 306970/2013-9), NSF (ref. TIN2013-40765-P, TIN2011-29520) and grant APVV-14-0013.

References

1. Beliakov, G., Pradera, A., Calvo, T.: Aggregation Functions: A Guide for Practitioners. Springer, Berlin Heidelberg (2007)
2. Grabisch, M., Marichal, J., Mesiar, R., Pap, E.: Aggregation Functions. Cambridge University Press, Cambridge (2009)

3. Amo, A., Montero, J., Biging, G., Cutello, V.: Fuzzy classification systems. Eur. J. Oper. Res. **156**(2), 495–507 (2004)
4. Bustince, H., Barrenechea, E., Calvo, T., James, S., Beliakov, G.: Consensus in multi-expert decision making problems using penalty functions defined over a cartesian product of lattices. Inf. Fusion **17**, 56–64 (2014)
5. Bustince, H., Pagola, M., Mesiar, R., Hüllermeier, E., Herrera, F.: Grouping, overlaps, and generalized bientropic functions for fuzzy modeling of pairwise comparisons. IEEE Trans. Fuzzy Syst. **20**(3), 405–415 (2012)
6. Bustince, H., Montero, J., Mesiar, R.: Migrativity of aggregation operators. Fuzzy Sets Syst. **160**(6), 766–777 (2009)
7. Bustince, H., Fernandez, J., Mesiar, R., Montero, J., Orduna, R.: Overlap functions. Nonlinear Anal. **72**(3–4), 1488–1499 (2010)
8. Alcala-Fdez, J., Alcala, R., Herrera, F.: A fuzzy association rule-based classification model for high-dimensional problems with genetic rule selection and lateral tuning. IEEE Trans. Fuzzy Syst. **19**(5), 857–872 (2011)
9. Jurio, A., Bustince, H., Pagola, M., Pradera, A., Yager, R.: Some properties of overlap and grouping functions and their application to image thresholding. Fuzzy Sets Syst. **229**, 69–90 (2013)
10. Wilkin, T., Beliakov, G.: Weakly monotone aggregation functions. Int. J. Intell. Sys. **30**, 144–169 (2015)
11. Bustince, H., Kolesárová, A., Fernandez, J., Mesiar, R.: Directional monotonicity of fusion functions. European Journal of Operational Research, in press.doi.10.1016/j.ejor.2015.01.018
12. Choquet, G.: Theory of capacitie. Ann. de l'Institut Fourier **5**, 131–295 (1953-1954)
13. Denneberg, D.: Representation of the choquet integral with the 6-additive Mbius transform. Fuzzy Sets Sys. **92**(2), 139–156 (1997)
14. Grabisch, M., Labreuche, C.: A decade of application of the choquet and sugeno integrals in multi-criteria decision aid. Ann. Oper. Res. **175**(1), 247–286 (2010)
15. Gilboa, I., Schmeidler, D.: Additive representations of non-additive measures and the choquet integral. Ann. Oper. Res. **52**(1), 43–65 (1994)
16. Lucca, G., Sanz, J., Dimuro, G. P., Bedregal, B., Mesiar, R., Kolesárová, A., Bustince, H.: Pre-aggregation functions: construction and an application. IEEE Transactions on Fuzzy Systems, accepted for publication
17. Barrenechea, E., Bustince, H., Fernandez, J., Paternain, D., Sanz, J.A.: Using the choquet integral in the fuzzy reasoning method of fuzzy rule-based classification systems. Axioms **2**(2), 208–223 (2013)
18. Bustince, H., Burillo, P., Soria, F.: Automorphisms, negations and implication operators. Fuzzy Sets Syst. **134**(2), 209–229 (2013)

Weighted Quasi-Arithmetic Mean on Two-Dimensional Regions and Their Applications

Yuji Yoshida[✉]

Faculty of Economics and Business Administration, University of Kitakyushu,
4-2-1 Kitagata, Kokuraminami, Kitakyushu 802-8577, Japan
yoshida@kitakyu-u.ac.jp

Abstract. This paper discusses a decision maker's attitude regarding risks, for example risk neutral, risk averse and risk loving in micro-economics by the convexity and concavity of utility functions. Weighted quasi-arithmetic means on two-dimensional regions are introduced, and some conditions on utility functions are discussed to characterize the decision maker's attitude. Risk premiums on two-dimensional regions are given and demonstrated. Some approaches to construct two-dimensional utilities from one-dimensional ones are given, and a lot of examples of weighted quasi-arithmetic means are shown.

1 Introduction

Weighted quasi-arithmetic means are important tools for subjective estimation of data in decision making such as management, artificial intelligence and so on, and they are also strongly related to utility functions in micro-economics (Fishburn [3]). Yoshida [9–11] has studied weighted quasi-arithmetic means of an interval by weighted aggregation operations where Kolmogorov [6] and Nagumo [7] studied the aggregation operators and Aczél [1] developed the theory regarding weighted aggregation. Yoshida [11] has discussed the relations between weighted quasi-arithmetic means on an interval and decision maker's attitude regarding risks.

For a continuous strictly increasing function $\varphi : [a, b] \mapsto (-\infty, \infty)$ as a decision maker's *utility function* and a continuous function $\omega : [a, b] \mapsto (0, \infty)$ as a *weighting function*, a *weighted quasi-arithmetic mean* on a closed interval $[a, b]$ is defined by

$$\varphi^{-1} \left(\int_a^b \varphi(x)\, \omega(x)\, dx \bigg/ \int_a^b \omega(x)\, dx \right). \tag{1.1}$$

Equation (1.1) is mathematically a *mean value* given by a real number $\mu (\in [a, b])$ satisfying

$$\varphi(\mu) \int_a^b \omega(x)\, dx = \int_a^b \varphi(x)\, \omega(x)\, dx \tag{1.2}$$

© Springer International Publishing Switzerland 2015
V. Torra and Y. Narukawa (Eds.): MDAI 2015, LNAI 9321, pp. 42–53, 2015.
DOI: 10.1007/978-3-319-23240-9_4

in the *mean value theorem for integration*. On the other hand, since φ is continuous and strictly increasing, decision maker's risk averse attitude is described as the following condition:

$$\varphi(E(X)) \geq E(\varphi(X)) \tag{1.3}$$

for all real valued random variables X, where $E(\cdot)$ denotes the expectation with some probability measure. Its equivalent representation with density function and normalization is

$$\varphi(\nu) \int_a^b \omega(x)\, dx \geq \int_a^b \varphi(x)\, \omega(x)\, dx \tag{1.4}$$

for all weighting functions ω, where $\nu (\in [a, b])$ is the risk neutral mean defined by

$$\nu = \int_a^b x\, \omega(x)\, dx \bigg/ \int_a^b \omega(x)\, dx. \tag{1.5}$$

From (1.2) and (1.4), decision maker's risk averse attitude is represented by $\mu \leq \nu$. On the other hand, (1.3) implies the concavity of the function φ. Therefore the following correspondence between the concavity of the function φ and weighted quasi-arithmetic means μ and ν holds [11]:

$$\varphi'' \leq 0 \iff \mu \leq \nu. \tag{1.6}$$

In this paper, we investigate weighted quasi-arithmetic means on two-dimensional regions, and we discuss whether this kind of relation (1.6) still holds or does not on two-dimensional regions.

In Sect. 2 we discuss a *decision maker's attitude regarding risks*, for example *risk neutral*, *risk averse* and *risk loving* in micro-economics by the convexity and concavity of utility functions. In Sect. 3 we introduce weighted quasi-arithmetic means on two-dimensional regions, and we discuss conditions on utility functions to characterize the decision maker's attitude. In Sect. 4, we demonstrate risk premiums on two-dimensional regions, which is one of important concepts for risk management in economics. In Sect. 5 we give a few approaches to construct two-dimensional utilities from one-dimensional utilities, and we show a lot of examples of weighted quasi-arithmetic means.

2   Risk Neutral, Risk Averse and Risk Loving

In this section, we discuss the convexity and concavity of utility functions on two-dimensional regions to characterize the decision maker's attitude regarding risks.

Let a two-dimensional space $\mathbb{R}^2 = (-\infty, \infty)^2$ and let a domain D be a non-empty open convex subset of $\mathbb{R}^2$. Let a utility f be a twice continuously differentiable (C^2-class) function on D which is strictly increasing, i.e. $f_x(x, y) > 0$ and $f_y(x, y) > 0$ for $(x, y) \in D$. We introduce concepts about a decision maker's attitude regarding risks with his utility f. Let (Ω, P) be a probability space, where Ω is a non-empty sample space and P is a non-atomic probability measure on Ω. Let $\mathcal{X}$ be a family of all real valued random variables $X : \Omega \mapsto \mathbb{R}$. A pair of random vectors is called a random vector in this paper. Let $\mathcal{X}(D) = \{\text{random vectors } (X, Y) : \Omega \mapsto D, \ X, Y \in \mathcal{X}\}$.

Definition 2.1 (Risk and decision making, [2–5]).

(i) Decision making with a utility function $f : D \mapsto \mathbb{R}$ is called *risk neutral* if

$$f(E(X), E(Y)) = E(f(X, Y)) \qquad (2.1)$$

for all random vectors $(X, Y) \in \mathcal{X}(D)$.

(ii) Decision making with a utility function $f : D \mapsto \mathbb{R}$ is called *risk averse* if

$$f(E(X), E(Y)) \geq E(f(X, Y)) \qquad (2.2)$$

for all random vectors $(X, Y) \in \mathcal{X}(D)$.

(iii) Decision making with a utility function $f : D \mapsto \mathbb{R}$ is called *risk loving* if

$$f(E(X), E(Y)) \leq E(f(X, Y)) \qquad (2.3)$$

for all random vectors $(X, Y) \in \mathcal{X}(D)$.

These decision maker's attitudes are related to the concavity and the convexity of his utility function f [9–11]. Hence we introduce the definitions of the concavity and the convexity of utility functions on two-dimensional regions [8].

Definition 2.2 (Concavity and convexity).

(i) A function $f : D \mapsto \mathbb{R}$ is called *concave* if

$$f((1 - \theta)x_1 + \theta x_2, (1 - \theta)y_1 + \theta y_2) \geq (1 - \theta)f(x_1, y_1) + \theta f(x_2, y_2) \quad (2.4)$$

for all $(x_1, y_1), (x_2, y_2) \in D$ and all real numbers θ satisfying $0 \leq \theta \leq 1$.

(ii) A function $f : D \mapsto \mathbb{R}$ is called *strictly concave* if

$$f((1 - \theta)x_1 + \theta x_2, (1 - \theta)y_1 + \theta y_2) > (1 - \theta)f(x_1, y_1) + \theta f(x_2, y_2) \quad (2.5)$$

for all $(x_1, y_1), (x_2, y_2) \in D$ satisfying $(x_1, y_1) \neq (x_2, y_2)$ and all real numbers θ satisfying $0 < \theta < 1$.

(iii) A function $f : D \mapsto \mathbb{R}$ is called *convex* if

$$f((1 - \theta)x_1 + \theta x_2, (1 - \theta)y_1 + \theta y_2) \leq (1 - \theta)f(x_1, y_1) + \theta f(x_2, y_2) \quad (2.6)$$

for all $(x_1, y_1), (x_2, y_2) \in D$ and all real numbers θ satisfying $0 \leq \theta \leq 1$.

(iv) A function $f : D \mapsto \mathbb{R}$ is called *strictly convex* if

$$f((1 - \theta)x_1 + \theta x_2, (1 - \theta)y_1 + \theta y_2) < (1 - \theta)f(x_1, y_1) + \theta f(x_2, y_2) \quad (2.7)$$

for all $(x_1, y_1), (x_2, y_2) \in D$ satisfying $(x_1, y_1) \neq (x_2, y_2)$ and all real numbers θ satisfying $0 < \theta < 1$.

The concavity and the convexity of utility functions are characterized with differentials as follows (Rockafellar [8]).

Lemma 2.1 *Let a utility $f : D \mapsto \mathbb{R}$ be a C^2-class function on D such that $f_x >$ 0 and $f_y > 0$ on D.*

(i) *The following (a)–(c) are equivalent:*
   (a) *f is concave (strictly concave respectively).*
   (b) *Its Hessian matrix*

$$H = \begin{pmatrix} f_{xx} & f_{xy} \\ f_{yx} & f_{yy} \end{pmatrix}$$

   is negative semi-definite (negative definite).
   (c) *f satisfies*

$$f_{xx} \leq 0, \ f_{yy} \leq 0 \ and \ |H| = f_{xx}f_{yy} - f_{xy}^2 \geq 0 \tag{2.8}$$
$$(f_{xx} < 0, \ f_{yy} < 0 \ and \ |H| = f_{xx}f_{yy} - f_{xy}^2 > 0) \tag{2.9}$$

   on D.
(ii) *The following (a')–(c') are equivalent:*
   (a') *f is convex (strictly convex respectively).*
   (b') *Its Hessian matrix*

$$H = \begin{pmatrix} f_{xx} & f_{xy} \\ f_{yx} & f_{yy} \end{pmatrix}$$

   is positive semi-definite (positive definite).
   (c') *f satisfies*

$$f_{xx} \geq 0, \ f_{yy} \geq 0 \ and \ |H| = f_{xx}f_{yy} - f_{xy}^2 \geq 0 \tag{2.10}$$
$$(f_{xx} > 0, \ f_{yy} > 0 \ and \ |H| = f_{xx}f_{yy} - f_{xy}^2 > 0) \tag{2.11}$$

   on D.

By Jensen's inequality, we obtain the following lemma from Definitions 2.1 and 2.2.

Lemma 2.2

(i) *If a utility function $f : D \mapsto \mathbb{R}$ is linear, i.e. $f(x,y) = \alpha x + \beta y + \gamma$ for $(x,y) \in D$ where real constants α, β, γ satisfying $\alpha > 0$ and $\beta > 0$, then decision making with the utility f is risk neutral.*
(ii) *If a utility function $f : D \mapsto \mathbb{R}$ is concave, then decision making with the utility f is risk averse.*
(iii) *If a utility function $f : D \mapsto \mathbb{R}$ is convex, then decision making with the utility f is risk loving.*

3   Weighted Quasi-Arithmetic Means on Two-Dimensional Regions

In this section, we introduce weighted quasi-arithmetic means on two-dimensional regions, and we discuss conditions on the decision maker's utility functions to characterize his attitude for risks based on the weighted quasi-arithmetic means. Let a domain D be a non-empty open convex subset of $\mathbb{R}^2$, and let a utility f be a C^2-class strictly increasing function on D. Now we take a weighting w as a density function of a random vector (X, Y) in Definition 2.1, where we assume w is a once continuously differentiable (C^1-class) positive valued function on D. Denote a family of rectangle regions by $\mathcal{R}(D) = \{R = I \times J \mid I \text{ and } J \text{ are bounded closed intervals and } R \subset D\}$. For a rectangle region $R \in \mathcal{R}(D)$, *weighted quasi-arithmetic means* on region R with utility f and weighting w are given by a subset $M_w^f(R)$ of R as follows.

$$M_w^f(R) = \left\{ (\tilde{x}, \tilde{y}) \in R \mid f(\tilde{x}, \tilde{y}) \iint_R w(x, y)\, dx\, dy = \iint_R f(x, y)w(x, y)\, dx\, dy \right\}. \quad (3.1)$$

Then we have $M_w^f(R) \neq \emptyset$ since f is continuous on R and

$$\min_{(\tilde{x}, \tilde{y}) \in R} f(\tilde{x}, \tilde{y}) \leq \iint_R f(x, y)w(x, y)\, dx\, dy \bigg/ \iint_R w(x, y)\, dx\, dy \leq \max_{(\tilde{x}, \tilde{y}) \in R} f(\tilde{x}, \tilde{y}).$$

Since $f_x > 0$ and $f_y > 0$ on R, there exists an implicit function ϕ which satisfies an equation

$$f(\tilde{x}, \phi(\tilde{x})) \iint_R w(x, y)\, dx\, dy = \iint_R f(x, y)w(x, y)\, dx\, dy, \quad (3.2)$$

and then ϕ is strictly decreasing since $\phi' = -\frac{f_x}{f_y} < 0$ on $M_w^f(R)$. Thus the set of weighted quasi-arithmetic means $M_w^f(R) = \{(x, y) \in R \mid y = \phi(x)\}$ becomes a continuous strictly decreasing curve segment on R. This curve is called *indifference curve* for utility function f in economics.

Lemma 3.1 *Let a rectangle region $R \in \mathcal{R}(D)$, and let a utility f be a C^2-class strictly increasing function on D. Let ϕ be an implicit function for (3.2). Then the following (i) and (ii) hold:*

(i) *If f is concave (strictly concave), then its implicit function ϕ is convex i.e. $\phi'' \geq 0$ (strictly convex i.e. $\phi'' > 0$ respectively).*

(ii) *If f is convex (strictly convex), then its implicit function ϕ is concave i.e. $\phi'' \leq 0$ (strictly concave i.e. $\phi'' < 0$ respectively).*

From Definition 2.1 we introduce the following concept depending on a rectangle region and a weighting function.

Definition 3.1 Let a rectangle region $R \in \mathcal{R}(D)$. Let a utility function $f : D \mapsto \mathbb{R}$ and let a weighting function $w : D \mapsto (0, \infty)$.

(i) Decision making with a utility f is called *risk neutral on R with weighting w* if

$$f(\overline{x}_R, \overline{y}_R) \iint_R w(x,y)\, dx\, dy = \iint_R f(x,y) w(x,y)\, dx\, dy, \qquad (3.3)$$

where we define a point $(\overline{x}_R, \overline{y}_R)$ on R by weighted means

$$\overline{x}_R = \iint_R x\, w(x,y)\, dx\, dy \Big/ \iint_R w(x,y)\, dx\, dy, \qquad (3.4)$$

$$\overline{y}_R = \iint_R y\, w(x,y)\, dx\, dy \Big/ \iint_R w(x,y)\, dx\, dy. \qquad (3.5)$$

(ii) Decision making with a utility f is called *risk averse on R with weighting w* if

$$f(\overline{x}_R, \overline{y}_R) \iint_R w(x,y)\, dx\, dy \geq \iint_R f(x,y) w(x,y)\, dx\, dy. \qquad (3.6)$$

(iii) Decision making with a utility f is called *risk loving on R with weighting w* if

$$f(\overline{x}_R, \overline{y}_R) \iint_R w(x,y)\, dx\, dy \leq \iint_R f(x,y) w(x,y)\, dx\, dy. \qquad (3.7)$$

Hence we investigate the weighted means (3.4) and (3.5) in Definition 3.1. Let a rectangle region $R \in \mathcal{R}(D)$. Let a weighting function $w : D \mapsto (0, \infty)$. From Lemma 2.2(i), we can give a risk neutral utility function $g : D \mapsto \mathbb{R}$ by a linear function: $g(x,y) = \alpha x + \beta y + \gamma$ for $(x,y) \in D$ with real constants α, β, γ satisfying $\alpha > 0$ and $\beta > 0$. Then its weighted quasi-arithmetic means are reduced to

$$M_w^g(R) = \{(x,y) \in R \mid \alpha(x - \overline{x}_R) + \beta(y - \overline{y}_R) = 0\}, \qquad (3.8)$$

where $(\overline{x}_R, \overline{y}_R)$ is defined by (3.4) and (3.5). In (3.8), it holds clearly that $(\overline{x}_R, \overline{y}_R) \in M_w^g(R)$ for linear risk neutral utility functions g with any real parameters α, β, γ satisfying $\alpha > 0$ and $\beta > 0$. Therefore $(\overline{x}_R, \overline{y}_R)$ is called an *invariant risk neutral point on R with weighting w*.

Example 3.1 Fix a domain $D = \mathbb{R}^2$ and a rectangle region $R = [0,1]^2$ and fix a weighting function $w = 1$. Then the invariant risk neutral point is $(\overline{x}_R, \overline{y}_R) = (\frac{1}{2}, \frac{1}{2})$. Hence we investigate the following three cases.

(i) (Strictly concave utility f). Take a utility function f as

$$f(x,y) = -x^2 - y^2 + 3x + 3y \qquad (3.9)$$

for $(x,y) \in \mathbb{R}^2$. Then $f_x > 0$ and $f_x > 0$ on R. We can easily check

$$f(\overline{x}_R, \overline{y}_R) \iint_R dx\, dy = \iint_R f(x,y)\, dx\, dy + \frac{1}{6}.$$

Thus the utility f is risk averse on R with weighting w. Its Hessian matrix is

$$H = \begin{pmatrix} f_{xx}(x,y) \; f_{xy}(x,y) \\ f_{xy}(x,y) \; f_{yy}(x,y) \end{pmatrix} = \begin{pmatrix} -2 \;\; 0 \\ 0 \; -2 \end{pmatrix}$$

and the determinant is $|H| = 4 > 0$. Therefore the utility f is strictly concave.

(ii) (Non-concave utility f). Take a utility function f as

$$f(x,y) = x^2 - 2y^2 + x + 5y \tag{3.10}$$

for $(x,y) \in \mathbb{R}^2$. Then $f_x > 0$ and $f_x > 0$ on R. We can easily check

$$f(\overline{x}_R, \overline{y}_R) \iint_R dx \, dy = \iint_R f(x,y) \, dx \, dy + \frac{1}{12}.$$

Thus the utility f is risk averse on R with weighting w. Its Hessian matrix is

$$H = \begin{pmatrix} f_{xx}(x,y) \; f_{xy}(x,y) \\ f_{xy}(x,y) \; f_{yy}(x,y) \end{pmatrix} = \begin{pmatrix} 2 \;\; 0 \\ 0 \; -4 \end{pmatrix}$$

and the determinant is $|H| = -8 < 0$. Therefore the utility f is not concave.

(iii) (Non-convex utility f). Take a utility function f as

$$f(x,y) = 2x^2 - y^2 + x + 3y \tag{3.11}$$

for $(x,y) \in \mathbb{R}^2$. Then $f_x > 0$ and $f_x > 0$ on R. We can easily check

$$f(\overline{x}_R, \overline{y}_R) \iint_R dx \, dy = \iint_R f(x,y) \, dx \, dy - \frac{1}{12}.$$

Thus the utility f is risk loving on R with weighting w. Its Hessian matrix is

$$H = \begin{pmatrix} f_{xx}(x,y) \; f_{xy}(x,y) \\ f_{xy}(x,y) \; f_{yy}(x,y) \end{pmatrix} = \begin{pmatrix} 4 \;\; 0 \\ 0 \; -2 \end{pmatrix}$$

and the determinant is $|H| = -8 < 0$. Therefore the utility f is not convex.

Example 3.1(ii) shows the concavity of utility f may not be a necessary condition for the risk averse on R, and Example 3.1(iii) also shows the convexity of utility f may not be a necessary condition for the risk loving on R. However it is possible to give the following necessary conditions for risk averse and risk loving (Fig. 1).

Theorem 3.1 *Let a utility f be a C^2-class strictly increasing function on D.*

(i) If decision making with utility f is risk averse on any rectangle region $R \in \mathcal{R}(D)$ with any C^1-class positive valued weighting function w on D, then it holds that

$$f_{xx} \le 0 \text{ and } f_{yy} \le 0 \quad \text{on } D. \tag{3.12}$$

(ii) If decision making with utility f is risk loving on any rectangle region $R \in \mathcal{R}(D)$ with any C^1-class positive valued weighting function w on D, then it holds that

$$f_{xx} \ge 0 \text{ and } f_{yy} \ge 0 \quad \text{on } D. \tag{3.13}$$

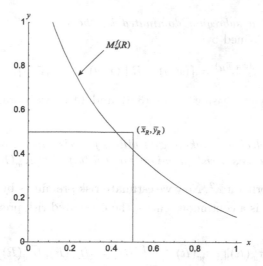

Fig. 1. Weighted quasi-arithmetic means $M_w^f(R)$ ($f(x, y) = -x^2 - y^2 + 3x + 3y$ and $w(x, y) = 1$ on $R = [0, 1]^2$)

4 Risk Premiums on Two-Dimensional Regions

Risk premiums are one of important concepts in financial theory. In this section we discuss risk premiums on two-dimensional regions. For this purpose, we introduce the following natural ordering on $\mathbb{R}^2$.

Definition 4.1 (A partial order $\preceq$ on $\mathbb{R}^2$). For two points $(\underline{x}, \underline{y}), (\overline{x}, \overline{y})(\in \mathbb{R}^2)$, an order $(\underline{x}, \underline{y}) \preceq (\overline{x}, \overline{y})$ implies $\underline{x} \leq \overline{x}$ and $\underline{y} \leq \overline{y}$.

Let a domain D be a non-empty open convex subset of $\mathbb{R}^2$. Let a utility f be a C^2-class strictly increasing function on D, and let a weighting w be a C^1-class positive valued function on D. We introduce the following concept from [5].

Definition 4.2 A vector $\pi_w^f(R) \in [0, \infty)^2$ satisfying the following equation is called a *risk premium for utility* f:

$$f((\overline{x}_R, \overline{y}_R) - \pi_w^f(R)) \iint_R w(x, y) \, dx \, dy = \iint_R f(x, y) w(x, y) \, dx \, dy. \qquad (4.1)$$

We denote the set of risk premiums satisfying (4.1) by the following $\Pi_w^f(R)$:

$$\Pi_w^f(R) = \{\pi_w^f(R) \mid (\overline{x}_R, \overline{y}_R) - \pi_w^f(R) \in M_w^f(R), \ \mathbf{0} \preceq \pi_w^f(R)\}, \qquad (4.2)$$

where $\mathbf{0}$ is the zero vector on $\mathbb{R}^2$.

Hence $\Pi_w^f(R)$ is also written as

$$\Pi_w^f(R) = \{(\overline{x}_R, \overline{y}_R) - (x, y) \mid (x, y) \in M_w^f(R) \cap R_-^{(\overline{x}_R, \overline{y}_R)}\}, \qquad (4.3)$$

where $R_-^{(\overline{x}_R, \overline{y}_R)}$ is *a subregion dominated by the invariant risk neutral point* $(\overline{x}_R, \overline{y}_R)$ which is defined by

$$R_-^{(\overline{x}_R, \overline{y}_R)} = \{(x,y) \in R \mid (x,y) \preceq (\overline{x}_R, \overline{y}_R)\}. \tag{4.4}$$

Since f is strictly increasing, from (3.6) and (4.1) we obtain the following theorem.

Theorem 4.1 *If decision making with utility f is risk averse on R with weighting w, then there exists a risk premium for utility f, i.e. $\Pi_w^f(R) \neq \emptyset$.*

Let $\|\cdot\|$ be a norm on $\mathbb{R}^2$. Now we estimate risk premiums by norm $\|\cdot\|$. Since $M_w^f(R) \cap R_-^{(\overline{x}_R, \overline{y}_R)}$ is a continuous curve, the estimated risk premiums become a closed interval:

$$\{\|\pi_w^f(R)\| \mid \pi_w^f(R) \in \Pi_w^f(R)\} = [\underline{\Pi}_w^f(R), \overline{\Pi}_w^f(R)], \tag{4.5}$$

where the maximum risk premium $\overline{\Pi}_w^f(R)$ and the minimum risk premium $\underline{\Pi}_w^f(R)$ are given from (4.3) as follows:

$$\overline{\Pi}_w^f(R) = \max_{(x,y) \in M_w^f(R) \cap R_-^{(\overline{x}_R, \overline{y}_R)}} \|(\overline{x}_R, \overline{y}_R) - (x,y)\|, \tag{4.6}$$

$$\underline{\Pi}_w^f(R) = \min_{(x,y) \in M_w^f(R) \cap R_-^{(\overline{x}_R, \overline{y}_R)}} \|(\overline{x}_R, \overline{y}_R) - (x,y)\|. \tag{4.7}$$

Hence we define sets of points, which are pairs of two-dimensional weighted quasi-arithmetic means, to attain the maximum risk premium and minimum risk premium as follows:

$$\overline{M}_w^f(R) = \arg \max_{(x,y) \in M_w^f(R) \cap R_-^{(\overline{x}_R, \overline{y}_R)}} \|(\overline{x}_R, \overline{y}_R) - (x,y)\|, \tag{4.8}$$

$$\underline{M}_w^f(R) = \arg \min_{(x,y) \in M_w^f(R) \cap R_-^{(\overline{x}_R, \overline{y}_R)}} \|(\overline{x}_R, \overline{y}_R) - (x,y)\|. \tag{4.9}$$

Using these tools, we can estimate risk premiums for risk averse utility. In Definition 2.1, risk neutral decision making is included in risk averse decision making and risk loving decision making. As a special case of Theorem 4.1, therefore, if decision making is risk neutral then the corresponding risk premium is $\Pi_w^f(R) = \{0\}$. Then $\overline{\Pi}_w^f(R) = \underline{\Pi}_w^f(R) = 0$ and $\overline{M}_w^f(R) = \underline{M}_w^f(R) = (\overline{x}_R, \overline{y}_R)$. The following Example 4.1 illustrates this concepts.

Example 4.1 We calculate risk premiums for Example 3.1(i). Take a domain $D = \mathbb{R}^2$ and a rectangle region $R = [0,1]^2$, and take a weighting function $w = 1$. Then the invariant risk neutral point is $(\overline{x}_R, \overline{y}_R) = (\frac{1}{2}, \frac{1}{2})$. Take a strictly concave increasing utility function f as (3.9). Then the utility f is risk averse on R with weighting w. From (4.3), we obtain risk premiums

$$\Pi_w^f(R) = \left\{ \left(\frac{1}{2} - x, \frac{1}{2} - y \right) \mid -x^2 - y^2 + 3x + 3y = \frac{7}{3}, 0 \le x \le \frac{1}{2}, 0 \le y \le \frac{1}{2} \right\}. \quad (4.10)$$

Take a norm $\|(x,y)\| = \sqrt{x^2 + y^2}$ for $(x,y) \in \mathbb{R}^2$. Then the estimated risk premiums becomes a closed interval:

$$\{\|\pi_w^f(R)\| \mid \pi_w^f(R) \in \Pi_w^f(R)\} = [\underline{\Pi}_w^f(R), \overline{\Pi}_w^f(R)] = \left[\frac{\sqrt{78}}{6} - \sqrt{2}, \frac{\sqrt{42}}{6} - 1 \right]. \quad (4.11)$$

Hence the maximum risk premium $\overline{\Pi}_w^f(R) = \frac{\sqrt{42}}{6} - 1 = 0.0801234\cdots$ is attained by two-dimensional weighted quasi-arithmetic means $\left(\frac{\sqrt{42}}{6} - 1, 0 \right), \left(0, \frac{\sqrt{42}}{6} - 1 \right)$ $\in \overline{M}_w^f(R)$, and the minimum risk premium $\underline{\Pi}_w^f(R) = \frac{\sqrt{78}}{6} - \sqrt{2} = 0.0577466\cdots$ is attained by a two-dimensional weighted quasi-arithmetic mean $\left(\frac{\sqrt{39}}{6} - 1, \frac{\sqrt{39}}{6} - 1 \right) \in \underline{M}_w^f(R)$.

5 Construction of Two-Dimensional Utilities from One-Dimensional Utilities

A lot of examples of utility functions on one-dimensional domains are known ([9–11]). Hence we give a few methods, which are easily checked from the definitions, to construct utility functions g on two-dimensional regions from utility functions on one-dimensional domains.

Lemma 5.1 *Let D be a non-empty open domain in $\mathbb{R}^2$, and let a rectangle region $R \in \mathcal{R}(D)$. Let I and J be closed sub-intervals of $\mathbb{R}$. Let g be a C^2-class concave (strictly concave) function on D. Let a pair of utilities $(\xi, \eta) : I \times J \mapsto D$ be C^2-class such that $\xi' > 0$ and $\xi'' \le 0$ on I and $\eta' > 0$ and $\eta'' \le 0$ on J. Then*

$$f(x,y) = g(\xi(x), \eta(y)) \quad (5.1)$$

is a C^2-class concave (strictly concave resp.) utility function on $I \times J$.

Corollary 5.1 *Let I and J be closed sub-intervals of $\mathbb{R}$. Let α and β be positive constants. Let two utilities $\xi : I \mapsto \mathbb{R}$ and $\eta : J \mapsto \mathbb{R}$ be C^2-class such that $\xi' > 0$ and $\xi'' \le 0$ on I and $\eta' > 0$ and $\eta'' \le 0$ on J. Then*

$$f(x,y) = \alpha\xi(x) + \beta\eta(y) \quad (5.2)$$

is a C^2-class concave utility function on $I \times J$.

Lemma 5.2 *Let D be a non-empty open domain in $\mathbb{R}^2$, and let I be a closed sub-interval of $\mathbb{R}$. Let $g : D \mapsto I$ be a C^2-class concave (strictly concave) utility function on D. Let a utility $\varphi : I \mapsto \mathbb{R}$ be C^2-class such that $\varphi' > 0$ and $\varphi'' \le 0$ on I. Then*

$$f(x,y) = \varphi(g(x,y)) \quad (5.3)$$

is a C^2-class concave (strictly concave) utility function on D.

Corollary 5.2 *Let a rectangle region $I \times J \in \mathcal{R}(D)$ and let α and β be positive constants. Let K be a closed sub-interval of $\mathbb{R}$ such that $K = \{\alpha x + \beta y \mid x \in I, y \in J\}$. Let a utility $\varphi : K \mapsto \mathbb{R}$ be C^2-class such that $\varphi' > 0$ and $\varphi'' < 0$ on K. Then*

$$f(x,y) = \varphi(\alpha x + \beta y) \tag{5.4}$$

is a C^2-class concave utility function on $I \times J$.

Example 5.1 In Table 1 we list up some economic utility functions φ on one-dimensional domains [10, 11], and then from (5.2) and (5.4) we can construct utility functions on two-dimensional regions by combining these functions. For example, from (5.2) and Table 1 we can obtain a utility function on two-dimensional domain $(0, \infty)^2$ by

$$f(x,y) = \alpha \log x + \beta \log y \tag{5.5}$$

for $(x, y) \in (0, \infty)^2$ with positive constants α and β. On the other hand from 5.4 and Table 1 we can give a utility function on two-dimensional domain $\mathbb{R}^2$ by

$$f(x,y) = 1 - e^{-(\alpha x + \beta y)} \tag{5.6}$$

for $(x, y) \in \mathbb{R}^2$ with positive constants α and β.

Table 1. Strictly concave utility functions φ on one-dimensional domains

Utility function, domain and parameters	$\varphi(x)$
Power utility $(0, \infty); 0 < \lambda < 1$	$\dfrac{x^\lambda}{\lambda}$
Logarithmic utility $(0, \infty); \lambda > 0$	$\lambda \log x$
Exponential utility $(-\infty, \infty); \lambda > 0$	$\dfrac{1 - e^{-\lambda x}}{\lambda}$
Quadratic utility $(0, \lambda); \lambda > 0$	$\lambda x - \dfrac{1}{2}x^2$
Sigmoid utility $(0, \infty); \lambda > 0$	$\dfrac{1}{1 + e^{-\lambda x}}$

Concluding Remark. Lemma 2.2 shows that the concavity of utility functions is a sufficient condition for the risk averse. However, in Example 3.1(ii) we found that the concavity of utility functions is not a necessary and sufficient condition for the risk averse. We need to find other conditions instead of the determinant condition for the Hessian in Lemma 2.1(i):

$$|H| = f_{xx} f_{yy} - f_{xy}^2 \geq 0 \tag{5.7}$$

on D.

References

1. Aczél, J.: On weighted mean values. Bull. Am. Math. Soc. **54**, 392–400 (1948)
2. Arrow, K.J.: Essays in the Theory of Risk-Bearing. Markham, Chicago (1971)
3. Fishburn, P.C.: Utility Theory for Decision Making. Wiley, New York (1970)
4. Gollier, G.: The Economics of Risk and Time. MIT Publishers, Cambridge (2001)
5. Eeckhoudt, L., Gollier, G., Schkesinger, H.: Economic and Financial Decisions under Risk. Princeton University Press, New Jersey (2005)
6. Kolmogoroff, A.N.: Sur la notion de la moyenne. Acad. Naz. Lincei Mem. Cl. Sci. Fis. Mat. Natur. Sez. **12**, 388–391 (1930)
7. Nagumo, K.: Über eine Klasse der Mittelwerte. Jpn. J. Math. **6**, 71–79 (1930)
8. Rockafellar, R.T.: Convex Analysis. Princeton University Press, Princeton (1970)
9. Yoshida, Y.: Aggregated mean ratios of an interval induced from aggregation operations. In: Torra, V., Narukawa, Y. (eds.) MDAI 2008. LNCS (LNAI), vol. 5285, pp. 26–37. Springer, Heidelberg (2008)
10. Yoshida, Y.: Quasi-arithmetic means and ratios of an interval induced from weighted aggregation operations. Soft Comput. **14**, 473–485 (2010)
11. Yoshida, Y.: Weighted quasi-arithmetic means and a risk index for stochastic environments. Int. J. Uncertainty Fuzziness Knowl. Based Syst. (IJUFKS) **16**(suppl), 1–16 (2011)

A Comparison of the GAI Model
and the Choquet Integral w.r.t. a k-ary Capacity

Christophe Labreuche[1](✉) and Michel Grabisch[2]

[1] Thales Research and Technology, Palaiseau, France
christophe.labreuche@thalesgroup.com
[2] Paris School of Economics, Université Paris I - Panthéon-Sorbonne,
Paris, France
michel.grabisch@univ-paris1.fr

Abstract. This paper proposes a comparison between a GAI model and the Choquet integral w.r.t. a k-ary capacity. We show that these two models are much closer than one would expect. Based on this comparison, we show a new result on the GAI models: any 2-additive GAI model can be rewritten in such a way that all utility terms in the GAI decomposition are non-negative and monotone. This is very important in practice since it allows reducing the number of monotonicity constraints to be enforced in the elicitation process, from an exponential number (of the number of attributes) to a quadratic number.

1 Introduction

Multi-Criteria Decision Making (MCDM) aims at representing the preferences of a Decision Maker (DM) regarding how to compare some options on the basis of their values on several attributes. The preferences of the DM can be projected to each attribute separately. Depending on the type of assumptions on these preferences over each attribute, two lines of MCDM model can be defined.

In the first one (called *attribute-decomposable*), the overall assessment of the options can be decomposed as an aggregation function applied to partial utility functions on each attribute. This representation implies some commensurability among criteria in the sense that the partial utility functions return an assessment in the common evaluation scale (e.g. [0,1] representing a satisfaction degree). The simplest model of this form uses the weighted sum model as an aggregation function. It is limited in the sense that it does not allow interaction among criteria. This has led to the use of the Choquet integral w.r.t. a capacity [11], or w.r.t. a k-ary capacity [13,14]. It has the ability to represent various important phenomena such as veto, favour, complementarity among criteria, among others.

In the second approach (called *additive-decomposable*), there are still some utility functions over the attributes, but it is not assumed that they return a commensurate evaluation, and one assumes some additivity in the overall utility. The best known model is the additive utility model. The GAI (Generalized Additive Independence) model has been designed as a generalization of the additive utility model [6,7] to allow interaction among attributes. It has been used

V. Torra and Y. Narakawa (Eds.): MDAI 2015, LNAI 9321, pp. 54–65, 2015.
DOI: 10.1007/978-3-319-23240-9_5

in AI [1]. In [4,9,10], the GAI model is learned based on standard gambles. In a configuration problem, a GAI model is elicited during the solution search using the expected value of information in [3] or minmax regret in [4]. In the OR community, a model called $UTA^{GMS}-INT$ very close to the GAI model has been proposed [16]. The model is learned using linear programming. Linear programming is also used in [2,20] to learn a GAI model.

We are especially interested in this paper in the GAI model. In the elicitation phase, one of the main challenges is to represent the monotonicity conditions, as the number of such conditions grows exponentially with the number of attributes.

We propose in this paper a comparison of the GAI model and k-ary capacities. Section 3 shows that when the attribute are discrete, a GAI model can be seen as a k-ary capacity. The discrete values of the attribute are mapped the reference elements $\{0, 1, \ldots, k\}$ used in a k-ary capacity. The concept of p-additivity, which is defined in Sect. 2, yield a GAI model which contains only utility terms depending on at most p attributes. This is similar to the concept of p additivity defined for capacities.

Section 4 addresses a problem that is crucial for the elicitation of a GAI model. A GAI model can be learned using linear programming, as described in [2,20] (see also [16] for $UTA^{GMS}-INT$). The linear constraints are the learning examples provided by the DM as well as the monotonicity conditions. These latter tell that the overall utility of the GAI model shall not decrease when the value on any value gets improved (according to the preference projected on each attribute separately). The main issue is that the number of such conditions growths exponentially with the number of attributes. This is already intractable with as few as 8 attributes. The main result of this paper shows that any 2-additive GAI model can be rewritten in such a way that all utility terms in the GAI decomposition are non-negative and monotone. This result implies that it is sufficient to enforce the monotonicity conditions only on the utility terms, which entails only a quadratic increase. This result is proved by turning a GAI model into a 2-additive k-ary capacity.

Note that there exists some similar reasoning in the context of capacities. The number of monotonicity constraints grows exponentially, even for a 2-additive capacity. It has been shown in [21] that the number of extreme points (vertices) of the set of 2-additive capacities is quadratic in the number of attributes. Hüllermeier and Fallah Tehrani used this property by writing any 2-additive capacity as a convex combination of these extreme points [18], which allows to have a quadratic number of unknowns and a quadratic number of monotonicity constraints.

Finally, Sect. 5 shows the link between a GAI model and the Choquet integral w.r.t. a k-ary capacity, when the attributes are intervals. We show that these two models are very similar, one using the multi-linear extension whereas the second uses the Lovász extension.

2   Basic Definitions

We are given a set of n attributes indexed by $N = \{1, \ldots, n\}$. Each attribute $i \in N$ is represented by a set X_i which can be discrete or continuous (an interval).

The alternatives are characterized by a value on each attribute, and are thus represented by an element in $X = X_1 \times \cdots \times X_n$. We assume that we are given a preference relation $\succsim_i$ over each attribute i. We denote by $\succ_i$ and $\sim_i$ the asymmetric and symmetric parts of $\succsim_i$ respectively. We aim to represent the overall assessment of a decision maker over alternatives

$$U : X \to \mathbb{R}. \tag{1}$$

According to partial preference $\succsim_i$, the best and worst elements in X_i are denoted $x_i^\top$ and $x_i^\perp$ respectively. We can assume w.l.o.g. that U returns values in $[0,1]$. Then we can enforce that

$$U(x_1^\top, \ldots, x_n^\top) = 1 \quad \text{and} \quad U(x_1^\perp, \ldots, x_n^\perp) = 0. \tag{2}$$

For $x, y \in X$ and $A \subseteq N$, we denote by X_A the set $\prod_{i \in A} X_i$, by x_A the restriction of x on attributes A, and by $(x_A, y_{-A}) \in X$ the compound alternative take value x_i for attribute i in A, and value y_i else.

Utility U is assumed to fulfil the following monotonicity conditions, which states that it should be consistent with each relation $\succsim_i$:

$$\forall x, y \in X \text{ with } y_i \succsim_i x_i \text{ for every } i \in N, \qquad U(y) \geq U(x) \tag{3}$$

There exist many different utility models of the form (1). In order to ease the elicitation process, we need to reduce the intrinsic complexity of model U. This can be used thanks to the concept of p-additivity, which generalizes 2-additivity [20]. Prior to that, given $x, y, z \in X$, the *discrete derivative* of U w.r.t. a subset $P \subseteq N$ at a triplet x, y, z is defined by

$$\Delta_P U(x_P, y_P, z_{-P}) = \sum_{T \subseteq P} (-1)^{|P|-|T|} U(y_T, x_{P \setminus T}, z_{-P}).$$

This generalizes the discrete derivative of on capacities defined by $\Delta_P \mu(S) = \sum_{T \subseteq P} (-1)^{|P|-|T|} \mu(S \cup T)$ for $S \subseteq N \setminus P$ and $\mu : \{0,1\}^N \to \mathbb{R}$ [8].

For instance,

$$\Delta_{\{i\}} U(x_i, y_i, z_{-i}) = U(y_i, z_{-i}) - U(x_i, z_{-i})$$
$$\Delta_{\{i,j\}} U(x_{i,j}, y_{i,j}, z_{-i,j}) = U(y_i, y_j, z_{-i,j}) - U(x_i, y_j, z_{-i,j})$$
$$- U(y_i, x_j, z_{-i,j}) + U(x_i, x_j, z_{-i,j})$$

We are now in a position to define p-additivity. It generalizes 2-additivity when $p = 2$ [20].

Definition 1. *Function U is said to be p-additive if for every $P \subseteq N$ with $|P| \leq p$, for every $x_P, y_P \in X_P$ and every $z_{-P}, t_{-P} \in X_{-P}$*

$$\Delta_P U(x_P, y_P, z_{-P}) = \Delta_P U(x_P, y_P, t_{-P}).$$

In the rest of this section, we describe two known models: (k-ary) capacities and the Choquet integral (Sect. 2.1), and the GAI model (Sect. 2.2).

2.1   (k-ary) Capacity and the Choquet Integral

Function U can take the form [19]

$$U(x) = F(u_1(x_1), \ldots, u_n(x_n)), \tag{4}$$

where $u_i : X_i \to \mathbb{R}$ is called the *utility function* (also called *value function*) on X_i and $F : \mathbb{R}^n \to \mathbb{R}$ is an *aggregation function*. Utility function u_i shall be consistent with $\succsim_i$ (i.e. $u_i(x_i) \geq u_i(y_i)$ whenever $x_i \succsim_i y_i$). A *criterion* is a preference over an attribute and corresponds to the pair $\langle X_i, u_i \rangle$. Hence function F aggregates the values on the n criteria. The Choquet integral is one of the most versatile aggregation function as it is able to capture various decision strategies representing interaction among criteria [5,11,15].

Capacity and the Choquet Integral.

Definition 2. *A fuzzy measure [23] or capacity [5] on N is a set function $\mu : 2^N \to \mathbb{R}$ satisfying (1) the monotonicity conditions: $\mu(A) \leq \mu(B)$ for every $A \subseteq B$, and (2) the normalization conditions: $\mu(\emptyset) = 0$, $\mu(N) = 1$.*

The Möbius transform (see e.g. [22]) of μ is defined by

$$m^\mu(A) = \sum_{B \subseteq A} (-1)^{|A \setminus B|} \mu(B). \tag{5}$$

A capacity is said to be p-additive if $m^\mu(A) = 0$ for every $A \subseteq N$ with $|A| > p$, and $m^\mu(A) \neq 0$ for at least one $A \subseteq N$ with $|A| = p$ [12]. The Choquet integral of $a \in \mathbb{R}^N$ w.r.t. capacity μ is defined by [5]

$$C_\mu(a) = \sum_{i=1}^{n} \left(a_{\tau(i)} - a_{\tau(i-1)} \right) \mu \left(\{ \tau(i), \cdots, \tau(n) \} \right), \tag{6}$$

where $a_{\tau(0)} := 0$ and τ is a permutation on N such that $a_{\tau(1)} \leq a_{\tau(2)} \leq \cdots \leq a_{\tau(n)}$.

k-ary Capacity and Choquet Integral.

The concept of a capacity contains 2 reference levels $\{0, 1\}$ on each criteria. It has been generalized to accommodate an arbitrary number of reference levels over each criterion. For $k \in \mathbb{N}_*$, we define

$$\mathcal{Q}_k(N) = \{0, 1, 2, \ldots, k\}^N. \tag{7}$$

and $\leq$ on $\mathcal{Q}_k(N)$ by

$$q \leq q' \quad \text{iff} \quad q_i \leq q'_i \ \forall i \in N. \tag{8}$$

We can now define k-ary capacities, where a usual capacity is a 1-ary capacity.

Definition 3 [13,14]. *A k-ary capacity on N is a function $v : \mathcal{Q}_k(N) \to \mathbb{R}$ satisfying (1) the monotonicity conditions:*

$$\forall q, q' \in \mathcal{Q}_k(N) \ s.t. \ q \leq q' \quad , \quad v(q) \leq v(q'), \tag{9}$$

and (2) the normalization conditions: $v(0, \ldots, 0) = 0$, $v(k, \ldots, k) = 1$.

Let $z \in \Omega := [0, k]^N$. The idea of the Choquet integral of z on a k-ary capacity is first to identify the 2^n nodes in $\mathcal{Q}_k(N)$ just around z (where the bottom element of these 2^n nodes will be denoted by q), then to look at the restriction of the k-ary capacity on these nodes (interpreted as a non-normalized capacity denoted by μ_q), and finally use the standard Choquet w.r.t. μ_q. Given z, we define $q \in \mathcal{Q}_k(N)$ by $q_i = \lfloor z_i \rfloor$ (the floor integer part of z_i). We also define a capacity given q by

$$\mu_q(S) = v((q+1)_S, q_{-S}). \tag{10}$$

Then the Choquet integral w.r.t. v at point z is defined by

$$\mathcal{C}_v(z) = \mathcal{C}_{\mu_q}(\phi) \tag{11}$$

$$\text{where} \quad \forall i \in N \qquad \phi_i = z_i - q_i \in [0, 1]. \tag{12}$$

Note that $z_i = k$, then $q_i = k - 1$. More generally, if $z_i \in \{0, \ldots, k\}$ then any $q_i \in \{z_i - 1, z_i, z_i + 1\} \cap \{0, \ldots, k\}$ will do, thanks to the *properly weighted* property of the Choquet integral.

2.2 GAI Model

General Model. The Generalized Additive Independence (GAI) model [1,6,7] takes the form of the sum of utilities over subsets of attributes:

$$U(x) = \sum_{S \in \mathcal{S}} u_S(x_S) \tag{13}$$

where $\mathcal{S}$ is a collection of subsets of N, and $u_S : X_s \to \mathbb{R}$. Set $\mathcal{S}$ contains all subsets of attributes that interact each other in the practical problem under study. Hence the additive model [19] is a particular case of the GAI model where $\mathcal{S}$ is composed of singletons only.

Case Where All Attributes Are Continuous. In order to determine all utility functions u_S, each attribute is discretized. For attribute $i \in N$, we keep only $\widehat{X}_i \subseteq X_i$ with $|\widehat{X}_i|$ finite in the learning phase. The unknowns of the GAI model are $\{u_S(z_S) : S \in \mathcal{S}, z_S \in \widehat{X}_S\}$, where $\widehat{X}_S = \prod_{i \in S} \widehat{X}_i$. The elements of $\widehat{X}_i$ are denoted by $a_i^0, a_i^1, \ldots, a_i^{m_i}$. In order to distinguish with the model u_S, we denote by

$$\widehat{u} := \{\widehat{u}_S(z_S) : S \in \mathcal{S}, z_S \in \widehat{X}_S\} \tag{14}$$

the set of all unknowns.

The value of u_S from $\widehat{u}_S$ is obtained by interpolation. In [20], a multi-linear interpolation is proposed: Consider $x_S \in X_S$. The following set

$$I = \{i \in S : x_i \notin \widehat{X}_i\}$$

contains all attributes in S not in the mesh $\widehat{X}_i$. For $i \in N$ we set

$$\underline{x}_i = \operatorname{argmax}\{z_i \in \widehat{X}_i \ : \ z_i \leq x_i\}$$
$$\overline{x}_i = \operatorname{argmin}\{z_i \in \widehat{X}_i \ : \ z_i \geq x_i\}$$

Note that $\overline{x}_i = \underline{x}_i$ iff $i \notin I$. Then [20]

$$u_S(x_S) = \sum_{A \subseteq I} \left[\prod_{i \in A} \frac{\overline{x}_i - x_i}{\overline{x}_i - \underline{x}_i} \times \prod_{i \in I \setminus A} \frac{x_i - \underline{x}_i}{\overline{x}_i - \underline{x}_i} \times \widehat{u}_S(\underline{x}_A, \overline{x}_{I \setminus A}, x_{S \setminus I}) \right], \quad (15)$$

where $(\underline{x}_A, \overline{x}_{I \setminus A}, x_{S \setminus I})$ is an alternative that is equal to $\underline{x}_k$ if $k \in A$, to $\overline{x}_k$ if $k \in I \setminus A$, and to x_k if $k \in S \setminus I$.

3   Link Between the GAI Model and a k-ary Capacity When All Attributes Are Discrete

We consider in this section the case where all attributes are discrete. Attribute X_i is denoted by $\{a_i^0, a_i^1, \ldots, a_i^{m_i}\}$. Without loss of generality, we assume that these elements are labelled in increasing order of preference, i.e. $a_i^{m_i} \succsim_i a_i^{m_i-1} \succsim_i \cdots \succsim_i a_i^0$.

We set $k = \max_{i \in N} m_i$. Let $\phi_i : X_i \to \{0, 1, 2, \ldots, k\}$ be given by

$$\phi_i(a_i^l) = l \qquad \forall l \in \{0, 1, 2, \ldots, k\}. \tag{16}$$

The conversion $\phi : X \to \mathcal{Q}_k(N)$ is defined by $\phi(x) = (\phi_1(x_1), \ldots \phi_n(x_n))$. Given a function U (see 1) that is monotone in the sense of (3) and that satisfies (2), we define v, for every $q \in \mathcal{Q}_k(N)$, by

$$v(q) = U(x) \qquad \text{if } \exists x \in X, \ \phi(x) = q \tag{17}$$

When the condition in (17) is not fulfilled, we extend v from $\phi(X)$ to $\mathcal{Q}_k(N)$ as follows:

$$v(q) = v(m_1, \ldots, m_n) \qquad \text{if } q \in \mathcal{Q}_k(N) \setminus \phi(X). \tag{18}$$

Yet v is a k-ary capacity thanks to (3). Hence we have constructed for any U a corresponding k-ary capacity.

Compared to a k-ary capacity, a GAI model appears only as a specialization where one assumes a special decomposition form (13). To push further the comparison, one could imagine introducing a decomposition (13) as in GAI models for a Choquet integral. Transposing this decomposition property to a simple capacity, one would say that a capacity μ is decomposable if it has the form

$$\mu(A) = \sum_{S \in \mathcal{S}, \, S \neq \emptyset} \mu_S(S \cap A)$$

Lemma 1. *For a fixed integer p, μ takes the form*

$$\mu(A) = \sum_{S \subseteq N \ S \neq \emptyset \ and \ |S| \leq k} \mu_S(S \cap A) \qquad (19)$$

iff μ is at most k-additive.

The following result can be shown.

Theorem 1. *U is p-additive if and only if there exists functions $u_A : X_A \to \mathbb{R}$, for every $A \subseteq N$ with $|A| \leq p$, such that U takes the form (13) with $\mathcal{S} = \{A \subseteq N, \ |A| \leq p\}$.*

It generalizes [20, Proposition 4], which is restricted to the case $p = 2$.

Then a p-additive GAI model is formally equivalent to a p-additive k-ary capacity.

Note that the inclusion-exclusion model proposes extensions of the p-additive Choquet integral using a t-norm instead of the minimum in the expression of the Choquet integral w.r.t. the Möbius coefficients [17].

4 A Much Cheaper Description of the Monotonicity Conditions When All Attributes Are Discrete

As in the previous section, we consider discrete attributes denoted by $X_i = \{a_i^0, \ldots, a_i^{m_i}\}$, where $a_i^{m_i} \succsim_i a_i^{m_i - 1} \succsim_i \cdots \succsim_i a_i^0$.

4.1 Complexity of the Monotonicity Conditions

We have described p-additive GAI models (see Definition 1). According to Theorem 1, a p-additive GAI model (a priori) contains all terms u_S with $|S| \leq p$. Then the unknowns of such a model are all terms $u_S(x_S)$, with $|S| \leq p$ and $x_S \in X_S$. Removing the empty set, we obtain the following number of unknowns for a p-additive GAI model:

$$\sum_{S \subseteq N, \ |S| \leq p, \ S \neq \emptyset} \prod_{i \in S} (m_i + 1). \qquad (20)$$

For instance, if $m_i = k$ for every i, then the number of unknowns becomes

$$\sum_{i=1}^{p} (k+1)^i \binom{n}{i}. \qquad (21)$$

Example 1. Consider $n = 10$ attributes and $k = 4$. The following table gives the number of unknowns, depending on the value of p.

p	1	2	3	4	5	6	7	8	9
# unknowns	50	1 175	16 175	147 425	934 925	4 216 175	13 591 175	31 169 300	50 700 550

Consider now $n = 20$ attributes and $k = 4$. The following table gives the number of unknowns, depending on the value of p.

p	1	2	3	4	5	6	7
# unknowns	100	4 850	147 350	3 175 475	51 625 475	657 250 475	6 713 500 475

p	8	9
# unknowns	55 920 531 725	3.83967E + 11

We see in the previous example that the number of unknowns increases extremely fast (exponentially) with p. For practical reasons, we often restrict ourselves to $p = 2$. We consider only 2-additive GAI models in the remaining of this section.

The unknowns are identified by asking the DM some training examples that are transformed into linear constraints on the unknowns [2, 20] (see also [16] for $UTA^{GMS} - INT$). One shall also add the monotonicity constraints on U – see (3). These conditions can be rewritten as follows

$$\forall i \in N \; \forall j_1 \in \{0, \ldots, m_1\} \ldots \forall j_{i-1} \in \{0, \ldots, m_{i-1}\} \forall j_i \in \{0, \ldots, m_i - 1\}$$
$$\forall j_{i+1} \in \{0, \ldots, m_{i+1}\} \ldots \forall j_n \in \{0, \ldots, m_n\} \tag{22}$$
$$U(a_1^{j_1}, \ldots, a_{i-1}^{j_{i-1}}, a_i^{j_i+1}, a_{i+1}^{j_{i+1}}, \ldots, a_n^{j_n}) \geq U(a_1^{j_1}, \ldots, a_{i-1}^{j_{i-1}}, a_i^{j_i}, a_{i+1}^{j_{i+1}}, \ldots, a_n^{j_n})$$

The number of elementary conditions contained in (22) is equal to

$$\sum_{i \in N} \left[m_i \times \prod_{j \in N \setminus \{i\}} (m_j + 1) \right]. \tag{23}$$

For instance, if $m_i = k$ for every i, then the number of monotonicity conditions becomes

$$n \times k \times (k + 1)^{n-1}. \tag{24}$$

The following example shows that this exponential increase becomes rapidly intractable with linear programming.

Example 2. Consider $k = 4$ and a 2-additive GAI model. The following chart gives the number of monotonicity constraint when n varies.

n	4	6	8	10	12	14
# of monotonicity constraints	2 000	75 000	2 500 000	78 125 000	2 343 750 000	68 359 375 000

n	16	18	20
# of monotonicity constraints	1.95313E + 12	5.49316E + 13	1.52588E + 15

Hence we are looking for a simpler monotonicity condition.

4.2   Representation Result of 2-additive GAI Models

In [16], the terms u_S can take both positive and negative signs (it is into the subtraction between two positive numbers) Let us start from the following example of a non-negative function $U(x_1, x_2)$ having a negative term:

$$U(x_1, x_2) = 2\,x_1 + x_2 - \max(x_1, x_2).\tag{25}$$

From the relation

$$\min(x_1, x_2) + \max(x_1, x_2) = x_1 + x_2,$$

(25) can be replaced by the equivalent expression:

$$U(x_1, x_2) = x_1 + \min(x_1, x_2).\tag{26}$$

In this illustrative example, the negative term has been replaced by a positive one. Moreover all terms are monotone. One wonders now if this process can be generalized to any 2-additive GAI model U. In other way, is it possible to transform any 2-additive GAI model in such a way that all terms u_S become non-negative and monotone?

The main theorem of this paper shows that the previous assertion is true.

Theorem 2. *Let us consider a 2-additive GAI model U that is monotone in the sense of (3) and that satisfies (2). Then there exists non-negative and monotone functions $u_{i,j} : X_i \times X_j \to [0,1]$ (for every $\{i,j\} \subseteq N$) and non-negative and monotone functions $u_i : X_i \to [0,1]$ (for every $i \in N$) such that for all $x \in X$*

$$U(x) = \sum_{i=1}^{n} u_i(x_i) + \sum_{\{i,j\} \subseteq N} u_{i,j}(x_i, x_j).\tag{27}$$

Due to space limitation, the proof of this theorem, which is quite long, is omitted.

Theorem 2 shows that, in order to ensure monotonicity of U, it is sufficient to enforce the monotonicity on each term u_S in the GAI decomposition. Then the number of monotonicity conditions is reduced to:

$$\sum_{i \in N} m_i + \sum_{\{i,j\} \subseteq N} (m_i(m_j + 1) + m_j(m_i + 1)).\tag{28}$$

For instance, if $m_i = k$ for every i, then the number of monotonicity conditions becomes

$$n\,k\Big[(n-1)(k+1) + 1\Big].\tag{29}$$

The number of constraint only growth quadratically with the number of criteria. The comparison between (24) and (29) is illustrated in the following example.

Example 3. Consider $k = 4$ and a 2-additive GAI model. The following chart give the number of monotonicity constraints.

n	4	6	8	10	12	14
# of monotonicity constraints with (3)	2 000	75 000	2 500 000	78 125 000	2 343 750 000	68 359 375 000
# of monotonicity constraints with Theorem 2	256	624	1 152	1 840	2 688	3 696

n	16	18	20
# of monotonicity constraints with (3)	$1.95313E + 12$	$5.49316E + 13$	$1.52588E + 15$
# of monotonicity constraints with Theorem 2	4 864	6 192	7 680

Thanks to Theorem 2, it is now possible to handle the monotonicity constraints in a linear programming solver.

5   Extension of k-ary Choquet Integral

We assume in this section that every attribute is continuous and described by an interval – see Sect. 2.2. In particular, the unknowns are the values of the utility terms u_S at a discretization $\widehat{X}_i = \{a_i^0, a_i^1, \ldots, a_i^k\}$ of X_i, for every i, with $a_i^0 < a_i^1 < \cdots < a_i^k$. For the sake of simplicity, we assume the same size k of discretization over each attribute.

We assume we are given the overall utility $\widehat{U}(x)$ for each element x in $\widehat{X}$. The two models (GAI, denoted by U^{GAI}, and the Choquet integral w.r.t. a k-ary capacity, denoted by $U^{k-\mathrm{ary}}$) interpolate differently from values $\{\widehat{U}(x),\ x \in \widehat{X}\}$. We have already analyzed in Sect. 3, whether a k-ary capacity can be decomposed in the same way as a GAI model, when attributes are discrete. We have seen in particular that a p-additive k-ary capacity corresponds to a p-additive GAI model and vice versa. The novelty when attributes are continuous is the interpolation. In our comparison of U^{GAI} and $U^{k-\mathrm{ary}}$, we focus thus on the interpolation power of these two models, without assuming a special decomposition of the GAI model. This means that we focus the analysis on one term u_S in the GAI model. Henceforth we fix for N a subset $S \in \mathcal{S}$ (we set $N = S$). Hence U^{GAI} is directly given by (15) with $S = N$. We clearly has for every $q \in \mathcal{Q}_p(N)$

$$U^{\mathrm{GAI}}(a_1^{q_1}, \ldots, a_n^{q_n}) = \widehat{U}(a_1^{q_1}, \ldots, a_n^{q_n}).$$

We assume, for every criterion i, that $a_i^k \succ_i \cdots \succ_i a_i^0$.

Let us describe model $U^{k-\mathrm{ary}}$ (use of a Choquet integral w.r.t. a k-ary capacity). After normalization by partial utility function u_i (see (4)), the criteria score shall lie in interval $[0, k]$. Moreover, utility function u_i shall map points $a_i^0, a_i^1, \ldots, a_i^k$ to $0, 1, \ldots, k$ respectively:

$$u_i(a_i^l) = l \qquad \forall l \in \{0, 1, 2, \ldots, k\}. \tag{30}$$

This is similar to (16). By assuming that u_i is a simple linear interpolation between any two successive points a_i^l and a_i^{l+1}, we obtain

$$u_i(x_i) = \begin{cases} 0 & \text{if } x_i \leq a_i^0 \\ \frac{x_i - a_i^l}{a_i^{l+1} - a_i^l}(l+1) + \frac{a_i^{l+1} - x_i}{a_i^{l+1} - a_i^l}l = l + \frac{x_i - a_i^l}{a_i^{l+1} - a_i^l} & \text{if } a_i^l \leq x_i < a_i^{l+1} \\ k & \text{if } x_i \geq a_i^k \end{cases} \quad (31)$$

The k-ary capacity v is defined for every $q \in \mathcal{Q}_p(N)$ by

$$v(q) = \widehat{U}(a_1^{q_1}, \dots, a_n^{q_n}).$$

Let $x \in X$. Setting operator $[\cdot]$ by $[x] \in X$ and $[x]_i = \max(\min(x_i, a_i^k), a_i^0)$, we have $u_i(x_i) = u_i([x])$. We define $q \in \mathcal{Q}_p(N)$ such that for all $i \in N$

$$q_i = \begin{cases} 0 & \text{if } x_i \leq a_i^0 \\ l & \text{if } a_i^l \leq x_i < a_i^{l+1} \\ k-1 & \text{if } x_i \geq a_i^{k-1} \end{cases} \quad (32)$$

Generalizing (12), we define function ϕ by

$$\forall i \in N \qquad \phi_i(x) = \min(u_i(x_i) - q_i, 1) = u_i([x]) - q_i \in [0,1]$$

$$= \begin{cases} 0 & \text{if } x_i \leq a_i^0 \\ \frac{x_i - a_i^l}{a_i^{l+1} - a_i^l} & \text{if } a_i^l \leq x_i < a_i^{l+1} \\ 1 & \text{if } x_i \geq a_i^k \end{cases} \quad (33)$$

Then the Choquet integral w.r.t. v for option x is given by (similar to (11) and (10))

$$U^{k-\mathrm{ary}}(x) = C_{\mu_q}(\phi) + v(q) \quad (34)$$

where the capacity μ_q is given by for every $S \subseteq N$

$$\mu_q(S) = v((q+1)_S, q_{N\setminus S}) - v(q). \quad (35)$$

The following result shows that $U^{k-\mathrm{ary}}$ is an interpolation.

Lemma 2. *For every* $q \in \mathcal{Q}_p(N)$

$$U^{k-\mathrm{ary}}(a_1^{q_1}, \dots, a_n^{q_n}) = \widehat{U}(a_1^{q_1}, \dots, a_n^{q_n}).$$

Hence U^{GAI} and $U^{k-\mathrm{ary}}$ return the same value on $\widehat{X}$. The only difference between U^{GAI} and $U^{k-\mathrm{ary}}$ is that U^{GAI} is the multi-linear extension of the values $\{\widehat{U}(x), \ x \in \widehat{X}\}$ (see (15)), whereas $U^{k-\mathrm{ary}}$ is the Lovász extension of the values $\{\widehat{U}(x), \ x \in \widehat{X}\}$ (see (34)).

References

1. Bacchus, F., Grove, A.: Graphical models for preference and utility. In: Conference on Uncertainty in Artificial Intelligence (UAI), pp. 3–10, Montreal, July 1995
2. Bigot, D., Fargier, H., Mengin, J., Zanuttini, B.: Using and learning GAI-decompositions for representing ordinal rankings. In: Workshop on Preference Learning, European Conference on Artificial Intelligence (ECAI), Montepellier, 27–31 August 2012

3. Braziunas, D., Boutilier, V.: Local utility elicitation in GAI models. In: Conference on Uncertainty in Artificial Intelligence (UAI), Edinburgh, July 2005
4. Braziunas, D., Boutilier, C.: Minimax regret based elicitation of generalized additive utilities. In: Proceedings of the Twenty-third Conference on Uncertainty in Artificial Intelligence (UAI-07), pp. 25–32, Vancouver (2007)
5. Choquet, G.: Theory of capacities. Ann. de l'Institut Fourier **5**, 131–295 (1953)
6. Fishburn, P.: Interdependence and additivity in multivariate, unidimensional expected utility theory. Int. Econ. Rev. **8**, 335–342 (1967)
7. Fishburn, P.: Utility Theory for Decision Making. Wiley, New York (1970)
8. Fujimoto, K., Kojadinovic, I., Marichal, J.-L.: Axiomatic characterizations of probabilistic and cardinal-probabilistic interaction indices. Games Econ. Behav. **55**, 72–99 (2006)
9. Gonzales, C., Perny, P.: GAI networks for utility elicitation. In: Proceedings of the 9th International Conference on the Principles of Knowledge Representation and Reasoning (KR), pp. 224–234 (2004)
10. Gonzales, C., Perny, P., Dubus, J.: Decision making with multiple objectives using GAI networks. Artif. Intell. J. **175**(7), 1153–1179 (2000)
11. Grabisch, M.: The application of fuzzy integrals in multicriteria decision making. European J. Oper. Res. **89**, 445–456 (1996)
12. Grabisch, M.: k-order additive discrete fuzzy measures and their representation. Fuzzy Sets Sys. **92**, 167–189 (1997)
13. Grabisch, M., Labreuche, C.: Capacities on lattices and k-ary capacities. In: International Conference of the Euro Society for Fuzzy Logic and Technology (EUSFLAT), Zittau, 10–12 September 2003
14. Grabisch, M., Labreuche, C.: Bipolarization of posets and natural interpolation. J. Math. Anal. Appl. **343**, 1080–1097 (2008)
15. Grabisch, M., Labreuche, C.: A decade of application of the Choquet and Sugeno integrals in multi-criteria decision aid. Ann. Oper. Res. **175**, 247–286 (2010)
16. Greco, S., Mousseau, V., Słowinski, R.: Robust ordinal regression for value functions handling interacting criteria. Eur. J. Oper. Res. **239**(3), 711–730 (2014)
17. Honda, A., Okamoto, J.: Inclusion-exclusion integral and its application to subjective video quality estimation. In: Hüllermeier, E., Kruse, R., Hoffmann, F. (eds.) IPMU 2010. CCIS, vol. 80, pp. 480–489. Springer, Heidelberg (2010)
18. Hüllermeier, E., Tehrani, A.F.: Efficient Learning of Classifiers Based on the 2-Additive Choquet Integral. In: Moewes, C., Nürnberger, A. (eds.) Computational Intelligence in Intelligent Data Analysis. SCI, vol. 445, pp. 17–29. Springer, Heidelberg (2013)
19. Keeney, R.L., Raiffa, H.: Decision with Multiple Objectives. Wiley, New York (1976)
20. Labreuche, C., Grabisch, M.: Use of the GAI model in multi-criteria decision making: inconsistency handling, interpretation. In: International Conference Of the Euro Society for Fuzzy Logic and Technology (EUSFLAT), Milano, (2013)
21. Miranda, P., Combarro, E., Gil, P.: Extreme points of some families of nonadditive measures. Euro. J. Oper. Res. **174**, 1865–1884 (2006)
22. Rota, G.: On the foundations of combinatorial theory I. theory of möbius functions. Zeitschrift für Wahrscheinlichkeitstheorie und Verwandte Gebiete **2**, 340–368 (1964)
23. Sugeno, M.: Theory of fuzzy integrals and its applications. Ph.D. thesis, Tokyo Institute of Technology (1974)

Estimating Unknown Values in Reciprocal Intuitionistic Preference Relations via Asymmetric Fuzzy Preference Relations

Francisco Chiclana[1]([⊠]), Raquel Ureña[2], Hamido Fujita[3],
and Enrique Herrera-Viedma[2]

[1] Centre for Computational Intelligence, Faculty of Technology,
De Montfort University, Leicester, UK
chiclana@dmu.ac.uk
[2] Department of Computer Science and Artificial Intelligence,
University of Granada, Granada, Spain
{raquel,viedma}@decsai.ugr.es
[3] Iwate Prefectural University, Takizawa, Iwate, Japan
issam@iwate-pu.ac.jp

Abstract. Intuitionistic preference relations are becoming increasingly important in the field of group decision making since they present a flexible and simple way to the experts to provide their preference relations, while at the same time allowing them to accommodate a certain degree of hesitation inherent to all decision making processes. In this contribution, we prove the mathematical equivalence between the set of asymmetric fuzzy preference relations and the set of reciprocal intuitionistic fuzzy preference relations. This result is exploited to tackle the presence of incomplete reciprocal intuitionistic fuzzy preference relation in decision making by developing a consistency driven estimation procedure via the corresponding equivalent incomplete asymmetric fuzzy preference relation.

Keywords: Intuitionistic preference relation · Asymmetric fuzzy preference relation · Consistency · Uninorm · Incomplete information

1 Introduction

Much research has been carried out in decision making with preferences modelled using fuzzy preference relations in comparison to using intuitionistic fuzzy preference relations. This is mainly to the longer existence of the former representation format of preferences but also to the increase computational complexity associated to the use of membership degree, non-membership degree and hesitation degree to model experts' subjective preferences with the latter representation format. Notice that in decision making, intuitionistic fuzzy preference relations are usually assumed to be reciprocal (Sect. 2).

In this paper the set of reciprocal intuitionistic fuzzy preference relations and the set of asymmetric fuzzy preference relations are proved to be mathematically

© Springer International Publishing Switzerland 2015
V. Torra and Y. Narakawa (Eds.): MDAI 2015, LNAI 9321, pp. 66–77, 2015.
DOI: 10.1007/978-3-319-23240-9_6

isomorphic. The importance of this result resides in that it can be exploited to use methodologies developed for fuzzy preference relations to the case of intuitionistic fuzzy preference relations and, ultimately, to overcome the computation complexity mentioned above and to extend the use of reciprocal intuitionistic fuzzy preference relations in decision making. Indeed, this result will allow us to take advantage of mature and well defined methodologies developed for fuzzy preference relations in an intuitionistic context while at the same time taking advantage of the flexibility of reciprocal intuitionistic fuzzy preference relations to model vagueness/uncertainty. In particular, in this paper we illustrate how this isomorphic equivalence is used to address the presence of incomplete reciprocal intuitionistic fuzzy preference relations in decision making.

Incomplete information, as a result from the incapability of experts to provide complete information about their preferences, may happens more frequently than expected because experts do not have a precise or sufficient level of knowledge of part of the problem, lack of time, difficulty to distinguish up to which degree one preference is better than other, or due to the presence of conflicting alternatives, among others [2,9]. In the literature, different approaches to deal with missing or incomplete information for the case of using fuzzy preference relations as the representation format of preferences have been extensively studied [24]. Most of the existing approaches are based on a methodology that 'builds' the matrix driven by the concept of consistency of information [1–5,10,12–14,16,17,25].

The case of incomplete intuitionistic fuzzy preference relations has also been addressed in literature in [25,26]. In both cases, a methodology driven by consistency was also adopted, although the way consistency of reciprocal preference relations was modelled was different. On the one hand, in [26] a straight forward transposition of the multiplicative consistency property for fuzzy preference relations was proposed for the case of reciprocal intuitionistic fuzzy preference relations, which later was proved to be incorrect [25] and publicly acknowledged by the authors that proposed it [27]. On the other hand, in [25] the concept of multiplicative consistency for reciprocal intuitionistic fuzzy preference relations was derived by formally extending the fuzzy preference relation multiplicative transitivity property via the use of both the *extension principle* [29] and *representation theorem* of fuzzy sets [28]. In this contribution, though, a different approach to incomplete reciprocal intuitionistic fuzzy preference relations is presented based on the aforementioned isomorphic mapping between the set of reciprocal intuitionistic fuzzy preference relations and the set of asymmetric fuzzy preference relations.

The rest of the paper is set out as follows: The first part presents the two mathematical frameworks for representing preferences (Sect. 2) and the basic concepts needed throughout the rest of the paper (Sect. 3). The second part of the paper demonstrates the isomorphism between the set of reciprocal intuitionistic fuzzy preference relations and the set of asymmetric fuzzy preference relations (Sect. 4) and its use to present a methodology to estimate missing values of reciprocal intuitionistic fuzzy preference relations (Sect. 5). The final part of the paper includes conclusions drawn form the results obtained (Sect. 6).

2   Preference Relations in Decision Making

The comparison of two elements of a set of feasible alternatives (X) by an expert can lead to the preference of one alternative to the other or to a state of indifference between them. Obviously, there is the possibility of an expert being unable to compare them. Two main mathematical models based on the concept of preference relation can be used in this context. In the first one, a preference relation is defined for each one of the above three possible preference states (preference, indifference, incomparability) [11], which is usually referred to as a preference structure on the set of alternatives [20]. The second one integrates the three possible preference states into a single preference relation [7]. In this paper, we focus on the second one as per the following definition:

Definition 1 (Preference Relation). *A preference relation P on a set X is a binary relation $\mu_P \colon X \times X \longrightarrow D$, where D is the domain of representation of preference degrees provided by the decision maker.*

For a set X of finite cardinality $(\#X = n)$ the following matrix representation of a preference relation P is used: $P = (p_{ij})$, with $p_{ij} = \mu_P(x_i, x_j)$ being interpreted as the degree or intensity of preference of alternative x_i over x_j $(i, j \in \{1, 2, \ldots, n\})$. The elements of P can be of a numeric or linguistic nature, i.e., could represent numeric or linguistic preferences, respectively [19]. The main types of numeric preference relations used in decision making are: crisp preference relations, additive preference relations, multiplicative preference relations, interval-valued preference relations and intuitionistic preference relations. In this contribution we focus on reciprocal intuitionistic fuzzy preference relations and their equivalence to a subclass of fuzzy preference relations, the asymmetric fuzzy preference relations.

2.1   Fuzzy Preference Relation

Recall that given a universal set U, with a generic element denoted by x, a fuzzy set X in U is a defined as a set of ordered pairs:

$$X = \big\{(x, \mu_X(x)) | x \in U\big\}$$

where $\mu_X \colon U \to [0, 1]$ is called the membership function of A and $\mu_X(x)$ represents the degree of membership of the element x in X. In this context, the degree of non-membership of the element x in X is normally defined as $\nu_X(x) = 1 - \mu_X(x)$, and as a consequence the following reciprocity property holds: $\mu_X(x) + \nu_X(x) = 1$. The reciprocal relationship between membership and non-membership makes the latter one unnecessary in the formulation as it can be derived from the former.

Definition 2 (Fuzzy Preference Relation). *A fuzzy preference relation $R = (r_{ij})$ on a finite set of alternatives X is a relation in $X \times X$ that is characterised by a membership function $\mu_R \colon X \times X \longrightarrow [0, 1]$.*

The following interpretation is assumed:

- $r_{ij} = 1$ indicates the maximum degree of preference for x_i over x_j
- $r_{ij} \in]0.5, 1[$ indicates a definite preference for x_i over x_j
- $r_{ij} = 1/2$ indicates indifference between x_i and x_j

When $r_{ij} + r_{ji} = 1$ $(\forall i, j \in \{1, \ldots, n\})$ is imposed we have a reciprocal fuzzy preference relation.

2.2 Intuitionistic Fuzzy Preference Relation

An intuitionistic fuzzy set X over a universe of discourse U is represented as [6]

$$X = \Big\{ \big(x, \langle \mu_X(x), \nu_X(x) \rangle \big) \big| x \in U \Big\}$$

where $\mu_X : U \longrightarrow [0, 1]$ and $\nu_X : U \longrightarrow [0, 1]$ verify

$$0 \leq \mu_X(x) + \nu_X(x) \leq 1 \quad \forall x \in U.$$

In this context, $\mu_X(x)$ and $\nu_X(x)$ are known as the degree of membership and degree of non-membership of x to X. Obviously, an intuitionistic fuzzy set becomes a fuzzy set when $\mu_X(x) = 1 - \nu_X(x)$ $\forall x \in U$. However, when there exists at least a value $x \in U$ for which $\mu_X(x) < 1 - \nu_X(x)$, an extra parameter known as the hesitancy degree is defined with intuitionistic fuzzy sets, $\tau_X(x) = 1 - \mu_X(x) - \nu_X(x)$, representing the amount of lacking information in determining the membership of x to X.

In [22], Szmidt and Kacprzyk defined the intuitionistic fuzzy preference relation as a generalisation of the concept of fuzzy preference relation.

Definition 3 (Intuitionistic Fuzzy Preference Relation). *An intuitionistic fuzzy preference relation B on a finite set of alternatives $X = \{x_1, \ldots, x_n\}$ is characterised by a membership function $\mu_B : X \times X \to [0, 1]$ and a non-membership function $\nu_B : X \times X \to [0, 1]$ such that*

$$0 \leq \mu_B(x_i, x_j) + \nu_B(x_i, x_j) \leq 1 \quad \forall (x_i, x_j) \in X \times X.$$

The value $\mu_B(x_i, x_j) = \mu_{ij}$ is interpreted as the certainty degree up to which x_i is preferred to x_j, while $\nu_B(x_i, x_j) = \nu_{ij}$ is interpreted as the certainty degree up to which x_i is non-preferred to x_j.

As with a fuzzy preference relation, an intuitionistic fuzzy preference relation is represented by a matrix $B = (b_{ij})$ with $b_{ij} = \langle \mu_{ij}, \nu_{ij} \rangle$ $\forall i, j = 1, 2, \ldots, n$. Obviously, when the hesitancy function is the null function we have that $\mu_{ij} + \nu_{ij} = 1$ $(\forall i, j)$ and the intuitionistic fuzzy preference relation $B = (b_{ij})$ is mathematically equivalent to the reciprocal fuzzy preference relation $R = (r_{ij})$, with $r_{ij} = \mu_{ij}$.

An intuitionistic fuzzy preference relation is referred to as reciprocal when the following additional conditions are imposed:

- $\mu_{ii} = \nu_{ii} = 0.5$ $\forall i \in \{1, \ldots, n\}$.
- $\mu_{ji} = \nu_{ij} \forall i, j \in \{1, \ldots, n\}$.

3   Consistency of Fuzzy Preferences

Consistency of fuzzy preference relations has been modelled using the notion of transitivity in the pairwise comparison among any three alternatives. If x_i is preferred to x_j ($x_i \succ x_j$) and this one to x_k ($x_j \succ x_k$) then alternative x_i should be preferred to x_k ($x_i \succ x_k$), which is normally referred to as *weak stochastic transitivity* [18]. Any property that guarantees the transitivity of the preferences is called a consistency property [8].

Different properties or conditions have been suggested as rational to be verified by a consistent fuzzy preference relation [8,15]: triangle condition, weak transitivity, max-min transitivity, max-max transitivity, restricted max-min transitivity, restricted max-max transitivity, additive transitivity, and multiplicative transitivity. The last two properties, proposed by Tanino in [23], are the most widely used in the context of incomplete information [8].

Definition 4 (Additive transitivity). *A fuzzy preference relation $R = (r_{ij})$ on a finite set of alternatives X is additive transitive if and only if*

$$(r_{ij} - 0.5) + (r_{jk} - 0.5) = r_{ik} - 0.5 \quad \forall i, j, k = 1, 2, \cdots, n$$

Additive transitivity for fuzzy preference relations is equivalent to Saaty's consistency property [21] for multiplicative preference relations [15]. However, it is also a fact that additive transitivity is in conflict with the $[0, 1]$ scale used for providing the preference values and therefore it is not appropriate to model consistency of fuzzy preference relations [8]. An alternative transitivity property for fuzzy preference relations to additive transitivity was also proposed by Tanino [23]:

Definition 5 (Multiplicative transitivity). *A fuzzy preference relation $R = (r_{ij})$ on a finite set of alternatives X is multiplicative transitive if and only if*

$$r_{ij} \cdot r_{jk} \cdot r_{ki} = r_{ik} \cdot r_{kj} \cdot r_{ji} \quad \forall i, k, j \in \{1, 2, \dots n\} \tag{1}$$

Multiplicative transitivity extends weak stochastic transitivity, and therefore extends the classical transitivity property of crisp preference relations.

The modelling of cardinal consistency of reciprocal fuzzy preference relations via a functional equation was proposed in [8], and it was proved that when such a function is almost continuous and monotonic (increasing) then it must be a representable uninorm. Furthermore, cardinal consistency with the conjunctive representable cross ratio uninorm

$$U(x, y) = \begin{cases} 0, & (x, y) \in \{(0, 1), (1, 0)\} \\ \dfrac{xy}{xy + (1 - x)(1 - y)}, & \text{otherwise} \end{cases} \tag{2}$$

is equivalent to Tanino's multiplicative transitivity property as per Definition 5. As any two representable uninorms are order isomorphic, it was concluded that multiplicative transitivity is the most appropriate property to model consistency of reciprocal fuzzy preference relations. This property is referred though as the *multiplicative consistency* property.

Multiplicative consistency property (1) allows to estimate the (fuzzy) preference value between a pair of alternatives (x_i, x_j) with $(i < j)$ using a different intermediate alternative x_k $(k \neq i, j)$, mr_{ij}^k, as

$$mr_{ij}^k = \frac{r_{ik} \cdot r_{kj} \cdot r_{ji}}{r_{jk} \cdot r_{ki}} \tag{3}$$

so long as the denominator is not zero. The value mr_{ij}^k is known as the partially multiplicative transitivity based estimated fuzzy preference value of the pair of alternatives (x_i, x_j) obtained using the intermediate alternative x_k.

The following points are noted:

- Expression (1) is always true when two of the three subindexes in $\{i, j, k\}$ are equal.
- When $k = i$ and $r_{ji} \neq 0$ then $mr_{ij}^i = r_{ij}$, while when $r_{ij} \neq 0$ then $mr_{ji}^i = r_{ji}$. Because $r_{ji} = 1 - r_{ij}$, then we have that: $r_{ji} \neq 0$ if and only if $r_{ij} \neq 1$. Thus, if $k = i$ and $(r_{ij}, r_{ji}) \notin \{(1,0), (0,1)\}$ we have $mr_{ij}^i = r_{ij}$ and $mr_{ji}^i = r_{ji}$.
- A similar reasoning and conclusion is obtained when $k = j$.
- Although it is possible to obtain the multiplicative transitivity based estimated fuzzy preference value of the pair of alternatives (x_i, x_j) when $k \in \{i, j\}$ and $(r_{ij}, r_{ji}) \notin \{(1,0), (0,1)\}$, it is also true that there is no indirect estimation process as described above.
- When the fuzzy preference value r_{ij} is unknown its estimation will automatically require that $k \notin \{i, j\}$.
- Finally, when $i = j$ we have by definition that $r_{ii} = 0.5$ and we would have $mr_{ii}^k = r_{ii}$ whenever $r_{ik} \notin \{(0,1), (1,0)\}$. Thus, this case will not be relevant when having incomplete information.

Thus, the global multiplicative transitivity based estimated value of the fuzzy preference value of the pair of alternatives (x_i, x_j) is defined as the following average of partially multiplicative transitivity based estimated values

$$mr_{ij} = \frac{\sum\limits_{k \in R_{ij}^{01}} mr_{ij}^k}{\# R_{ij}^{01}};$$

where $R_{ij}^{01} = \{k \neq i, j | (r_{ik}, r_{kj}) \notin R^{01}\}$, $R^{01} = \{(1,0), (0,1)\}$, and $\# R_{ij}^{01}$ is the cardinality of R_{ij}^{01}.

Given a fuzzy preference relation, $R = (r_{ij})$, its multiplicative transitivity based fuzzy preference relation, $MR = (mr_{ij})$, can be constructed. If R is multiplicative transitive then (1) holds $\forall i, j, k$, and we have

$$r_{ij} = \frac{r_{ik} \cdot r_{kj} \cdot r_{ji}}{r_{jk} \cdot r_{ki}};$$

whenever $k \in R_{ij}^{01}$. Consequently, $mr_{ij}^k = r_{ij}$ $\forall i, j, k \in R_{ij}^{01}$ and therefore it is $r_{ij} = mr_{ij}$ $\forall i, j$. The following alternative definition of multiplicative transitivity for fuzzy preference relations is justified.

Definition 6 (Multiplicative Consistency). *A fuzzy preference relation* $R = (r_{ij})$ *is multiplicative consistent if and only if* $R = MR$.

The similarity value between the fuzzy preference relation, R, and its multiplicative transitivity based fuzzy preference relation, MR is defined as follows:

$$CL = \frac{\sum_{i,j=1;\ i\neq j}^{n} CL_{ij}}{n \cdot (n-1)}.$$

where $CL_{ij} = 1 - d(r_{ij}, mr_{ij})$ $\forall i, j$, and $d(r_{ij}, mr_{ij})$ represents the distance between the values r_{ij} and mr_{ij}. We have the following:

- If R is multiplicative consistent then it is $r_{ij} = mr_{ij}$ $\forall i, j$. Consequently, $d(r_{ij}, mr_{ij}) = 0$ $\forall i, j$, i.e. $CL = 1$.
- If $CL = 1$ then it is $\sum_{i,j=1,i\neq j}^{n} CL_{ij} = n \times (n-1)$, and because $CL_{ij} \in [0,1]$ then $CL_{ij} = 1$ $\forall i \neq j$, i.e. $r_{ij} = mr_{ij}$ $\forall i \neq j$. Finally, when $i = j$ we have $mr_{ii}^{k} = r_{ii} = 0.5$ whenever $r_{ik} \notin \{(0,1),(1,0)\}$, and therefore $mr_{ii} = 0.5$ $\forall i$. Thus, we have $r_{ij} = mr_{ij}$ $\forall i, j$, i.e. $R = MR$, and R is multiplicative consistent.

This proves that a fuzzy preference relation R is multiplicative consistent if and only if $CL = 1$, and therefore provides a characterisation of multiplicative consistency of a fuzzy preference relation based on its similarity value to its multiplicative transitivity based fuzzy preference relation.

4   Reciprocal Intuitionistic Fuzzy Preference Relations and Asymmetric Fuzzy Preference Relations

Let denote with $\mathcal{B}$ the set of reciprocal intuitionistic fuzzy preference relations:

$$\mathcal{B} = \Big\{ B = (b_{ij}) | \forall ij :\ b_{ij} = \langle \mu_{ij}, \nu_{ij} \rangle,\ \mu_{ij}, \nu_{ij} \in [0,1],$$

$$\mu_{ii} = \nu_{ii} = 0.5\, \mu_{ij} = \nu_{ji},\ 0 \leq \mu_{ij} + \nu_{ij} \leq 1 \Big\} \quad (4)$$

and with $\mathcal{R}$ the set of fuzzy preference relations

$$\mathcal{R} = \Big\{ R = (r_{ij}) | \forall ij :\ r_{ij} \in [0,1] \Big\}$$

Let $f : [0,1] \times [0,1] \longrightarrow [0,1]$ be the following function $f(x_1, x_2) = x_1$. We can define the following mapping, $F : \mathcal{B} \longrightarrow \mathcal{R}$, between the set of reciprocal intuitionistic fuzzy preference relations, $\mathcal{B}$, and the set of fuzzy preference relations, $\mathcal{R}$,

$$R = F(B) = (f(b_{ij})) = (\mu_{ij}).$$

We have:

– Function F is well defined, i.e. given $B \in \mathcal{B}$ it is true that $f(B) \in \mathcal{R}$.
– Function F is an injection. Indeed, let $B_1 = (b_{ij}^1)$ and $B_2 = (b_{ij}^2)$ be two reciprocal intuitionistic fuzzy preference relations such that $F(B_1) = F(B_2)$. Then we have that

$$f(b_{ij}^1) = f(b_{ij}^2) \Leftrightarrow \mu_{ij}^1 = \mu_{ij}^2 \ \forall i, j.$$

Because $\mu_{ij}^1 = \nu_{ji}^1$ and $\mu_{ij}^2 = \nu_{ji}^2$ then it is obvious that

$$\nu_{ij}^1 = \nu_{ij}^2 \ \forall i, j.$$

Therefore we have that

$$b_{ij}^1 = \langle \mu_{ij}^1, \nu_{ij}^1 \rangle = \langle \mu_{ij}^2, \nu_{ij}^2 \rangle = b_{ij}^2 \ \forall i, j.$$

Consequently, it is concluded that

$$B_1 = B_2.$$

– Function F is not a surjection as not all fuzzy preference relations $R \in \mathcal{R}$ verify $0 \le r_{ij} + r_{ji} \le 1$. Thus the range of function function f is the set of asymmetric fuzzy preference relations.

Summarising:
The set of reciprocal intuitionistic fuzzy preference relations is isomorphic to the set of asymmetric fuzzy preference relations.

5   Estimating Unknown Values in Incomplete Reciprocal Intuitionistic Fuzzy Preference Relations

It is assumed that for incomplete reciprocal intuitionistic fuzzy preference relations, given a pair of alternatives (x_i, x_j) for which b_{ij} is not known, both membership and non-memberships will be unknown. Due to reciprocity, we have that if b_{ij} is not known then b_{ji} is also not known.

If B is an incomplete reciprocal intuitionistic fuzzy preference relation, then $R = F(B)$ will be an incomplete asymmetric fuzzy preference relation. However, the missing preference value r_{ij} $(i \ne j)$ cannot be partially estimated, using an intermediate alternative x_k, via expression (1) because r_{ji} is also unknown. In these cases we use expression (2). Thus the missing preference value $r_{ij}(i \ne j)$ can be partially estimated, using an intermediate alternative x_k, with the value:

$$cr_{ij}^k = \begin{cases} 0, & (r_{ik}, r_{kj}) \in \{(0,1), (1,0)\} \\ \dfrac{r_{ik} \cdot r_{kj}}{r_{ik} \cdot r_{kj} + (1 - r_{ik}) \cdot (1 - r_{kj})}, & \text{Otherwise.} \end{cases} \tag{5}$$

The global multiplicative transitivity based estimated value, cr_{ij}, is defined as:

$$cr_{ij} = \frac{\sum\limits_{k \in R_{ij}^{01}} cr_{ij}^{k}}{\# R_{ij}^{01}}$$

where $H_{ij}^{01} = \{k \in R_{ij}^{01} | (i, j) \in MV \,\&\, (i, k), (k, j) \in EV\}$; MV is the set of pairs of different alternatives for which the fuzzy preference degree is unknown or missing; EV is the set of pairs of different alternatives with known fuzzy preference values.

The iterative procedure to complete reciprocal fuzzy preference relations developed in [14] can be applied to complete R and, consequently, to complete B as the following example illustrates.

Example 1. Let $X = \{x_1, x_2, x_3, x_4\}$ be a set of alternatives evaluated by a decision maker against a particular criterion using the following incomplete reciprocal intuitionistic fuzzy preference relation [25]:

$$B = \begin{pmatrix} \langle 0.50, 0.50 \rangle & \langle 0.40, 0.30 \rangle & x & x \\ \langle 0.30, 0.40 \rangle & \langle 0.50, 0.50 \rangle & \langle 0.50, 0.40 \rangle & x \\ x & \langle 0.40, 0.50 \rangle & \langle 0.50, 0.50 \rangle & \langle 0.30, 0.40 \rangle \\ x & x & \langle 0.40, 0.30 \rangle & \langle 0.50, 0.50 \rangle \end{pmatrix}$$

The associated incomplete asymmetric fuzzy preference relation is:

$$R = \begin{pmatrix} 0.5 & 0.4 & - & - \\ 0.3 & 0.5 & 0.5 & - \\ - & 0.4 & 0.5 & 0.3 \\ - & - & 0.4 & 0.5 \end{pmatrix}$$

Step 1: The set of elements that can be estimated at this stage are:

$$EMV_1 = \{(1, 3), (2, 4), (3, 1), (4, 2)\}.$$

The computation of the estimated values cr_{13} and cr_{31} requires the intermediate alternative $k = 2$, for which we have

$$cr_{13}^2 = \frac{r_{12} \cdot r_{23}}{r_{12} \cdot r_{23} + (1 - r_{12}) \cdot (1 - r_{23})} = \frac{0.4 \cdot 0.5}{0.4 \cdot 0.5 + 0.6 \cdot 0.5} = 0.4,$$

and

$$cr_{31}^2 = \frac{r_{32} \cdot r_{21}}{r_{32} \cdot r_{21} + (1 - r_{32}) \cdot (1 - r_{21})} = \frac{0.4 \cdot 0.3}{0.4 \cdot 0.3 + 0.6 \cdot 0.7} = 0.22.$$

The computation of the estimated values cr_{24} and cr_{42} is done using intermediate alternative $k = 3$

$$cr_{24}^3 = \frac{r_{23} \cdot r_{34}}{r_{23} \cdot r_{34} + (1 - r_{23}) \cdot (1 - r_{34})} = \frac{0.5 \cdot 0.3}{0.5 \cdot 0.3 + 0.5 \cdot 0.7} = 0.3,$$

and

$$cr_{42}^3 = \frac{r_{43} \cdot r_{32}}{r_{43} \cdot r_{32} + (1 - r_{43}) \cdot (1 - r_{32})} = \frac{0.4 \cdot 0.4}{0.4 \cdot 0.4 + 0.6 \cdot 0.6} = 0.31.$$

After the estimation process is applied, we have:

$$R = \begin{pmatrix} 0.5 & 0.4 & \mathbf{0.4} & - \\ 0.3 & 0.5 & 0.5 & \mathbf{0.3} \\ \mathbf{0.22} & 0.4 & 0.5 & 0.3 \\ - & \mathbf{0.31} & 0.4 & 0.5 \end{pmatrix}$$

Step 2: The remaining unknown elements can be estimated at this stage, $EMV_2 = \{(1, 4), (4, 1)\}$. The computation process of the estimated values are:

$$cr_{14}^2 = \frac{r_{12} \cdot r_{24}}{r_{12} \cdot c_{24} + (1 - c_{12}) \cdot (1 - c_{24})} = \frac{0.4 \cdot 0.3}{0.4 \cdot 0.3 + 0.6 \cdot 0.7} = 0.22;$$

$$cr_{14}^3 = \frac{r_{13} \cdot r_{34}}{r_{13} \cdot r_{34} - (1 - r_{13}) \cdot (1 - r_{34})} = \frac{0.4 \cdot 0.3}{0.4 \cdot 0.3 + 0.6 \cdot 0.7} = 0.22;$$

$$cr_{14} = \frac{cr_{14}^2 + cr_{14}^3}{2} = 0.22.$$

$$cr_{41}^2 = \frac{r_{42} \cdot r_{21}}{r_{42} \cdot c_{21} + (1 - c_{42}) \cdot (1 - c_{21})} = \frac{0.31 \cdot 0.3}{0.31 \cdot 0.3 + 0.69 \cdot 0.7} = 0.16;$$

$$cr_{41}^3 = \frac{r_{43} \cdot r_{31}}{r_{43} \cdot r_{31} - (1 - r_{43}) \cdot (1 - r_{31})} = \frac{0.4 \cdot 0.22}{0.4 \cdot 0.22 + 0.6 \cdot 0.78} = 0.16;$$

$$cr_{41} = \frac{cr_{41}^2 + cr_{41}^3}{2} = 0.16.$$

The following completed asymmetric fuzzy preference relation R is obtained:

$$R = \begin{pmatrix} 0.5 & 0.4 & \mathbf{0.4} & \mathbf{0.22} \\ 0.3 & 0.5 & 0.5 & \mathbf{0.3} \\ \mathbf{0.22} & 0.4 & 0.5 & 0.3 \\ \mathbf{0.16} & \mathbf{0.31} & 0.4 & 0.5 \end{pmatrix}$$

The complete reciprocal intuitionistic fuzzy preference relation is:

$$B = F^{-1}(R) = \begin{pmatrix} \langle 0.50, 0.50 \rangle & \langle 0.40, 0.30 \rangle & \mathbf{\langle 0.40, 0.22 \rangle} & \mathbf{\langle 0.22, 0.16 \rangle} \\ \langle 0.30, 0.40 \rangle & \langle 0.50, 0.50 \rangle & \langle 0.50, 0.40 \rangle & \mathbf{\langle 0.30, 0.31 \rangle} \\ \mathbf{\langle 0.22, 0.40 \rangle} & \langle 0.40, 0.50 \rangle & \langle 0.50, 0.50 \rangle & \langle 0.30, 0.40 \rangle \\ \mathbf{\langle 0.16, 0.22 \rangle} & \mathbf{\langle 0.31, 0.30 \rangle} & \langle 0.40, 0.30 \rangle & \langle 0.50, 0.50 \rangle \end{pmatrix}$$

Notice that the completed reciprocal intuitionistic fuzzy preference relation obtained coincides with the one in [25], where there was a typo in b_{41} (b_{14}) that appeared as $\langle 0.19, 0.22 \rangle$ ($\langle 0.22, 0.19 \rangle$) instead of the correct one shown here.

6 Conclusion

The set of asymmetric fuzzy preference relations is isomorphic to the set of reciprocal intuitionistic fuzzy preference relations. This result is important because it allows to use methodologies developed for fuzzy preference relations to the case of intuitionistic fuzzy preference relations and, ultimately, to overcome their associated computation complexity and to extend the use of reciprocal intuitionistic fuzzy preference relations in decision making. Indeed, this result has been exploited here to address the issue of incomplete reciprocal intuitionistic fuzzy preference relations in decision making.

Acknowledgments. This work has been developed with the financing of the Andalusian Excellence research project TIC-5991 and FEDER funds in the Spanish National research project TIN2013-40658-P. Raquel Ureña would like to acknowledge the support received by the mobility grant program awarded by the University of Granada's International Office. Prof. Francisco Chiclana and Prof. Hamido Fujita would like to acknowledge the support provided by the University of Granada 'Strengthening through Short-Visits' (Ref. GENIL-SSV 2015) programme.

References

1. Alonso, S., Cabrerizo, F., Chiclana, F., Herrera, F., Herrera-Viedma, E.: Group decision making with incomplete fuzzy linguistic preference relations. Int. J. Intell. Syst. **24**(2), 201–222 (2009a)
2. Alonso, S., Chiclana, F., Herrera, F., Herrera-Viedma, E.: A Learning Procedure to Estimate Missing Values in Fuzzy Preference Relations Based on Additive Consistency. In: Torra, V., Narukawa, Y. (eds.) MDAI 2004. LNCS (LNAI), vol. 3131, pp. 227–238. Springer, Heidelberg (2004)
3. Alonso, S., Chiclana, F., Herrera, F., Herrera-Viedma, E., Alcalá-Fdez, J., Porcel, C.: A consistency-based procedure to estimate missing pairwise preference values. Int. J. Intell. Syst. **23**(2), 155–175 (2008)
4. Alonso, S., Herrera-Viedma, E., Chiclana, F., Herrera, F.: Individual and social strategies to deal with ignorance situations in multi-person decision making. Int. J. Inf. Technol. Decis. Making **8**(2), 313–333 (2009b)
5. Alonso, S., Herrera-Viedma, E., Chiclana, F., Herrera, F.: A web based consensus support system for group decision making problems and incomplete preferences. Inf. Sci. **180**(23), 4477–4495 (2010)
6. Atanassov, K.T.: Intuitionistic fuzzy sets. Fuzzy Sets Syst. **20**(1), 87–96 (1986)
7. Bezdek, J., Spillman, B., Spillman, R.: A fuzzy relation space for group decision-theory. Fuzzy Sets Syst. **1**(4), 255–268 (1978)
8. Chiclana, F., Herrera-Viedma, E., Alonso, S., Herrera, F.: Cardinal consistency of reciprocal preference relations: a characterization of multiplicative transitivity. IEEE Trans. Fuzzy Syst. **17**(1), 14–23 (2009)
9. Ebenbach, D.H., Moore, C.: Incomplete information, inferences, and individual differences: the case of environmental judgments. Organ. Behav. Hum. Decis. Process. **81**(1), 1–27 (2000)
10. Fedrizzi, M., Giove, S.: Incomplete pairwise comparison and consistency optimization. Eur. J. Oper. Res. **183**(1), 303–313 (2007)

11. Fishburn, P.: Utility theory for decision making. Krieger, Melbourne (1979)
12. Genc, S., Boran, F.E., Akay, D., Xu, Z.: Interval multiplicative transitivity for consistency, missing values and priority weights of interval fuzzy preference relations. Inf. Sci. **180**(24), 4877–4891 (2010)
13. Herrera-Viedma, E., Alonso, S., Chiclana, F., Herrera, F.: A consensus model for group decision making with incomplete fuzzy preference relations. IEEE Trans. Fuzzy Syst. **15**(5), 863–877 (2007a)
14. Herrera-Viedma, E., Chiclana, F., Herrera, F., Alonso, S.: Group decision-making model with incomplete fuzzy preference relations based on additive consistency. IEEE Trans. Syst. Man Cybern. Part B Cybern. **37**(1), 176–189 (2007b)
15. Herrera-Viedma, E., Herrera, F., Chiclana, F., Luque, M.: Some issues on consistency of fuzzy preference relations. Eur. J. Oper. Res. **154**(1), 98–109 (2004)
16. Lee, L.-W.: Group decision making with incomplete fuzzy preference relations based on the additive consistency and the order consistency. Expert Syst. Appl. **39**(14), 11666–11676 (2012)
17. Liu, X., Pan, Y., Xu, Y., Yu, S.: Least square completion and inconsistency repair methods for additively consistent fuzzy preference relations. Fuzzy Sets Syst. **198**(1), 1–19 (2012)
18. Luce, R.D., Suppes, P.: Preferences, utility and subject probability. In: Handbook of Mathematical Psychology, New York, vol. 3 (1965)
19. Pérez-Asurmendi, P., Chiclana, F.: Linguistic majorities with difference in support. Appl. Soft Comput. **18**, 196–208 (2014)
20. Roubens, M., Vincke, P.: Preference modeling. Springer, Berlin (1985)
21. Saaty, T.L.: The Analytic Hierarchy Process. McGraw-Hill, New York (1980)
22. Szmidt, E., Kacprzyk, J.: Using intuitionistic fuzzy sets in group decision making. Control and Cybern. **31**(4), 1037–1053 (2002)
23. Tanino, T.: Fuzzy preference orderings in group decision making. Fuzzy Sets Syst. **12**, 117–131 (1984)
24. Ureña, R., Chiclana, F., Morente-Molinera, J., Herrera-Viedma, E.: Managing incomplete preference relations in decision making: a review and future trends. Inf. Sci. **302**, 14–32 (2015)
25. Wu, J., Chiclana, F.: Multiplicative consistency of intuitionistic reciprocal preference relations and its application to missing values estimation and consensus building. Knowl.-Based Syst. **71**, 187–200 (2014)
26. Xu, Z., Cai, X., Szmidt, E.: Algorithms for estimating missing elements of incomplete intuitionistic preference relations. Int. J. Intell. Syst. **26**(9), 787–813 (2011)
27. Xu, Z., Liao, H.: A survey of approaches to decision making with intuitionistic fuzzy preference relations. Knowledge-Based Systems, 2014, doi: 10.1016/j.knosys.2014.12.034 (2015 in press)
28. Zadeh, L.A.: Fuzzy sets. Inf. Control **8**(3), 338–357 (1965)
29. Zadeh, L.A.: The concept of a linguistic variable and its application to approximate reasoning-i. Inf. Sci. **8**, 199–249 (1975)

Handling Risk Attitudes for Preference Learning and Intelligent Decision Support

Camilo Franco$^{(\boxtimes)}$, Jens Leth Hougaard, and Kurt Nielsen

IFRO, Faculty of Science, Copenhagen University, Frederiksberg, Denmark
cf@ifro.ku.dk

Abstract. Intelligent decision support should allow integrating human knowledge with efficient algorithms for making interpretable and useful recommendations on real world decision problems. Attitudes and preferences articulate and come together under a decision process that should be explicitly modeled for understanding and solving the inherent conflict of decision making. Here, risk attitudes are represented by means of fuzzy-linguistic structures, and an interactive methodology is proposed for learning preferences from a group of decision makers (DMs). The methodology is built on a multi-criteria framework allowing imprecise observations/measurements, where DMs reveal their attitudes in linguistic form and receive from the system their associated type, characterized by a preference order of the alternatives, together with the amount of consensus and dissention existing among the group. Following on the system's feedback, DMs can negotiate on a common attitude while searching for a satisfactory decision.

Keywords: Interval mutlicriteria · Fuzzy-linguistic structures · Human-system interaction · Consensus-dissention · Social decision making

1 Introduction

Uncertainty is naturally present in real-world decision problems. In fact, uncertainty is always present in human evaluations, measurements and judgments, which represent the available information that has to be dealt with for gaining relevant knowledge and making decisions. Under this view, support is required to give decision makers (DMs) useful and insightful feedback for arriving at satisfactory solutions. Based on multi-criteria decision modeling (see e.g. [5,13,22]), in particular the Weighted Overlap Dominance (WOD) procedure [13] which deals with imprecise (interval) data problems, we address the specific challenge of handling risk decision attitudes for *intelligent decision support* (see e.g. [6,10,16,27,28]).

The decision support system (DSS) process dynamics that will be examined throughout this paper is illustrated in Fig. 1, being composed by three main phases, namely INFO, WOD and IACT:

© Springer International Publishing Switzerland 2015
V. Torra and Y. Narakawa (Eds.): MDAI 2015, LNAI 9321, pp. 78–89, 2015.
DOI: 10.1007/978-3-319-23240-9_7

1. INFO. All the available information is introduced into the system, consisting in a fixed set of alternatives, a given set of interval-valued criteria with their respective weights, and the risk attitudes of DMs.
2. WOD. For every DM, alternatives are ordered according to their weighted multi-dimensional interval scores, obtaining for every pair of alternatives either a dominance/outranking or an indifference relation.
3. IACT. The system learns the type of every DM according to an associated preference order, measuring the amount of consensus and dissention among types, so DMs can negotiate/rectify their attitudes, restarting the process at INFO while searching for a satisfactory/optimal solution. The process stops when no further consensus can be reached.

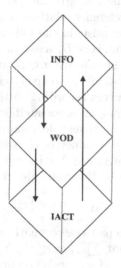

Fig. 1. The DSS process dynamics

Focusing on WOD, it is stated that one alternative outranks/dominates another one if there is *sufficient evidence* for affirming so, otherwise they are considered to be indifferent [7, 13]. Under this approach, the *verification* of sufficient evidence is examined in relation to the *risk attitude* of the DM (as it will be examined in detail in Sect. 4). Hence, the inherent conflict of the multicriteria problem, associated to the *incomparability* [20] among alternatives, can be explained by learning the different attitudinal types of DMs, like opposing postures (sources of disagreement) which have to come closer together for finding a *social decision*. In this way, the objective of this paper is to establish a decision support methodology that builds useful and reliable knowledge from the linguistic interaction with DMs, aiding their negotiation process while searching for results with greater *coherence* among them, maximizing group consensus by reducing (pairwise) minimal dissention among types.

In order to do so, this paper is organized as follows. Section 2 offers an outline of the WOD inference process as it was originally presented in [13]. Section 3 introduces fuzzy-lingustic structures, presenting the preliminary concepts that are used in Sect. 4 for modeling risk attitudes. In Sect. 5 the methodology for learning the types of DMs is explained, and in Sect. 6 the DSS human-system interaction is summarized under Algorithm 3, producing decision support while searching for an agreement on the social solution. Finally there are some notes and comments concerning open problems for future research.

2   Inferring Preferences from Imprecise Data

The WOD procedure [13] allows coping with the natural imprecision of real life observations and measurements, as given by interval values. This procedure makes use of criteria weights and risk attitude parameters to make sense of the interval data, identifying the preference relations holding among the alternatives. In short, the WOD procedure consists in the following.

Consider a set of decision makers D, a set of alternatives N and a set of criteria C, such that for every alternative $a \in N$ and criterion $i \in C$ there is a lower and upper bounded valuation, respectively given by $x_{ai}^L, x_{ai}^U \in [0,1]$, such that $x_{ai}^L \leq x_{ai}^U$, scoring alternatives according to the characteristic property of the criterion. Every criterion has an associated weight expressing its relative importance, given by $w_i \in \mathbb{R}^+$, and every decision maker $e \in D$ has a subjective decision attitude represented by parameters $\beta_e \in [0,1]$ and $\gamma_e \in \mathbb{R}^+$.

Therefore, for every alternative $a \in N$, the suitability of a regarding the set of criteria C, $|C| = m$, is given by the multi-dimensional (hyper) cube,

$$c_a = \left[x_{a1}^L, x_{a1}^U \right] \times \cdots \times \left[x_{am}^L, x_{am}^U \right]. \tag{1}$$

Based on this information, a pairwise comparison process is developed among alternatives $a, b \in N$, such that $\sum_{i=1}^m w_i x_{ai}^U \geq \sum_{i=1}^m w_i x_{bi}^U$. According to the amount of overlap between c_a and c_b, the WOD procedure infers the preference relation holding among a and b. There are three kinds of overlap, namely *no overlap*, *partial overlap* and *complete overlap*. In the case of *no overlap*, such that

$$\sum_{i=1}^m w_i x_{ai}^L > \sum_{i=1}^m w_i x_{bi}^U, \tag{2}$$

it certainly holds that a dominates b, which is represented by the *outranking relation* $\succ$, such that

$$a \succ b. \tag{3}$$

On the other hand, if there is *partial overlap*, such that

$$\sum_{i=1}^m w_i x_{ai}^L > \sum_{i=1}^m w_i x_{bi}^L \tag{4}$$

and

$$\sum_{i=1}^m w_i x_{ai}^U > \sum_{i=1}^m w_i x_{bi}^U, \tag{5}$$

then it holds that,

$$a \succ b \;\Leftrightarrow\; P(a,b) > \beta. \tag{6}$$

Here $P(a,b)$ expresses a proxy for the likelihood that alternative a in fact dominates alternative b, due to the possibility that some point in (or randomly taken from) c_a can be greater than another point from c_b (see [13] for a specific example on how to estimate such proxy). This likelihood has to be higher than β in order for a to outrank b. Otherwise, if

$$P \leq \beta, \tag{7}$$

then both alternatives are said to be *indifferent*, such that

$$a \sim b. \tag{8}$$

Lastly, if there is *complete overlap*, such that

$$\sum_{i=1}^{m} w_i x_{ai}^L < \sum_{i=1}^{m} w_i x_{bi}^L \tag{9}$$

and

$$\sum_{i=1}^{m} w_i x_{ai}^U > \sum_{i=1}^{m} w_i x_{bi}^U, \tag{10}$$

then it holds that,

$$a \succ b \;\Leftrightarrow\; G(a,b) > \gamma, \tag{11}$$

where $G(a,b)$ expresses the likelihood that any point belonging to c_a is greater than any other point in c_b (see again [13] for more details). Hence, if $G(a,b)$ is greater than γ, it holds that $a \succ b$. Otherwise, it either holds that $b \succ a$ or $a \sim b$ if it is respectively verified that $G(a,b)$ is less than or equal to γ.

Notice that the indifference relation of the WOD procedure, due to the interval nature of data, does not hold as a transitive or equivalence relation. Therefore, the outranking order assigned on N is *semi-transitive*, such that for every $a,b,c \in N$ it holds that $a \succ b, b \succ c \not\succ c \succ a$ (see again [13] but also [7]).

Under this framework, the parameters β and γ denote risk thresholds for establishing an outranking relation, such that their meaning is being modeled in direct relation to a crisp number. On the other hand, acknowledging the general character of words, concepts and perceptions, it is necessary to take a closer look at the correspondence between DMs' risk attitudes and their numerical translation/estimation. Thus, a given attitude should at least refer to a set of values, which under an explicit semantic structure, allows incorporating the gradualness and generality of its numerical estimation.

In order to undertake computations with attitudes under the DSS (see again Fig. 1), the estimation of linguistic values for β and γ can be examined through the computing with words and perceptions paradigm (see [30–32], but also [18,23]). Thus, the following analysis is based on the intuition that *language* is the means to represent the subjective thinking process and the relation between perception and reality, enhancing the interaction with technology and the affective (decision-wise) states of DMs.

The complete procedure for the articulation of binary preference relations is specified under the WOD Algorithm 1. In the following section fuzzy-linguistic structures are introduced, which will be later used for undertaking a linguistic modelization of the attitudes explaining the β and γ parameters.

Algorithm 1. WOD algorithm

Input: For every $a \in N$ and $i \in C$, the hyper cubes c_a, the criteria weights w_i and for every $e \in D$, the risk attitude parameters β_e and γ_e.
Output: For every $e \in D$, a preference order on N.
$(WOD - 1)$ For every $a, b \in N$, establish an outranking or indifference relation according to (1)-(11).

---

3 Fuzzy-Linguistic Structures

Fuzzy logic [30, 31] allows representing the meaning of words and concepts, examining human reasoning through natural/ordinary language. Under this approach, commonly known as the Computing with Words paradigm [9, 11, 14, 32], words are taken as linguistic terms that are susceptible of being represented by fuzzy sets. Thus, through their associated membership functions, the meaning of fuzzy sets is supported by a particular structure maintaining a specific order among them (see e.g. [6, 17, 18]). Such structure is here referred to as a *fuzzy-linguistic structure*.

Addressing the general character of words, and in particular of gradable predicates that are susceptible of verification *up to a certain degree*, fuzzy sets are an appropriate tool for designing the means of such verification process. In this way, a fuzzy set representing a linguistic term can be regarded as containing a *core* and a *support*, such that its core is the subset of the universe U where the term is known to hold true, while its co-support consists in the subset of U where it is known that it does not hold true. Hence, there is a space in between the core and the co-support that can be gradually filled in by a continuous and monotone transition (in fact, the specific form of this transition is a matter of design [24, 25]), representing the intensity in which the elements of U verify the meaning of the fuzzy set. Thus, the elements belonging to the core are considered to have absolute intensity, while the ones belonging to the co-support have null intensity.

For a general valuation scale L, the membership function $\mu : U \to L$ can be expressed as an ordered quadruple of the ordinates $(\mu^1, \mu^2, \mu^3, \mu^4)$, such that the interval $[\mu^1, \mu^4]$ stands as the support and the interval $[\mu^2, \mu^3]$ stands as the core of the fuzzy set. So, for any pair of consecutive linguistic terms $l_t, l_{t+1} \in L$, respectively represented by μ_{l_t} and $\mu_{l_{t+1}}$, the order relation $\trianglelefteq$ is defined such that $\mu_{l_t} \trianglelefteq \mu_{l_{t+1}}$ holds only if $\mu^3_{l_t} \leq \mu^1_{l_{t+1}}$ and $\mu^4_{l_t} \leq \mu^2_{l_{t+1}}$. Now, *fuzzy-linguistic structures* can be defined as follows.

Definition 1. *Given a set of different and consecutive linguistic labels $L = \{l_1, l_2, ..., l_T\}$, where each label $l_t \in L$, $t = 1, 2, ..., T$, is represented by means of a fuzzy set with a membership function given by $\mu_{l_t} = (\mu_{l_t}^1, \mu_{l_t}^2, \mu_{l_t}^3, \mu_{l_t}^4)$, a fuzzy-linguistic structure is such that for any pair of consecutive labels $l_t, l_{t+1} \in L$, it holds that $\mu_{l_t} \trianglelefteq \mu_{l_{t+1}}$.*

In this way, a fuzzy-linguistic structure contains the reference ordered set of linguistic terms, such that l_1 and l_T are respectively the minimum and maximum objects of the structure. This approach can be further developed to handle words in a manner that is more approximate to natural language and its use of gradable predicates, taking into consideration *linguistic modifiers* and *linguistic aggregation operators*, following the initial proposal of [18] (see also [6]).

Linguistic modifiers can be defined as unary functions $M : L \rightarrow L$, such that their effect on the meaning of the terms can be either *compressing* or *expanding* [18]. A compressing M is such that for any $l_t \in L$, it holds that $M(l_t) \subset l_t$, while an expanding M is such that $l_t \subset M(l_t)$. Some examples for compressing M can be "very"-l_t, "strictly"-l_t or "strongly"-l_t, while for an expanding M, they can be "around"-l_t, "almost"-l_t or "roughly"-l_t.

For example, given a linguistic term $l_t \in L$ represented by means of the membership function μ_{l_t}, and given an averaging operator k, a *compressing* M, denoted by CM, is such that

$$CM(\mu_{l_t}) = (k(\mu_{l_t}^1, \mu_{l_t}^2), \mu_{l_t}^2, \mu_{l_t}^3, k(\mu_{l_t}^3, \mu_{l_t}^4)), \tag{12}$$

and an *expanding* M, denoted by EM, is such that

$$EM(\mu_{l_t}) = (\mu_{l_t}^1, k(\mu_{l_t}^1, \mu_{l_t}^2), k(\mu_{l_t}^2, \mu_{l_t}^3), \mu_{l_t}^4). \tag{13}$$

On the other hand, linguistic aggregation operators allow using the existing linguistic labels to generate new labels, such that new terms can appear *in between* any pair of consecutive terms, while maintaining the order among the linguistic components of the structure [6,18]. In this way, a new term can arise *in between* any pair $l_t, l_{t+1} \in L$, by means of an operator specifically designed for the inclusion of new linguistic labels.

Definition 2. *Given a fuzzy-linguistic structure, the in between linguistic aggregation operator is a mapping $LA : L^2 \rightarrow L$ such that for any pair of consecutive terms $l_t, l_{t+1} \in L$ and their associated fuzzy sets, it holds that $CM(\mu_{l_t}) \trianglelefteq LA(\mu_{l_t}, \mu_{l_{t+1}}) \trianglelefteq CM(\mu_{l_{t+1}})$.*

In this way, given two averaging operators k_1, k_2, such that for any pair of elements $u_1, u_2 \in U$ it holds that $k_1(u_1, u_2) \leq k_2(u_1, u_2)$, LA can be taken as in the following example, previously undertaking the compression of the consecutive terms, as in $\nu_{l_t} = CM(\mu_{l_t})$ and $\nu_{l_{t+1}} = CM(\mu_{l_{t+1}})$,

$$LA(l_t, l_{t+1}) = (\nu_{l_t}^3, k_1(\nu_{l_t}^3, \nu_{l_{t+1}}^2), k_2(\nu_{l_t}^3, \nu_{l_{t+1}}^2), \nu_{l_{t+1}}^2). \tag{14}$$

Under the general framework of fuzzy-linguistic structures, the design of different examples for M and LA can be further developed, including more linguistic terms and modifiers that preserve the order relation among every pair

$l_t, l_{t+1} \in L$, while enhancing the granularity of L as much as required (see [6, 18]). Its application for the representation and measurement of risk attitudes will be explored next.

4 Measuring Risk Attitudes with Fuzzy Linguistic Structures

Based on fuzzy-linguistic structures, the β and γ risk attitudes can be modelized and incorporated in the articulation of preferences under the WOD interactive decision process. The incorporation of attitudes is particularly relevant for decision support under imprecision, where attitudes play a central role (see e.g. [28, 29], but also [7]). In this sense, examining the meaning of risk as a concept which is used by DMs, the attitude towards risk can be measured on a linguistic scale built from the two opposite categories of *aversion* and *proneness* (see [1, 15, 19] for a general view on the evaluation of attitudes under different bipolar evaluation spaces).

As it has been examined in Sect. 2 and the DSS process dynamics of Fig. 1, attitudes guide the articulation of preferences through the interaction between the system and the group of DMs. In particular, attitudes towards risk refer to the amount of evidence needed to affirm an outranking relation for every pair $a, b \in N$, such that $a \succ b$ ($\succ$), instead of having that $a \sim b$ ($\sim$) or even that $b \succ a$ ($\succ^{-1}$), the latter only for the case of complete overlap and the parameter γ.

Therefore, high values of β correspond with a low risk attitude, because an outranking relation will only hold if there is a high amount of evidence existing in favor of $\succ$. In this way, β is defined over a scale with minimum element 0, denoting *high* risk, and a maximum element $K = 1$, denoting *low* risk, with an indeterminate space of *medium* risk consisting of being *in between* high and low risk attitudes (see Fig. 2). So, if β is close to 0, the attitude towards risk is considered to be of *risk proneness*, and if β is close to 1, then the attitude is considered to be of *risk aversion*, being the middle attitude regarded as *risk neutrality*. Notice that here neutrality refers to a middle attitude (as in [8, 19]), although a linearity between extreme and neutral attitudes may not necessarily hold (see e.g. [15, 17]).

On the other hand, on the contrary to the partial overlap case of β, γ refers to the three possibilities of obtaining $\succ$, $\sim$ or the inverse relation $\succ^{-1}$, where every time that $\succ$ does not hold, it reciprocally holds that $\succeq^{-1}$, such that $\succeq = \langle \succ, \sim \rangle$. Hence, γ is measured over a scale with a minimum element 0, denoting *high* risk for affirming $\succ$ (or inversely, low risk for affirming $\succ^{-1}$), and a maximum element $K \in \mathbb{R}^+$, denoting *low* risk for affirming $\succ$ (or inversely, high risk for affirming $\succ^{-1}$). Thus, there is some space for a *medium* state of risk consisting of being *in between* high and low risk attitudes (see again Fig. 2), where low values of γ denote a *risk prone* attitude, high values denote a *risk averse* attitude, and intemediate values denote a *risk neutral* attitude.

Overall, the risk attitude parameters β and γ refer to the measurement of attitudes with respect to three basic components, namely *proneness*, *neutrality*

Fig. 2. Measuring risk attitudes for affirming $\succ$ on a commonl linear scale for β and γ

and *aversion*, ordered according to a specific structure that holds among them. Acknowledging the general character of words, those terms naturally refer to a region or interval of the numerical scale, suggesting their correspondence with a set of numbers instead of a correspondence with a unique number. Even more, adjacent terms suggest a gradual intersection between them, where e.g. diminishing intensities of risk proneness may coincide with increasing intensities of risk neutrality.

In consequence, a risk attitude R can be measured with respect to a basic fuzzy-linguistic structure L^R, composed of at least the two opposite and most extreme linguistic labels (l_1, l_T) of proneness (l_1) and aversion (l_T), such that,

$$L^R = \{l_1 = prone, l_T = averse\}. \tag{15}$$

Based on this basic structure, the meaning of the terms can be modified, where it is possible for the decision maker to express linguistic grades of risk by attaching different words to the terms, such as "very"or "strictly" in the case of the compressing modifiers CM, or of "roughly"or "around" in the case of expanding EM. Besides, with the use of aggregation operators, such as the *in between* operator LA, new terms can emerge from any pair of consecutive terms, enabling the decision maker to create and use a new term for valuing attitudes. For example, the first new term consists in being neither "prone"nor "averse", but "in between"them, denoting the state of $l_2 = risk\ neutrality$ (see Fig. 3, where the opposite terms l_1 and l_T compress, making room for l_2). Following the same line of reasoning, the decision maker can be as specific as required, e.g. being "in between neutral and prone" or "roughly strongly-averse".

As a result, the attitude towards risk for DMs can be expressed by some (M or LA) modifed term in L^R, assigning linguistic values for computing with β and γ parameters. In this way, for every ordinate of the fuzzy set representing a given attitude, the WOD phase infers an order, so in the next phase the different types of DMs can be identified, as it will be examined in the next section.

5   Learning Types for Decision Support

Following the decision process, the system computes a preference order for every DM according to their attitudes. As it has been pointed out in Sect. 2, the outranking order resulting from Algorithm 1 is a semi-transitive one, such that a definite procedure can be used to further refine it and learn a weak order or ranking.

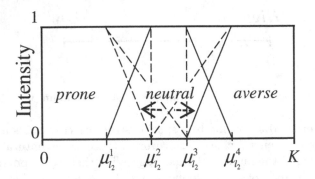

Fig. 3. Emergence of the middle term denoting a "neutral" attitude

Alternatives are ranked according to their *relevance* [7], taking into account the amount and the importance of the alternatives that they outrank. In this way, for every alternative $a \in N$, the relevance of a is given by $\sigma(a)$, such that,

$$\sigma(a) = s_a + \sum_{\forall b \in S_a} s_b, \tag{16}$$

where S_a is the set of alternatives that are dominated/outranked by a, and $s_a = |S_a|$.

The procedure for learning types of DMs is summarized under Algorithm 2. First, for every DM $e \in D$, there is a linguistic term denoting e's attitude, given by $l_t^e \in L^R$. The system then computes an (outranking semi-transitive) order on N, resulting from the WOD evaluation of every ordinate of the membership function representing l_t^e. All four ordinates are then aggregated into an overall ranking by means of (16). Having identified all the rankings that follow from the information given by DMs, the system returns the set of types Θ explaining the different attitudes.

Thus, different attitudes can obtain the same characteristic order, implying that the *type* of a DM can be completely described by a unique order and all the attitudes associated to it.

Algorithm 2. Learning Types (LT) algorithm

Input: For every $e \in DM$, the linguistic value l_t^e denoting their risk attitude.
Output: The set of types Θ.
$(LT - 1)$ Compute the WOD algorithm (1) for every ordinate of μ_{l_t}.
$(LT - 2)$ Aggregate the outranking orders associated to l_t^e by means of the relevance ranking operator (16).
$(LT - 3)$ Assign to every non-equivalent ranking a distinct type $q \in \Theta$.

Therefore, types $q \in \Theta$ are completely described by a ranking ρ^q of N and their associated risk attitudes $\{l_t\}_{\rho^q}$. Once the types of DMs are known, the system can offer support for resolving conflict among them, aiming at reducing the number of incomparable alternatives as it will be explained in the next section. The goal of the system focuses on using the linguistic interaction with and within DMs for maximizing consensus and arriving at a social satisfactory solution.

6   Intelligent Decision Support

The interactive human-system dynamics of Fig. 1 can now be addressed under the setting described by Algorithm 2, such that the decision process is guided towards reducing discrepancies among DMs. The system aids in identifying the predominant types and suggesting negotiation paths to arrive at an agreement or socially acceptable solution maximizing consensus.

Given the set of types θ, the group consensus is measured by the *general consensus index* $CI = 1/|\Theta|$. Complementing the information on consensus, *dissention degrees* are introduced here to measure the distances among pairs of types.

In this way, for every $a \in N$ and $q, q' \in \Theta$, the system computes the position of a in rankings ρ^q and $\rho^{q'}$, denoted respectively by ρ_a^q and $\rho_a^{q'}$, and obtains the overall dissention degree $ds(q, q')$, such that,

$$ds(q, q') = \sum_{\forall a \in N} dist(\rho_a^q, \rho_a^{q'}), \tag{17}$$

where *dist* represents a given distance measure (see e.g. [2]), like e.g. the 1-norm distance,

$$dist(\rho_a^q, \rho_a^{q'}) = |\rho_a^q - \rho_a^{q'}|. \tag{18}$$

The decision process aims at maximizing consensus (see Algorithm 3), based on the previous calculation of dissention degrees among all the different pairs of types. Thus, the system identifies all pairs $q, q' \in \Theta$ with minimal dissention, so DMs can look for an agreement among the nearest types, negotiating a common attitude that increases the general consensus index CI. In consequence, under the complete DSS process dynamics, attitudes not only guide the articulation of preferences through the interaction between the system and the group of DMs, but also (and under the same linguistic form) guide the negotiation among the different DMs.

Algorithm 3. Minimal dissention (MD) algorithm

Input: For every $e \in D$, the risk attitudes associated to e.
Output: All pairs $q, q' \in \Theta$ with minimal dissention.
$(MD - 1)$ For every $e \in D$, learn the type for e according to the LT-algorithm, identify the pairs $q, q' \in \Theta$ with minimal dissention and repeat for every new input until $CI = 1$ or no further negotiation is possible (CI remains constant for a fixed number of iterations).

---

7 Conclusions

A DSS methodology has been provided with the purpose of aiding the consensus/negotiation process between different DMs. The system infers individuals' preferences from their attitudes towards risk, learns the predominant types of DMs and measures the dissention and consensus among them. DMs can use the knowledge generated by the system to search for a satisfactory decision or identify the source of the conflict making it impossible to arrive at a unique (optimal) social solution.

It remains for further research to test and implement the DSS dynamic process in a real-case scenario, exploring the difficulties that may emerge in a real negotiation process. From a theoretical standpoint, the estimation for the likelihood of dominance between intervals/hypercubes remains to be explored in more detail, as well as the ranking and consensus procedures, which should be compared with other techniqes found in literature (see e.g. [3,4,12,21,26]).

References

1. Cacioppo, J.T., Berntson, G.G.: The affect system, architecture and operating characteristics - current directions. Psycological Sci. **8**, 133–137 (1999)
2. Chiclana, F., Tapia-Garcia, J.M., del Moral, M.J., Herrera-Viedma, E.: A statistical comparative study of different similarity measures of consensus in group decision making. Inf. Sci. **221**, 110–123 (2013)
3. Cook, W.D., Kress, M.: A data envelopment model for aggregating preference rankings. Manag. Sci. **36**, 1302–1310 (1990)
4. Emond, E.J., Mason, D.W.: A new rank correlation coefficient with application to the consensus ranking problem. J. Multi-Criteria Decis. Anal. **11**, 17–28 (2002)
5. Fodor, J., Roubens, M.: Fuzzy Preference Modelling and Multicriteria Decision Support. Kluwer Academic Publishers, Dordrecht (1994)
6. Franco, C.A.: On the analytic hierarchy process and decision support based on fuzzy-linguistic preference structures. Knowl.-Based Syst. **70**, 203–211 (2014)
7. Franco, C., Hougaard, J.L., Nielsen K.: Ranking alternatives based on imprecise multi-criteria data and pairwise overlap dominance relations. MSAP Working Papers Series (2014)
8. Grabisch, M., Greco, S., Pirlot, M.: Bipolar and bivariate models in multi-criteria decision analysis: descriptive and constructive approaches. Int. J. Intel. Syst. **23**, 930–969 (2008)
9. Herrera, F., Alonso, S., Chiclana, F., Herrera-Viedma, E.: Computing with words in decision making: foundations, trends and prospects. Fuzzy Optim. Decis. Making **8**, 337–364 (2009)
10. Herrera, F., Herrera-Viedma, E., Verdegay, J.L.: A model of consensus in group decision making under linguistic assessments. Fuzzy Sets Syst. **78**, 73–87 (1996)
11. Herrera, F., Martínez, L.: A 2-tuple fuzzy linguistic representation model for computing with words. IEEE Trans. Fuzzy Syst. **8**, 746–752 (2000)
12. Herrera-Viedma, E., Cabrerizo, F.J., Kacprzyk, J., Pedrycz, W.: A review of soft consensus models in a fuzzy environment. Inf. Fusion **17**, 4–13 (2014)
13. Hougaard, J.L., Nielsen, K.: Weighted overlap dominance - a procedure for interactive selection on multidimensional interval data. Appl. Math. Model. **35**, 3958–3969 (2011)

14. Kacprzyk, J., Zadrozny, S.: Computing with words in decision making through individual and collective linguistic choice rules. Int. J. Uncertainty, Fuzziness and Knowl.-Based Syst. **9**, 89–102 (2001)
15. Kaplan, K.: On the ambivalence-indifference problem in attitude theory and measurement: a suggested modification of the semantic differential technique. Psychol. Bull. **77**, 361–372 (1972)
16. Martínez, L., Ruan, D., Herrera, F.: Computing with words in decision support systems: An overview on models and applications. Int. J. Comput. Intell. Syst. **3**, 382–395 (2010)
17. Montero, J., Gómez, D., Bustince, H.: On the relevance of some families of fuzzy sets. Fuzzy Sets Syst. **158**, 2429–2442 (2007)
18. Moraga, C., Trillas, E.: A Computing with Words path from fuzzy logic to natural language. In: Proceedings IFSA-EUSFLAT 2009, pp. 687–692 (2009)
19. Osgood, Ch., Suci, G., Tannenbaum, P.: The Measurement of Meaning. University of Illinois Press, Urbana (1958)
20. Roy, B.: The outranking approach and the foundation of ELECTRE methods. Theor. Decis. **31**, 49–73 (1991)
21. Tastle, W.J., Wierman, M.J., Dumdum, U.R.: Ranking ordinal scales using the consensus measure. Issues Inf. Syst. **6**, 96–102 (2005)
22. Le Téno, J.F., Mareschal, B.: An interval version of PROMETHEE for the comparison of building products' design with ill-defined data on environmental quality. Eur. J. Oper. Res. **109**, 522–529 (1998)
23. Trillas, E.: On a model for the meaning of predicates - A naïve approach to the genesis of fuzzy sets. Stud. Fuzziness Soft Comput. **243**, 175–205 (2009)
24. Trillas, E., Guadarrama, S.: Fuzzy representations need a careful design. Int. J. General Syst. **39**, 329–346 (2010)
25. Trillas, E., Moraga, C.: Reasons for a careful design of fuzzy sets. In: Proceedings EUSFLAT **2013**, pp. 140–145 (2013)
26. Wierman, M.J., Tastle, W.J.: Consensus and dissention: theory and properties. In: Proceedings NAFIPS **2005**, pp. 75–79 (2005)
27. Wu, J., Chiclana, F.: Visual information feedback mechanism and attitudinal prioritisation method for group decision making with triangular fuzzy complementary preference relations. Inf. Sci. **279**, 716–734 (2014)
28. Yager, R.R.: Fuzzy modeling for intelligent decision making under uncertainty. IEEE Trans. Syst. Man Cybern. **30**, 60–70 (2000)
29. Yager, R.R.: An approach to ordinal decision making. Int. J. Approximate Reasoning **12**, 237–261 (1995)
30. Zadeh, L.A.: The concept of a linguistic variable and its application to approximate reasoning-I. Inf. Sci. **8**, 199–249 (1975)
31. Zadeh, L.A.: Fuzzy logic and approximate reasoning. Synthese **30**, 407–428 (1975)
32. Zadeh, L.A.: From computing with numbers to computing with words. From manipulation of measurements to manipulation of perceptions. IEEE Trans. Circ. Syst. **45**, 105–119 (1999)

Representation of Ordinal Preferences over Infinite Products

Marta Cardin[(✉)]

Department of Economics, Università Ca' Foscari Venezia, 30121 Venice, Italy
mcardin@unive.it

Abstract. In many decision processes data aggregation is required. In many models the need arises to aggregate data of varying dimension while aggregation operators are considered for a fixed number of arguments. In many contexts inputs to be aggregated are of a qualitative nature. This paper analyzes the evaluation of sequences of ordinal input and of variable length. We consider various axioms against which different ranking methods can be compared.

Keywords: Preference structure · Archimedean preference · Average utility · Ordered weighted maximum · Step-based preference structure

1 Introduction

Aggregation operators are mathematical functions that are used to fuse information of several inputs in a single outcome which is a very common need in particular in the area of artificial intelligence for decision making(see [12] for a general background on aggregation theory). Real-valued non-additive measures and their associated integrals are widely used aggregation operators. There are many situations where inputs to be aggregated are qualitative and numerical values are used by convenience. The aim of this paper is to generalize some well known aggregation functionals in a purely ordinal context. In this case only maximum and minimum are used for aggregation of different inputs.

Moreover in many cases aggregation operators are considered for a fixed number of arguments that is too restrictive in many important models (see examples in Sect. 2). In [5–7] and [11] are investigated extended aggregation operators satisfying a property called *arity-monotonicity* and a similar approach is considered in [4].

Let us consider a non empty set X and a relation $>$ on X expressing a cardinal or ordinal evaluation. We use a linear scale with a neutral point with "good" values above and "bad" values below. We address the problem of representability of orders defined in a infinite dimensional subset of X . The most classical representation is the numerical representability through real-valued functions, but we propose a general model that consider elements with different priority and so we do not assume an Archimedean axiom. We study an additive representation of a general preference structure without assuming an Archimedean property.

© Springer International Publishing Switzerland 2015
V. Torra and Y. Narakawa (Eds.): MDAI 2015, LNAI 9321, pp. 90–99, 2015.
DOI: 10.1007/978-3-319-23240-9_8

This property is necessary for a real valued representations but in many cases it is not empirically clear. The use of non-real representation is a strength of a model. Considering evaluation of scientific research, in many case it can be assumed that it is not possible to compensate an high level research product with some low level research products (e.g. a monograph with some papers). A research product can be evaluated by some citation indices, by considering different bibliometric database (ISI, Scopus or Google Scholar) but we have to consider also different evaluations practices. In a more general approach when we consider humanities and social science we have to evaluate different kind of products like monographs, edited volumes, chapters and translations.

In this paper we focus our attention on additive preference structure and we study step-based structure generalizing the approach in [8]. The paper will be set out as follows. In Sect. 2 we briefly mention some basic concepts, we provide the necessary definitions and we introduce some examples. Sect. 3 formulate our characterization of additive symmetric preference structure while Sect. 4 is devoted to step-based symmetric preference structures that are preference structures represented by an order weighted maximum while in Sect. 5 we briefly discuss possible directions for future work.

2 Notations and Settings

In this section we give some basic notations and terminology and we introduce our evaluation model from an axiomatic point of view. Some interpretations of our model are presented.

2.1 Definitions

First we recall some definitions. $\mathbb{N}$ is the set of positive integers. A *weak order* $\succeq$ on an non-empty set A is a transitive and complete binary relation on A. A *total order* (also called *linear order*) is an antisymmetric weak order. The notations $\succ, \preceq, \prec, \sim$ are as usual. So $\succ, \sim$ stand for the asymmetric and the symmetric parts of $\succeq$, respectively, i.e.,

$$a \succ b \text{ if and only if } \text{not } (b \succeq a), \quad a \sim b \text{ if and only if } (a \succeq b \text{ and } b \succeq a)$$

where a, b are elements of A. We assume that $(X, >)$ is a bipolar scale that is a totally ordered set with a prescribed element $\mathbb{O}$ separating the positive evaluations $x > \mathbb{O}$ from the negative evaluation $x < \mathbb{O}$. The non empty set X can be finite or infinite. The most obvious example of a *bipolar scale* is the real line and the most simple example is the set $I = \{-, \mathbb{O}, +\}$.

For every $n \in \mathbb{N}$ we consider the set X^n and we define $\mathbf{X} = \bigcup_{n \in} X^n$. Then $\mathbf{X}$ is the set of finite sequences of any length.

If $x = (x_1, \ldots, x_n) \in X^n$ and $y = (y_1, \ldots, y_m) \in X^m$ the *concatenation* (x, y) denotes $(x, y) = (x_1, \ldots, x_n, y_1, \ldots, y_m)$. Moreover $1x = x$ and $2x = (x, x)$ and $nx = (x, (n-1)x)$. We identify elements in X with corresponding sequences of length 1.

The main goal of this paper is to study weak order on the space $\mathbf{X}$. Let $\succeq$ be a weak order on $\mathbf{X}$ where the statement $a \succeq b$ is interpreted as meaning that a is at least as good as b. We assume that $\succeq$ is *increasing* i.e. such that

$$\text{if} \quad x, y \in X^n, \quad x_i \geq y_i \quad \text{for every } i, 1 \leq i \leq n \quad \text{then} \quad x \succeq y.$$

Moreover we suppose that $\succeq$ is *symmetric* weak order on $\mathbf{X}$ where symmetry holds if

$$(x_1, \ldots, x_n) \sim (x_{\pi(1)}, \ldots, x_{\pi(n)})$$

for every permutation π of $N = \{1, \ldots, n\}$.

A *symmetric preference structure* is a triple $(\mathbf{X}, >, \succeq)$ with the properties considered above.

2.2 Examples

We briefly apply the considered model to different fields.

(i) *Social Choice.* We consider social decisions where different individuals evaluate alternatives rather than rank them as in the Arrovian framework.

We are interested in different evaluating situations in which the set of voters as well as the the set of alternatives might vary. We consider $N = \{1, \ldots, n\}$ voters that are allowed to express their opinion on the candidates, social alternatives or proposals by assigning to each alternative an element of the set X. Voters can assign negative and positive evaluations to candidates and we can consider numerical evaluations, ordinal evaluations and also the case in which a voter can approve, disapprove or being neutral with respect to an alternative if for example $X = \{-, \mathbb{0}, +\}$. As a result every alternative is characterized by an element $x \in \mathbf{X}$ and we have to rank any pair of alternatives according to their label vectors. In this case symmetry is often called anonymity and so we assume that every individual is endowed with the same voting power as it is used in usual voting procedures. It is important to note that in our framework when X is different from $\{-, \mathbb{0}, +\}$ voters are able to express intensity of their preference.

(ii) *Measurement of Quality.* When we consider measurement of quality in particular perceived quality data are often only available at an ordinal level. University students can be viewed as "experts" evaluating university courses giving information about the quality of the didactics. In this case the set $N = \{1, \ldots, n\}$ represents a set of of individuals, X is the set of evaluations and the collective evaluation of course is represented by an element $x \in \mathbf{X}$.

(iii) *Research Evaluation.* Two factors mainly characterize the scientific production of a research unit (not only a individual author but also as a journal or a university department): the number of scientific papers and the impact or the importance of the considered papers. Different metrics can be introduced for evaluating the quality of a research product. The citation count is often considered the prevalent measure of research quality, but the meaning of citations can be ambiguous, thus citation-based statistics are not as objective as sometimes affirmed. At present, there are a huge number of indexes that have been proposed to evaluate scientists' research output and many of them are based also on citation statistics.

These research metrics in some cases are ambiguous also because a paper may be cited for positive or negative reasons. In a more general approach we can evaluate a paper with a negative or positive label and then a scholar with n papers can be represented by a vector $(x_1, \ldots, x_n)$ where x_i is the evaluation of the i-th paper. The symmetry property implies that the impact or the importance of a publication is independent on the impact of the preceding publications and so there is no reputation effect.

(iv) *Risk Measurement*. During the last decades the field of risk measurement has become of great importance to the finance and insurance management and, at the same time, it leads to interesting mathematical problems.

Risk measures constitute an important and widely studied tool and different families of risk measures have been proposed in the literature.

The paper that lays the foundations of the axiomatic approach in defining a risk measure is [2]. In 2007 Heyde et al. [13] introduced the natural risk statistics that are risk measures depending on data (se also [1]). The natural risk statistics are associated with a finite sample and satisfy a more general subadditivity assumption then that of classical coherent risk measures, and are robust, thus particularly suitable for external risk measurement. We assume that the behavior of a random loss is represented by a collection of data observation $x = (x_1, x_2, \ldots, x_n) \in \mathbb{R}^n$ (could be empirical or subjective or both). In this case obviously $X = \mathbb{R}$ and also in that case we have positive and negative elements in X. A risk statistics ρ is a mapping from the data in $\mathbb{R}^n$ to a numerical value in $\mathbb{R}$. One drawback of these risk measures is their dependence on the space dimension n. It is important to construct natural risk statistics defined for data samples of all sizes. In [3] it is introduced a class of data-based risk measures on the space of infinite sequences. However when data are available only at an ordinal scale, as is common for non financial companies, a quantitative approach is not possible.

2.3   Representation of a Symmetric Preference Structure

If $(X, \succeq)$ and $(Y, \succeq)$ are weak orders we say that $(X, \succeq)$ is representable in $(Y, \succeq)$ if there exists a map $f \colon X \to Y$ such that $f(x) \succeq f(x')$ if and only if $x \succeq x'$.

Traditionally the literature on (utility) representations deals with *real*-valued functions. It is also well known that a real representations impose restrictive conditions on the weak order and that there are many contexts in which such conditions are not satisfied (see for example [14]).

The interest for alternative representations is witnessed by the fact that in modeling multidimensional preferences lexicographic structure arise quite naturally. In this paper utility functions with values in a *ordered Abelian group* that is a triple $(G, +, >)$ where G is a set , $+$ is a commutative group operation with neutral element 0 and $>$ is a total order compatible with group operation i.e. such that if a, b are elements of G and $a > 0$ then $a + b > b$.

An Abelian group G is *Archimedean* if for $a, b \in G$ with $a > 0$ there exists $r \in \mathbb{N}$ such that $ra \geq b$.

3   Additive Representation of Independent Symmetric Preference Structure

We formalize some axioms that are sufficient for the construction of an additive representation of our preference structure.

We consider the *independence* axiom defined by

$$\text{if} \quad x \in X^n, y \in X^m, \quad a \in X, \quad \text{then} \quad x \succeq y \quad \text{if and only if} \quad (a, x) \succeq (a, y).$$

The axiom of independence considers also elements of different length. In our model an independence condition implies that inserting an extra common coordinate at any place does not change preference. Obviously we can insert any number of common coordinates by repeated application.

Independence axioms are undoubtedly strong assumptions but in many contexts are seen as essential conditions. In the framework of research evaluation independence means that if two authors share the same paper then their relative position does not depend on the impact or the importance of the common paper.

We can note that many bibliometric indeces as the well-known h-index ([15], [18])do not satisfy independence.

The following result provides necessary and sufficient conditions for a additive representation of a preference structure on a Abelian ordered group.

Proposition 1. *A symmetric preference structure* $(\mathbf{X}, >, \succeq)$ *satisfies independence axiom if and only if there exists an ordered abelian group* $(G, +, >)$ *and an increasing function* $f \colon \mathbf{X} \to G$ *with* $f(\mathbb{O}) = 0$ *such that if* $x \in X^n, y \in X^m$ *then*

$$x \succeq y \quad \text{if and only if} \quad \sum_{i=1}^{n} f(x_i) \geq \sum_{i=1}^{m} f(y_i).$$

Proof. It is easy to note that an additive weak order defined by a function $f \colon \mathbf{X} \to G$ where $(G, +, >)$ is an ordered Abelian group defines a symmetric preference structure that satisfies independence axiom.

We prove the reverse implication by considering the proof of Theorem 1 of [17]. We assume that $I = \mathbb{N}$ and we identify the elements of $X^n \subseteq \mathbf{X}$ with elements of X with infinite $\mathbb{O}$-components with $x_i = \mathbb{O}$ for every $i > n$. So the weak order in $\mathbf{X}$ can be extended to a strictly finitary transitive binary relation $\succeq$ in X . The relation $\succeq$ satisfy the assumptions of Theorem 1 in [17]. As in the proof of Theorem 1 in [17] we define a representation function with values in a free Abelian group by defining element o in [17] by $o = \mathbb{O}$.

If we refer to the components of a linearly ordered Abelian group Proposition 1 proves that any preference structure is based on classes of merit and that the classes are lexicographically ordered. Moreover in our general model we can consider the possibility of infinite or infinitesimal differences between the elements in $\mathbf{X}$. We assume that there are elements with different priority and that many elements with lower priority cannot compensate an element with higher priority. Archimedean axioms are necessary whenever one want to obtain a numerical representation.

Let us say that a symmetric preference structure $(\mathbf{X}, >, \succeq)$ is *Archimedean*

if for every $x \in X^n, x \succeq \mathbb{O}$ and $y \in X^m$ there exists $r \in \mathbb{N}$ such that $rx \succeq y$.

Proposition 2. *A symmetric preference structure* $(\mathbf{X}, >, \succeq)$ *is Archimedean and satisfy independence axiom if and only if there exists an increasing function* $f \colon \mathbf{X} \to \mathbb{R}$ *with* $f(\mathbb{O}) = 0$ *and such that if* $x \in X^n, y \in X^m$ *then*

$$x \succeq y \quad \text{if and only if} \quad \sum_{i=1}^{n} f(x_i) \geq \sum_{i=1}^{m} f(y_i).$$

Proof. If there exists a real-valued representation function for a symmetric preference structure $(\mathbf{X}, >, \succeq)$ it is straightforward to prove that $(\mathbf{X}, >, \succeq)$ is an Archimedean structure.

Conversely let $(\mathbf{X}, >, \succeq)$ be a symmetric preference structure that satisfy independence and Archimedean axioms. By Proposition 1 there exists an ordered Abelian group $(G, +, >)$ and an increasing function $f \colon \mathbf{X} \to G$ with $f(\mathbb{O}) = 0$ and such that if $x \in X^n, y \in X^m$ then $x \succeq y$ if and only if $\sum_{i=1}^{n} f(x_i) \geq \sum_{i=1}^{m} f(y_i)$. By the proof of Theorem 1 of [17] the range $f(\mathbf{X})$ is a subgroup of the group G. If $a, b \in f(\mathbf{X})$ wih $a > 0$ then $a = \sum_{i=1}^{n} f(x_i)$ and $b = \sum_{i=1}^{m} u(y_i)$. Let $x = (x_1, \ldots, x_n)$ and $y = (y_1, \ldots, y_m)$. By Archimedean property of $\mathbf{X}$ there exists $r \in \mathbb{N}$ such that $rx \succeq y$ since $x > \mathbb{O}$ and so it is easy to prove that $ra \succeq b$. We note also that by Hölder's theorem every Archimedean totally ordered group is order-isomorphic to a subgroup of the additive group of real numbers with the natural order and then we can assume that the function f is real-valued.

Now we consider average utility as in [16] that is characterized by the following axioms of weak separability, weak archimedeaness and replication equivalence. The *weak separability* axiom is satisfied when

if $\quad x, y \in X^n, \quad a \in X, \quad$ then $\quad x \succeq y \quad$ if and only if $\quad (a, x) \succeq (a, y),$

while the *weak Archimedean axiom* holds when

if $\quad x, y \in X^n, v, w \in X^m, x \succeq y$ there exists $r \in \mathbb{N}$ such that $(rx, v) \succeq (ry, w).$

The *replication equivalence* axiom holds

if for every $\quad x \in \mathbf{X} \quad n \in \mathbb{N} \quad x \sim nx.$

The next proposition [16] characterizes average utility.

Proposition 3. *A symmetric preference structure* $(\mathbf{X}, >, \succeq)$ *is Archimedean and satisfy weak independence axiom, weak Archimedeaness and replication equivalence axiom if and only if there exists an increasing function* $f \colon \mathbf{X} \to \mathbb{R}$ *with* $f(\mathbb{O}) = 0$ *and such that if* $x \in X^n, y \in X^m$ *then*

$$x \succeq y \quad \text{if and only if} \quad \frac{\sum_{i=1}^{n} f(x_i)}{n} \geq \frac{\sum_{i=1}^{m} f(x_i)}{m}.$$

Proof. We refer to the proof of Theorem 7 in [16] where the element $\mathbb{O}$ plays the role of neutral element for concatenation operation. Our weak Archimedeaness is Archimedean axiom in [16] while our weak independence is joint independence in [16] .

4 Step-Based Preference Structure

In [8] Chambers and Miller introduced from an axiomatic point of view a class of measures of scholarly influence that they called step-based indices. The axioms that characterize this class are satisfied by the h-index and also by some other natural indices such as the maximum-index, the i10-index and the publication count. These axioms are defined by lattice operations and are linked to maxitive and minitive properties in Aggregation Theory(see [9] and [10]). We generalize the approach in [8] considering evaluation in a totally ordered set and characterizing our step-based ordinal structure by weaker axioms.

Throughout this section we consider only positive evaluations and so we have that $x \geq \mathbb{O}$ for every $x \in X$. We assume also that an element $x \in X^n$ is equivalent to an element $x \in X^m$, $m > n$ with $m - n$ $\mathbb{O}$ - elements.

For $x = (x_1, \ldots, x_n) \in X^n$ and $y \in X$ we write

$$x /_i y = (x_1, \ldots, x_{i-1}, y, x_{i+1}, \ldots, x_n)$$

and we say that a symmetric preference structure $(\mathbf{X}, >, \succeq)$ is *size-bounded* if for every $n \in \mathbb{N}$ there exists $u_n \in \mathbb{N}$ such that

$$x /_i y = (x_1, \ldots, x_{i-1}, u_n, x_{i+1}, \ldots, x_n) \succeq x$$

for very $1 \leq i \leq n$. Then for every $x \in X^n$ we have $n u_n \succeq x$.

If $x = (x_1, \ldots, x_n) \in X^n$ we denote by $(\cdot)$ a permutation on $N = \{1, \ldots, n\}$ which arranges the elements of the vector by increasing values that is $x_{(1)} \geq x_{(2)} \geq, \ldots, \geq x_{(n)}$ and if $x, y \in X^n$ we define

$$x \vee_s y = (x_{(1)} \vee y_{(1)}, \ldots, x_{(n)} \vee y_{(n)}), \quad x \wedge_s y = (x_{(1)} \wedge y_{(1)}, \ldots, x_{(n)} \wedge y_{(n)})$$

A preference structure $(\mathbf{X}, >, \succeq)$ satisfy *S-maxitivity*(symmetric maxitivity) axiom

$$\text{if} \quad x, y \in X^n \quad \text{and} \quad x \succeq y \quad \text{then} \quad x \vee_s y \sim x$$

and satisfy the *S-minitivity* (simmetric minitivity) axiom

$$\text{if} \quad x, y \in X^n \quad \text{and} \quad x \succeq y \quad \text{then} \quad x \wedge_s y \sim y.$$

The following result characterizes our ordinal step-based symmetric preference structures.

Proposition 4. *Let* $(\mathbf{X}, >, \succeq)$ *be a symmetric and size-bounded preference structure that satisfies S-maxitivity and S-minitivity axioms. Then there exists a totally ordered set* $(Q, >)$, *a sequence of functions* $f_i \colon X \to Q$ *and a sequence* w_i *in* Q *such that if* $x \in X^n, y \in X^m$ *then*

$$x \succeq y \quad \text{if and only if} \quad \bigvee_{1 \leq i \leq n} f_i(x_{(i)}) \wedge w_i \geq \bigvee_{1 \leq i \leq m} f_i(y_{(i)}) \wedge w_i$$

Proof. Since $(\mathbf{X}, \succeq)$ is linearly ordered we can consider the set of equivalence class $\mathbf{X}/\sim$ that can be bijectively mapped in a qualitative scale $(Q, >)$.

Let $f_i(x)$ the equivalence class of ix for every $x \in X$ and w_i the equivalence class of iu_i. It is important to note that the equivalence class $f_i(x)$ does not depend on n if we consider the element ix as an element of X^n with $n - i\mathbb{O}$ - components. The same remark is applicable also for w_i. We consider an element $x \in X^n$ and we assume that $x_{(1)} > x_{(2)} > \ldots > x_{(n)}$ since the proof in the general case is similar. We can prove that

$$x = \bigvee_{1 \leq i \leq n} ix_{(i)} \wedge iu_i$$

where $(iu_i, 0, \ldots, 0)$ is an element of X^n. The equivalence class of $ix_{(i)}$ is $f_i(x_{(i)})$ by definition while the equivalence class of $(iu_i, 0, \ldots, 0)$ is w_i so by S-minitivity the equivalence class of $ix_{(i)} \wedge iu_i$ is $f_n(x_{(i)}) \wedge w_i$. Moreover by S-maxitivity the equivalence class of the element x is

$$\bigvee_{1 \leq i \leq n} f_n(x_{(i)}) \wedge w_i.$$

We have introduced and characterized ordered weighted maximum aggregation operators in our framework. As in the well-known case of the h-index (see [18]) the considered preference structure is one for which there is an increasing set of steps and the evaluations is determined by the best step that an element x achieves. that is the value n for which is assumed the maximum value. If the maximum value is $f_n(x_{(n)}) \wedge w_n$ the element $x \in X^n$ reaches the first step, if the maximum value is $f_{n-1}(x_{(n-1)}) \wedge w_{n-1}$ the element $x \in X^n$ achieves the second step and so on.

5 Concluding Remarks

We have defined and axiomatically characterized preference relations on sequences of variable length. We consider positive and negative scores and we do not consider only real-valued representation as is usually done.

Then we do not assume an Archimedean axiom and we consider an ordinal context. A symmetry property characterizes our preference relations and in our framework we study compensative and non-compensative aggregation operators. This paper has tried to point out some potential application fields for our classes of aggregation operators.

An obvious topic for future research is to analyze other functionals for sequences of variable length. Moreover sometimes we need to evaluate objects with a scale that is not totally ordered and then we have to consider also the case of sequences with values in a complete lattice.

References

1. Ahmed, S., Filipović, D., Svintland, G.: A note on natural risk statistics. Oper. Res. Lett. **36**, 662–664 (2008)
2. Artzner, P., Delbaen, F., Eber, J.M., Heath, D.: Coherent measures of risk. Math Financ **9**(3), 203–228 (1999)
3. Assa, H., Morales, M.: Risk measures on the space of infinite sequences. Math. Finan. Econ. **2**, 253–275 (2010)
4. Beliakov, G., James, S.: Stability of weighted penalty-based aggregation functions. Fuzzy Sets Syst. **226**, 1–18 (2013)
5. Cena, A., Gagolewski, M.: OM3: ordered maxitive, minitive, and modular aggregation operators. axiomatic analysis under arity-dependence (I). In: Bustince, H., Fernandez, J., Mesiar, R., Calvo, T. (eds.) Aggregation Functions in Theory and in Practise. AISC, vol. 228, pp. 93–103. Springer, Heidelberg (2013)
6. Cena, A., Gagolewski, M.: OM3: ordered maxitive, minitive, and modular aggregation operators. a simulation study (II). In: Bustince, H., Fernandez, J., Mesiar, R., Calvo, T. (eds.) Aggregation Functions in Theory and in Practise. AISC, vol. 228, pp. 107–118. Springer, Heidelberg (2013)
7. Cena, A., Gagolewski, M.: OM3: Ordered maxitive, minitive, and modular aggregation operators axiomatic and probabilistic properties in an arity-monotonic setting. Fuzzy Set Syst. **264**, 138–159 (2015)
8. Chambers, C.P., Miller, A.D.: Scholarly influence. J. Econ. Theory **151**(1), 571–583 (2014)
9. Couceiro, M., Marichal, J.-L.: Axiomatizations of quasi-polynomial functions on bounded chains. Aequationes Math **396**, 195–213 (2009)
10. Couceiro, M., Marichal, J.-L.: Axiomatizations of Lovász extensions of pseudo-Boolean functions. Fuzzy Set Syst. **181**, 28–38 (2011)
11. Gagolewski, M., Grzegorzewski, P.: Arity-monotonic extended aggregation operators. In: Hüllermeier, E., Kruse, R., Hoffmann, F. (eds.) IPMU 2010. CCIS, vol. 80, pp. 693–702. Springer, Heidelberg (2010)
12. Grabisch, M., Marichal, J.L., Mesiar, R., Pap, E.: Aggregation Functions. Encyclopedia of Mathematics and its Applications. Cambridge University Press, Cambridge (2009)
13. Heyde C.C., Kou S.G., Peng X.H.: What is a good external risk measure: Bridging the gaps between robustness, subadditivity and insurance risk measures (2007). (preprint)
14. Herden, G., Mehta, G.B.: The Debreu gap lemma and some generalizations. J. Math. Econom. **40**, 747–769 (2004)
15. Hirsch, J.E.: An index to quantify an individual's scientific research output. Proc. Natl. Acad. Sci. U.S.A. **102**, 16569–16572 (2005)

16. Kothiyal, A., Spinu, V., Wakker, P.P.: Average utility maximization: a preference foundation. Oper. Res. **62**(1), 207–218 (2014)
17. Pivato, M.: Additive representation of separable preferences over infinite products. Theory Dec. **77**, 31–83 (2014)
18. Torra, V., Narukawa, Y.: The h-index and the number of citations: two fuzzy integrals. IEEE Trans. Fuzzy Syst. **16**, 795–797 (2008)
19. Yager, R.R.: On ordered weighted averaging aggregation operators in multicriteria decision making. IEEE Trans. Syst. Man Cybern. **18**(1), 183–190 (1988)

Clustering and Similarity

Spherical k-Means++ Clustering

Yasunori Endo[✉] and Sadaaki Miyamoto

Faculty of Engineering, Information and Systems, University of Tsukuba,
Tennodai 1-1-1, Tsukuba, Ibaraki 305-8573, Japan
{endo,miyamoto}@risk.tsukuba.ac.jp

Abstract. k-means clustering (KM) algorithm, also called hard c-means clustering (HCM) algorithm, is a very powerful clustering algorithm [1,2], but it has a serious problem of strong initial value dependence. To decrease the dependence, Arthur and Vassilvitskii proposed an algorithm of k-means++ clustering (KM++) algorithm on 2007 [3]. By the way, there are many case that each object is allocated on an unit sphere, e.g. text clustering. Dhillon and Modha proposed the primitive spherical k-means clustering algorithm to classify such objects on 2007 [4] and Honik, Kober, and Buchta proposed new spherical k-means clustering (SKM) algorithm on 2012 [5]. However, both of the algorithms also have the same problem of initial value dependence as KM. Therefore, the paper discuss the following points: (1) the dissimilarity of SKM is extended to satisfy the triangle inequality, and (2) spherical k-means++ clustering (SKM++) algorithm which works well for the problem is proposed. The paper shows that the effectiveness of SKM++ is theoretically guaranteed.

1 Introduction

Recently, information from large-scaled social data sets has great effect on many aspects of society. We can mention recommendation systems as an example. When a person uses online markets, the recommendation system estimates commodities that he prefers from his purchase history and show the commodities on the display.

Some data mining tools to retrieve useful information from such social data sets play very important role in such systems. One of the most representative tool is spherical k-means clustering (SKM) algorithm by Honik, Kober, and Buchta on 2012 [5] based on k-means clustering (KM) [1,2] and the primitive spherical k-means clustering [4]. The cosine correlation is used as the dissimilarity between each datum and the cluster center in the SKM algorithm. Therefore it can be considered that all data are on the unit sphere, that is, the norm of the data handled by the algorithm is normalized to one. It is sufficient because many data in the social data sets are normalized when we retrieve useful information from the social data set. The SKM algorithm is very useful and it implemented on some powerful software, e.g. R of a free software environment for statistical computing and graphics.

© Springer International Publishing Switzerland 2015
V. Torra and Y. Narakawa (Eds.): MDAI 2015, LNAI 9321, pp. 103–114, 2015.
DOI: 10.1007/978-3-319-23240-9_9

However, both of SKM and the primitive spherical k-means clustering have a serious problem, that is, strong initial value dependence (i.v.d.). Therefore, this paper shows an algorithm which works well for the i.v.d. problem, called spherical k-means++ clustering (SKM++) algorithm. This work is inspired by k-means++ clustering (KM++) algorithm by Arthur and Vassilvitskii on 2007 [3].

First, we extend the dissimilarity between data in SKM to satisfy the triangle inequality. Second, clustering results by the SKM with the extended dissimilarity is equivalent to the original SKM. This fact is necessary to construct SKM++. Third, we show a way to select initial values in the clustering process, and prove that the way decreases the i.v.d. of SKM. Forth, we show that i.v.d. of SKM++ is theoretically estimated as a half of KM++.

2   Preparation

Let n be the number of object and $x \in \Re^p$ be each object. Without loss of generality, we can assume that $\|x\| = 1$ to simplify the discussion. $X = \{x\}$ means a set of objects. Let c, $v \in \Re^p$, and V be the number of clusters, a cluster center, and a set of cluster centers, respectively. $C^v = \{\arg\min_x d(x, v) \mid x \in X\}$ means a cluster with a cluster center v. Moreover, let C_i ($i = 1, \ldots, c$), $v_i \in \Re^p$, and $V^* = \{v_i \in \Re^p\}$ be the i-th optimal cluster, a center of C_i (the optimal solution), and a set of v_i (a set of optimal solutions).

2.1   Spherical k-Means Clustering

Spherical k-means clustering (SKM) is a very useful tool to classify the data whose norms are normalized as one. In this case, all data are allocated on the unit sphere. One of the most representative example is text mining. Now text mining is paid a lot of attention as an important methodology to analyze online data, e.g. social network service (SNS). Therefore, It is no exaggeration to say that SKM is more important than k-means clustering (KM).

SKM algorithm is constructed to minimize the following objective function:

$$J_{\text{SKM}}(V) = \sum_{v \in V} \sum_{x \in C^v} d_1(x, v).$$

$$(d_1(x, v) = 1 - \langle x, v \rangle)$$

Minimization of $J_{\text{SKM}}(V)$ is equivalent to maximization of $\sum_{v \in V} \sum_{x \in C^v} \langle x, v \rangle$ from

$$J_{\text{SKM}}(V) = \sum_{v \in V} \sum_{x \in C^v} d_1(x, v) = \sum_{v \in V} \sum_{x \in C^v} (1 - \langle x, v \rangle) = |X| - \sum_{v \in V} \sum_{x \in C^v} \langle x, v \rangle.$$

Algorithm 1. Spherical k-Means Clustering (SKM)

SKM1. Give the initial value of V.

SKM2. Allocate each $x \in X$ to a cluster C whose cluster center v is closest to x than other cluster centers as follows:

$$C^v = \{x \mid d_1(x, v) = d_1(x, V), v \in V\}.$$

SKM3. Update each cluster center v as follows:

$$v = \text{mean}(C^v) = \frac{\sum_{x \in C^v} x}{\left\| \sum_{x \in C^v} x \right\|}.$$

The operator is called Fischer mean.

SKM4 If a given stop criterion satisfies, finish the algorithm. Otherwise, go back to **SKM2**.

---

3 Extension of Dissimilarity in SKM

We call an algorithm α-SKM in which the dissimilarity d_1 of SKM is extended as follows:

$$d(x, v) = \alpha - \langle x, v \rangle. \quad (\alpha \geq 1)$$

That is, α-SKM is the algorithm using d instead of d_1 and d_1 is a special case of d. The objective function is as follows:

$$J(V) = \sum_{v \in V} \sum_{x \in C^v} d(x, v) = |X|\alpha - \sum_{v \in V} \sum_{x \in C^v} \langle x, v \rangle.$$

Minimization of J_α-SKM(V) is equivalent to maximization of $\sum_{v \in V} \sum_{x \in C^v} \langle x, v \rangle$, and finally, α-SKM is equivalent to SKM. Therefore, the α-SKM algorithm is the same as SKM and the following discussion for α-SKM can be applicable to SKM.

At a glance, it looks like the extension is meaningless. However, Lemma 1 shows that d satisfies the triangle inequality when $\alpha \geq 3/2$, and the fact plays very important role for considering SKM++.

Lemma 1. d satisfies the triangle inequality when $\alpha \geq 3/2$.

Proof. Let a, b, and c ($0 \leq a, b, c < \pi$) be angles between x and y, y and z, and x and z, respectively. Then we get

$$d(x, z) \leq d(x, y) + d(y, z) \Leftrightarrow \alpha - \langle x, z \rangle \leq \alpha - \langle x, y \rangle + \alpha - \langle y, z \rangle$$
$$\Leftrightarrow \langle x, y \rangle + \langle y, z \rangle - \langle x, z \rangle \leq \alpha$$
$$\Leftrightarrow \cos(a) + \cos(b) - \cos(c) \leq \alpha.$$

x, y, and z are all on the unit sphere so that a, b, and c satisfy the triangle inequality, that is, $a + b \geq c$. Thus this problem is rewritten as following an optimization problem:

$$\text{maximize} \quad \cos(a) + \cos(b) - \cos(c),$$
$$\text{subject to} \quad 0 \leq a, b, c < \pi, \ c \leq a + b.$$

Let's consider the following four cases:

1. $0 \leq a, b < \pi/2$.
2. $\pi/2 \leq a, b < \pi$.
3. $0 \leq a \leq \pi/2$, $\pi/2 \leq b \leq \pi$, and $\pi/2 \leq a + b \leq \pi$.
4. $0 \leq a \leq \pi/2$, $\pi/2 \leq b \leq \pi$, and $\pi \leq a + b \leq (3/2)\pi$.

In the first case,

$$c \leq a + b \Rightarrow \cos(a + b) \leq \cos(c)$$

from $0 \leq c < \pi$. Let $g(a, b) = \cos(a) + \cos(b) - \cos(a + b)$. From

$$\frac{\partial g(a, b)}{\partial a} = -\sin(a) + \sin(a + b) = 2\cos(a + \frac{b}{2})\sin(\frac{b}{2})$$

and $0 \leq a + b/2 \leq (3/4)\pi$, we know that g takes the maximum value $h(b)$:

$$h(b) = g(\frac{\pi}{2} - \frac{b}{2}, b) = 2\sin(\frac{b}{2}) + \cos(b)$$

when $a + b/2 = \pi/2$. From

$$\frac{\partial h(b)}{\partial b} = \cos(\frac{b}{2}) - \sin(b) = 2\cos(\frac{3}{4}b + \frac{\pi}{4})\cos(\frac{b}{4} + \frac{\pi}{4}),$$

and $\pi/4 \leq (3/4)b + \pi/4 < (5/8)\pi$ and $\pi/4 \leq b/4 + \pi/4 < (3/8)\pi$, we know that $h(b)$ takes the maximum value when $(3/4)b + \pi/4 = \pi/2$, that is, $b = \pi/3$. Finally,

$$\cos(a) + \cos(b) - \cos(c) \leq g(a, b) \leq \frac{3}{2}$$

when $a = b = \pi/3$ under the consideration of the symmetry of a and b.

In the second case, $c \leq \pi$ from $\pi/2 \leq a, b < \pi$ and $c \leq a + b$. Therefore, $\max \cos(a) = \max \cos(b) = 0$ and $\min \cos(c) = -1$. Finally,

$$\cos(a) + \cos(b) - \cos(c) \leq 1 < \frac{3}{2}.$$

In the third case,

$$c \leq a + b \Rightarrow \cos(a + b) \leq \cos(c)$$

from $0 \leq c \leq a + b \leq \pi$. From the same discussion of the first case, the lemma holds true.

In the forth case, $\cos(a)$ and $\cos(b)$ are monotonic decrease, then $\cos(a) + \cos(b)$ is also monotonic decrease. Thus, $\cos(a)+\cos(b)$ is maximum when $a+b = \pi$ from $\pi \leq a+b \leq (3/2)\pi$, and the maximum value is $\cos(a) + \cos(\pi - a) = 0$. On the other hand, $\cos(c) > -1$ from $c \leq a+b \leq (3/2)\pi$. Finally,

$$\cos(a) + \cos(b) - \cos(c) < 1.$$

From the above discussion, the lemma holds true.   Q.E.D.

From the above lemma, we can assume that $\alpha \geq 3/2$.

4   Analysis

4.1   Preliminary Step

Our aim is to show that SKM++ decreases the initial value dependence in comparison with SKM. We try to analyze it by a similar flow of Ref. [6].

Lemma 2. *For any cluster C and any object z,*

$$\sum_{x \in C} d(x, z) - \sum_{x \in C} d(x, \operatorname{mean}(C)) = \left\| \sum_{x \in C} x \right\| d(\operatorname{mean}(C), z) - \left\| \sum_{x \in C} x \right\| (\alpha - 1).$$

Proof

$$\sum_{x \in C} d(x, z) - \sum_{x \in C} d(x, \operatorname{mean}(C)) = \sum_{x \in C} \langle x, \operatorname{mean}(C) \rangle - \sum_{x \in C} \langle x, z \rangle$$

$$= \left\| \sum_{x \in C} x \right\| d(\operatorname{mean}(C), z) - \left\| \sum_{x \in C} x \right\| (\alpha - 1).$$

Q.E.D.

Lemma 3. *Let z be an object selected randomly from an arbitrary cluster C. For the expectation $E(J(C, z))$ of the value of objective function $J(C, z) = \sum_{x \in C} d(x, z)$, the following relation holds true:*

$$E(J(C, z)) = \left(1 + \frac{\| \sum_{x \in C} x \|}{|C|} \right) J(C) - \left\| \sum_{x \in C} x \right\| (\alpha - 1).$$

Here $J(C)$ means the value of the objective function with a cluster C and $J(C) = J(C, \operatorname{mean}(C)) = \sum_{x \in C} d(x, \operatorname{mean}(C))$.

Proof

$$E(J(C,z)) = \frac{1}{|C|} \sum_{z \in C} J(C,z)$$

$$= \frac{1}{|C|} \sum_{z \in C} \left(\sum_{x \in C} d(x, \text{mean}(C)) + \left\| \sum_{x \in C} x \right\| d(\text{mean}(C), z) - \left\| \sum_{x \in C} x \right\| (\alpha - 1) \right)$$

$$= \left(1 + \frac{\left\| \sum_{x \in C} x \right\|}{|C|} \right) J(C) - \left\| \sum_{x \in C} x \right\| (\alpha - 1).$$

Q.E.D.

Lemma 4. *Let z be an object selected randomly from an arbitrary cluster C. For the expectation $E(J(C,z))$ of the value of objective function $J(C,z)$, the following relation holds true:*

$$E(J(C,z)) \le 2J(C).$$

Proof. It holds true from Lemma 3 and the following relation:

$$\|x\| = 1 \Rightarrow \left\| \sum_{x \in C} x \right\| \le |C|.$$

Q.E.D.

Lemma 5. *For any objects x and y, and any set Z, if $\alpha \ge 3/2$ the following relation holds true:*

$$D(x, Z) \le d(x, y) + D(y, Z).$$

Here $D(x, Z)$ means the dissimilarity between a point x and a set Z and $D(x, Z) = \min_{z \in Z} d(x, z)$.

Proof. We define $z^{\dagger} = \arg\min_{z \in Z} d(x, z)$, and $z^{\ddagger} = \arg\min_{z \in Z} d(y, z)$.
When $z^{\dagger} = z^{\ddagger}$, we get the following relation from Lemma 1:

$$D(x, Z) = d(x, z^{\dagger}) \le d(x, y) + d(y, z^{\dagger})$$
$$= d(x, y) + D(y, Z).$$

When $z^{\dagger} \ne z^{\ddagger}$, we get the following relation from Lemma 1:

$$D(x, Z) = d(x, z^{\dagger}) \le d(x, z^{\ddagger})$$
$$\le d(x, y) + d(y, z^{\ddagger})$$
$$= d(x, y) + D(y, Z).$$

Q.E.D.

Lemma 6. *Cluster centers are selected according to the following process:*

Step 1. *Let the number of iterations $t = 1$. Select $v \in X$ at random and let the initial set of cluster centers $V^t = \{v\}$.*

Step 2. *Select x with the probability proportional to $D(x, V^t)$ as a new cluster center. Let $V^{t+1} = V^t \cup \{x\}$.*

Step 3. *$t := t + 1$. If $t = k$, finish the algorithm. Otherwise, go back to Step 2.*

We assume that the t-th iteration has finished ($t < k$) and Let V^t be a set of cluster centers selected until the end of t-th iteration, and $z \in C_i$ be the next selected cluster center. That is, $V^{t+1} = V^t \cup \{z\}$. For the conditional expectation of $J(C_i, V^t \cup \{z\}) = J(C_i, V^{t+1})$,

$$E(J(C_i, V^{t+1}) \mid V^t, \{z \in C_i\}) \leq 4J(C_i).$$

Proof. The probability to select z is represented as $\frac{D(z, V^t)}{\sum_{z \in C_i} D(z, V^t)} = \frac{D(z, V^t)}{J(C_i, V^t)}$ because z is selected with the probability proportional to $D(z, V^t)$ and $z \in C_i$. Moreover we get

$$E(J(C_i, V^{t+1}) \mid V^t, \{z \in C_i\}) = \sum_{z \in C_i} \frac{D(z, V^t)}{J(C_i, V^t)} \sum_{x \in C_i} \min\{J(x, V^t), d(x, z)\}$$

from

$$J(C_i, V^{t+1}) = J(C_i, V^t \cup \{z\}) = \sum_{x \in C_i} \min\{J(x, V^t), d(x, z)\}.$$

Here we know that the following relation holds true from Lemma 5:

$$D(z, V^t) \leq d(z, x) + D(x, V^t)$$

because $\alpha \geq 3/2$. Thus we get the following relation:

$$\sum_{x \in C_i} D(z, V^t) \leq \sum_{x \in C_i} d(z, x) + \sum_{x \in C_i} D(x, V^t)$$
$$= J(C_i, z) + J(C_i, V^t).$$

Therefore,

$$D(z, V^t) \leq \frac{1}{|C_i|} \left(J(C_i, z) + J(C_i, V^t) \right).$$

Finally, we get the following relation from the above relation and Lemma 4:

$$E(J(C_i, V^{t+1}) \mid V^t, \{z \in C_i\})$$
$$\leq \sum_{z \in C_i} \frac{\frac{1}{|C_i|} \left(\sum_{x \in C_i} J(C_i, z) + \sum_{x \in C_i} D(x, V^t) \right)}{J(C_i, V^t)} \sum_{x \in C_i} \min\{J(x, V^t), d(x, z)\}$$
$$= \frac{2}{|C_i|} \sum_{z \in C_i} J(C_i, z) = 2E(J(C_i, z)) \leq 4J(C_i).$$

Q.E.D.

4.2 Main Step

Here, we introduce the following symbols. Let $H^t = \{i \mid 1 \leq i \leq k, \; C_i \cup V^t \neq \phi\}$, $U^t = \{i \mid 1 \leq i \leq k\} \backslash H^t$, and $W^t = t - |H^t|$ be a set of "hit" clusters at the end of t-th iteration, a set of "uncovered" clusters at the end of t-th iteration, and the number of "wasted" iterations at the end of t-th iteration, respectively.

First, Lemma 7 for H^t holds true.

Lemma 7. *For any* $t \leq k$,

$$E\left(\sum_{i \in H^t} J(C_i, V^t) \right) \leq 4J^*$$

Here J^* *means the optimal value of the objective function* $J(V)$ *and* $J^* = \sum_{i=1}^{c} \sum_{x \in C_i} d(x, v_i)$.

Proof. Let C_t be a cluster selected at the t-th iteration. The following relation holds true from $V^0 = \phi$ and Lemma 4:

$$E(J(C_1, z_1)) = E(C_1, V^1 \mid V^0, \{z_1 \in C_1\}) \leq 2J(C_1) \leq 4J(C_1).$$

Therefore, we get the following relation from Lemma 6:

$$E\left(\sum_{i \in H^t} J(C_i, V^t) \right) = E(J(C_1, V^1) \mid V^0, \{z_1 \in C_1\}) + \ldots + E(J(C_t, V^t) \mid V^{t-1}, \{z_t \in C_t\})$$

$$\leq 4 \sum_{i=1}^{t} J(C_i) \leq 4 \sum_{i=1}^{k} J(C_i) = 4J^*.$$

$$\text{Q.E.D.}$$

Next, we consider the following function for U^t:

$$\Theta^t = \frac{W^t \sum_{i \in U^t} J(C_i, V^t)}{|U^t|}$$

Let assume that the t-th iteration has finished. We can consider the following two cases for a cluster center selected at the $(t+1)$-th iteration:

1. The cluster center is selected from U^t.
2. The cluster center is selected from H^t.

The desirable case is former.

For the former case, the following lemma holds true.

Lemma 8. *Let* $R^t = \{q \mid q$ *is an index of the cluster which is randomly selected from* U^r *at the* $r - th$ *iteration,* $r = 1, \ldots, t\}$ *and we assume that a cluster center is selected from* C_j *at the* $(t+1)$*-th iteration. Then, we get the following relation:*

$$E(\Theta^{t+1} - \Theta^t \mid R^t, \{j \in U^t\}) \leq 0.$$

Proof. If $j \in U^t$, we get $H^{t+1} = H^t \cup \{j\}$, $W^{t+1} = W^t$, and $U^{t+1} = U^t \backslash \{j\}$. Thus,

$$
\begin{aligned}
\Theta^{t+1} &= \frac{W^{t+1} \sum_{i \in U^{t+1}} J(C_i, V^{t+1})}{|U^{t+1}|} \\
&= \frac{W^t \left(\sum_{i \in U^t} J(C_i, V^t) - J(C_j, V^t) \right)}{|U^t| - 1}.
\end{aligned}
$$

If j is randomly selected from U^t, we get the following relation:

$$
\begin{aligned}
E(J(C_j, V^t) \mid R^t, \{j \in U^t\}) &= \sum_{j \in U^t} \frac{J(C_j, V^t)}{\sum_{i \in U^t} J(C_i, V^t)} J(C_j, V^t) \\
&\geq \frac{1}{|U^t|} \sum_{i \in U^t} J(C_i, V^t)
\end{aligned}
$$

from Cauchy-Schwarz inequality and the fact that the probability to select C_j from clusters whose indices belong to U^t is as follows:

$$
\frac{J(C_j, V^t)}{\sum_{i \in U^t} J(C_i, V^t)}.
$$

Hence,

$$
\begin{aligned}
E(\Theta^{t+1} \mid R^t, \{j \in U^t\}) &= E\left(\frac{W^{t+1} \sum_{i \in U^{t+1}} J(C_i, V^t)}{|U^{t+1}|} \mid R^t, \{j \in U^t\} \right) \\
&= \frac{W^t \sum_{i \in U^t} J(C_i, V^t)}{|U^t|} = \Theta^t.
\end{aligned}
$$

Finally, we get the following relation:

$$
E(\Theta^{t+1} \mid R^t, \{j \in U^t\}) - \Theta^t = E(\Theta^{t+1} - \Theta^t \mid R^t, \{j \in U^t\}) \leq 0.
$$

Q.E.D.

For the latter case, the following lemma holds true.

Lemma 9. *If $j \in H^t$,*

$$
\Theta^{t+1} - \Theta^t = \frac{\sum_{i \in U^t} J(C_i, V^t)}{|U^t|}.
$$

Proof. If $j \in H^t$, we get $H^{t+1} = H^t$, $W^{t+1} = W^t + 1$, and $U^{t+1} = U^t$. Thus,

$$
\begin{aligned}
\Theta^{t+1} - \Theta^t &= \frac{W^{t+1} \sum_{i \in U^{t+1}} J(C_i, V^{t+1})}{|U^{t+1}|} - \frac{W^t \sum_{i \in U^t} J(C_i, V^t)}{|U^t|} \\
&= \frac{\sum_{i \in U^t} J(C_i, V^t)}{|U^t|}.
\end{aligned}
$$

Q.E.D.

From the above two cases, the following lemma holds true.

Lemma 10. *For any t $(0 \leq t \leq k - 1)$,*

$$E(\Theta^{t+1} - \Theta^t \mid R^t) \leq \frac{\sum_{i \in H^t} J(C_i, V^t)}{k - t}.$$

Proof. From Lemmas 8 and 9, we get the following relation:

$$E(\Theta^{t+1} - \Theta^t \mid R^t) = \frac{\sum_{i \in H^t} J(C_i, V^t)}{J(V^t)} E(\Theta^{t+1} - \Theta^t \mid R^t, \{j \in H^t\})$$

$$+ \frac{\sum_{i \in U^t} J(C_i, V^t)}{J(V^t)} E(\Theta^{t+1} - \Theta^t \mid R^t, \{j \in U^t\})$$

$$\leq \frac{\sum_{i \in H^t} J(C_i, V^t)}{k - t}.$$

Q.E.D.

From the above lemmas, we can derive the following theorem.

Theorem 1. *If V is selected by the process in Lemma 6, the following relation holds true:*

$$E(J(V)) \leq 4(\ln k + 2)J^*.$$

Proof. From the following property of Θ^t:

$$\lim_{t \to k} \Theta^t = \sum_{i \in U^k} J(C_i)$$

$$\sum_{i \in H^k} J(C_i, V^k) + \sum_{i \in U^k} J(C_i, V^k) = J(V),$$

and Lemmas 7 and 10,

$$E(J(V)) = E\left(\sum_{i \in H_k} J(C_i, V)\right) + E\left(\sum_{i \in U_k} J(C_i, V)\right)$$

$$\leq 4(\ln k + 2)J^*.$$

Q.E.D.

5   Spherical k-Means++

From the above theoretical discussion, we can construct the following algorithm by substituting d for d_1 in SKM as dissimilarity and introducing the process in Lemma 6 as selection of initial values. We show the SKM++ algorithm in Algorithm 2.

Algorithm 2. Spherical k-Means++ Clustering (SKM++)

SKM++1. Give the initial values V by the following process:

   SKM++1-1.

   Step 1. Set the number of iteration as $t = 1$. Select $v \in X$ at random, and let the initial set of cluster centers $V^t = \{v\}$.

   SKM++1-2. Select x with the probability of $\frac{D(x, V^t)}{J(V^t)}$. Let x be a new cluster center and $V^{t+1} = V^t \cup \{x\}$.

   SKM++1-3. Let $t := t + 1$. If $t = k$, finish the process. Otherwise, go back to **SKM++1-2.**

SKM++2. Allocate each $x \in X$ to a cluster C^v whose cluster center v is closest to x than other cluster centers as follows:

$$C^v = \{x \mid d(x, v) = d(x, V), v \in V\}$$

SKM++3. Update each cluster center v as follows:

$$v = \text{mean}(C^v) = \frac{\sum_{x \in C^v} x}{\|\sum_{x \in C^v} x\|}.$$

SKM++4. If a given stop criterion satisfies, finish the algorithm. Otherwise, go back to **SKM2++.**

Theorem 1 shows that the maximum value of the expectation of the objective function of SKM is $4(\ln k + 2)$ times as large as the optimal value of the objective function of SKM by the SKM++ algorithm. On the other hand, it is proved that the maximum value of the expectation of the objective function of KM is $8(\ln k + 2)$ times as large as the optimal value of the objective function of KM by the KM++ algorithm.

It means that i.v.d. of SKM++ is theoretically estimated as a half of KM++.

6   Conclusion

This paper show a new clustering algorithm SKM++. SKM++ is very useful for the problem of initial value dependence. Importance of SKM will increase because the scale of data on the Internet will make large and a lot of such data allocate on the unit sphere. From such a viewpoint, it is expected that SKM++ is more emphasized than KM++.

Acknowledgment. This work has partly been supported by JSPS KAKENHI Grant Numbers 26330270 and 26330271.

References

1. Steinhaus, H.: Sur la division des corps matériels en parties. Bulletin de l'Académie Polonaise des Sci. **4**(12), 801–804 (1957)
2. MacQueen, J.B.: Some methods for classification and analysis of multivariate observations. In: Proceedings of the 5th Berkeley Symposium on Mathematical Statistics and Probability, Statistics, vol. 1, pp. 281–297. University of California Press (1967)
3. Arthur, D., Vassilvitskii, S.: k-means++: the advantages of careful seeding, In: Proceedings of the Eighteenth Annual ACM-SIAM symposium on Discrete algorithms, pp. 1027–1035. Society for Industrial and Applied Mathematics, Philadelphia (2007)
4. Dhillon, I.S., Modha, D.S.: Concept decompositions for large sparse text data using clustering. Mach. Learn. **42**, 143–175 (2001)
5. Hornik, K., Feinerer, I., Kober, M., Buchta, C.: Spherical k-Means Clustering, vol. 50(10), September 2012
6. Dasgupta, S.: Lecture 3 – Algorithms for k-means clustering (2013). http://cseweb.ucsd.edu/dasgupta/291-geom/kmeans.pdf

On Possibilistic Clustering Methods Based on Shannon/Tsallis-Entropy for Spherical Data and Categorical Multivariate Data

Yuchi Kanzawa[✉]

Shibaura Institute of Technology, Koto, Tokyo 135-8548, Japan
kanzawa@sic.shibaura-it.ac.jp

Abstract. In this paper, four possibilistic clustering methods are proposed. First, we propose two possibilistic clustering methods for spherical data — one based on Shannon entropy, and the other on Tsallis entropy. These methods are derived by subtracting the cosine correlation between an object and a cluster center from 1, to obtain the object-cluster dissimilarity. These methods are derived from the proposed spherical data methods by considering analogies between the spherical and categorical multivariate fuzzy clustering methods, in which the fuzzy methods' object-cluster similarity calculation is modified to accommodate the proposed possibilistic methods. The validity of the proposed methods is verified through numerical examples.

Keywords: Possibilistic clustering · Spherical data · Categorical multivariate data

1 Introduction

Fuzzy c-means (FCM), proposed by Bezdek [1], is the most popular algorithm for performing fuzzy clustering on linear data. FCM is fuzzified through its membership in the hard c-means (HCM) objective function [2]. Other HCM fuzzification methods include entropy-regularized FCM (eFCM) [3] and Tsallis entropy-based FCM (tFCM) [4].

FCM and its variants are useful clustering methods; however, their memberships do not always correspond well to the degree of belonging of the data. To address this weakness of FCM, Krishnapuram and Keller [5] proposed a possibilistic c-means (PCM) algorithm that uses a possibilistic membership function. Krishnapuram and Keller [6], and Ménard et al. [4] proposed other possibilistic clustering techniques that employ Shannon entropy and Tsallis entropy, respectively. In this study, these two methods are respectively referred to as entropy-regularized PCM (ePCM) and Tsallis-entropy-regularized PCM (tPCM).

All the aforementioned clustering methods are designed for linear data. In other application domains, linear data clustering methods may yield poor results. For example, information retrieval applications show that cosine similarity is a more accurate similarity measure for clustering text documents than Euclidean

© Springer International Publishing Switzerland 2015
V. Torra and Y. Narakawa (Eds.): MDAI 2015, LNAI 9321, pp. 115–128, 2015.
DOI: 10.1007/978-3-319-23240-9_10

distortion of dissimilarity [7]. Such domains require spherical data use, and only consider the directions of the unit vectors. In particular, spherical K-means [8] and its fuzzified variants [9–13] are designed to process spherical data. However, a possibilistic approach for clustering spherical data has not been proposed in the literatures; this was a motivation for this work. The spherical clustering methods that correspond to eFCM and tFCM are denoted as eFCS and tFCS in this paper.

In recent studies [13–17], various fuzzy clustering methods have been proposed for categorical multivariate data (FCCM). In these methods, a categorical multivariate dataset is provided in the form of a cross-classification table, contingency table, or co-occurrence matrix. Because the optimization problems [13,15] are similar to spherical clustering, these FCCM methods can be extended into possibilistic clustering, which was another motivation for this work. The method described in [15] is referred to as entropy-regularized FCCM (eFCCM), and the method described in [13] is referred to as Tsallis entropy-regularized FCCM (tFCCM), in order to distinguish these methods in this paper.

In this study, four possibilistic clustering methods are proposed — two for spherical data and two for categorical multivariate data. First, we propose the possibilistic clustering methods for spherical data: entropy-regularized possibilistic clustering for spherical data (ePCS) and Tsallis entropy-regularized possibilistic clustering for spherical data (tPCS). These methods are derived by subtracting the cosine correlation between an object and a cluster center from 1, to obtain the object-cluster dissimilarity; this value is used in place of the squared Euclidean distance between an object and an cluster center, which is commonly used in conventional linear data methods. Second, we propose two possibilistic clustering methods for categorical multivariate data: entropy-regularized possibilistic clustering for categorical multivariate data (ePCCM) and Tsallis entropy-regularized possibilistic clustering for categorical multivariate data (tPCCM). These methods are derived from the proposed spherical data methods (ePCS and tPCS) by considering analogies between the fuzzy methods for spherical data and categorical multivariate data; here, the object-cluster similarity calculation in the fuzzy methods is modified to accommodate the proposed possibilistic methods. The validity of the proposed methods is verified through numerical examples.

The rest of this paper is organized as follows. In Sect. 2, the notation and the conventional methods are introduced. Section 3 presents the proposed methods, and Sect. 4 provides some numerical examples. Section 5 contains our concluding remarks.

2   Preliminaries

2.1   Notation, Fuzzy c-Means, and Its Variants

Let $X = \{x_k \in \mathbb{R}^p \mid k \in \{1, \cdots, N\}\}$ be a dataset of p-dimensional points, referred to as linear data. The membership of x_k that belongs to the i-th cluster is denoted by $u_{i,k}$ ($i \in \{1, \cdots, C\}, k \in \{1, \cdots, N\}$) and the set of $u_{i,k}$ is denoted by u, which is also known as the partition matrix. The cluster center set is denoted by $v = \{v_i \mid v_i \in \mathbb{R}^p, i \in \{1, \cdots, C\}\}$. The squared Euclidean distance between the k-th datum and the i-th cluster center is denoted by $d_{i,k} = \|x_k - v_i\|_2^2$.

One approach for membership fuzzification is to regularize the objective function of HCM by introducing a regularization term with a positive parameter λ into the objective function. This approach was successfully implemented by Miyamoto and Mukaidono [3]. Using the entropy term, the entropy-regularized FCM (eFCM) is defined as

$$\underset{u,v}{\text{minimize}} \sum_{i=1}^{C} \sum_{k=1}^{N} u_{i,k} d_{i,k} + \lambda^{-1} \sum_{i=1}^{C} \sum_{k=1}^{N} u_{i,k} \log(u_{i,k}) \tag{1}$$

$$\text{subject to } \sum_{i=1}^{C} u_{i,k} = 1. \tag{2}$$

Ménard adopted Tsallis entropy [20] instead of Shannon entropy to perform fuzzy clustering, and proposed tFCM [4] defined as

$$\underset{u,v}{\text{minimize}} \sum_{i=1}^{C} \sum_{k=1}^{N} u_{i,k}^m d_{i,k} + \frac{\lambda^{-1}}{m-1} \sum_{i=1}^{C} \sum_{k=1}^{N} [u_{i,k}^m - u_{i,k}] \tag{3}$$

subject to Eq. (2).

2.2 Possibilistic Clustering

To improve the fidelity of fuzzy clustering, Krishnapuram and Keller [5] relaxed the constrained condition Eq. (2), which yielded a possibilistic membership function. The memberships for a certain cluster and its cluster center are released from constraint Eq. (2), and are obtained independent of these for other clusters. Hereafter, we only consider cases in which $C = 1$, where the number 1 signifies that only one cluster is searched for at a time. With this setting, cluster fusion [18] is useful, given that many cluster centers become nearer to each other as iteration proceeds; the distance between two clusters frequently approaches zero. The cluster fusion is described in the following algorithm:

Algorithm 1

1. Select a subset of objects as initial cluster centers. It is possible to select all objects: $C = N$; $v_i = x_i$ ($i \in \{1, \cdots, C\}$).
2. Perform possibilistic clustering, and obtain C cluster centers.
3. Merge cluster centers that have negligible distances between them.

Krishnapuram and Keller [6], and Ménard [4] proposed possibilistic clustering methods using entropy, defined as

$$\underset{u,v}{\text{minimize}} \sum_{k=1}^{N} u_{1,k} d_{1,k} + \lambda^{-1} \sum_{k=1}^{N} u_{1,k} \log(u_{1,k}) - \lambda^{-1} \sum_{k=1}^{N} u_{1,k}, \tag{4}$$

$$\underset{u,v}{\text{minimize}} \sum_{k=1}^{N} u_{1,k}^m d_{1,k} + \frac{\lambda^{-1}}{m-1} \sum_{k=1}^{N} (u_{1,k}^m - u_{1,k}) - \lambda^{-1} \sum_{k=1}^{N} u_{1,k}. \tag{5}$$

These two methods are referred to as ePCM and tPCM because the usual (Shannon) entropy and Tsallis entropy are employed in these methods, respectively. The optimal solutions for the membership and cluster center are described as

$$u_{1,k} = \exp(-\lambda d_{1,k}), \tag{6}$$

$$v_1 = (\sum_{k=1}^{N} u_{1,k} x_k) / (\sum_{k=1}^{N} u_{1,k}) \tag{7}$$

for ePCM, and

$$u_{1,k} = (1 - \lambda (1-m) d_{1,k})^{\frac{1}{1-m}}, \tag{8}$$

$$v_1 = (\sum_{k=1}^{N} u_{1,k}^m x_k) / (\sum_{k=1}^{N} u_{1,k}^m) \tag{9}$$

for tPCM. These equations are alternatively iterated during the second step in Algorithm 1. Ménard denoted the third term in Eqs. (4) and (5) as a possibilistic constraint, and showed that these two methods were derived by adding this constraint to the eFCM and tFCM objective functions in Eqs. (1) and (3).

2.3   Fuzzy Clustering for Spherical Data

If objects are on the unit hypersphere, $1 - x_k^\mathsf{T} v_i$ can be used as the dissimilarity between an object x_k and a cluster center v_i. Such objects are referred to as spherical data. Two methods that correspond to Eqs. (1) and (3) are obtained for the following optimization problems:

$$\underset{u,v}{\text{minimize}} \sum_{i=1}^{C} \sum_{k=1}^{N} u_{i,k}(1 - x_k^\mathsf{T} v_i) + \lambda^{-1} \sum_{i=1}^{C} \sum_{k=1}^{N} u_{i,k} \log(u_{i,k}), \tag{10}$$

$$\underset{u,v}{\text{minimize}} \sum_{i=1}^{C} \sum_{k=1}^{N} u_{i,k}^m(1 - x_k^\mathsf{T} v_i) + \frac{\lambda^{-1}}{m-1} \sum_{i=1}^{C} \sum_{k=1}^{N} [u_{i,k}^m - u_{i,k}], \tag{11}$$

respectively, subject to Eq. (2) and

$$\|v_i\|_2 = 1, \tag{12}$$

referred to as eFCS [9] and tFCS [13], respectively. It is shown in [19] that eFCS optimization problem in Eq. (10) can be equivalently described as the following maximizing problem

$$\underset{u,v}{\text{maximize}} \sum_{i=1}^{C} \sum_{k=1}^{N} u_{i,k} x_k^\mathsf{T} v_i - \lambda^{-1} \sum_{i=1}^{C} \sum_{k=1}^{N} u_{i,k} \log(u_{i,k}). \tag{13}$$

However, to the best of our knowledge, a possibilistic approach to spherical clustering has not yet been investigated.

2.4 Fuzzy Clustering for Categorical Multivariate Data

Assume that for datasets $X = \{x_k \mid k \in \{1, \dots, N\}\}$ and $Y = \{y_\ell \mid \ell \in \{1, \dots, M\}\}$, the co-occurrence information between x_k and y_ℓ, $R_{k,\ell}$ is given. R is the matrix whose (k, ℓ)-th element is $R_{k,\ell}$. We refer to X and Y as the row and column datasets, respectively, because the k-th row of R represents the similarities between x_k and y_ℓ, and the ℓ-th column of R represents the similarities between y_ℓ and x_k. The membership of datum x_k belonging to the i-th cluster is denoted by $u_{i,k}$. The (i, k)-th element of matrix u is denoted by $u_{i,k}$, and u satisfies the constraint in Eq. (2). The membership of datum y_ℓ belonging to the i-th cluster is denoted by $w_{i,\ell}$. The (i, ℓ)-th element of matrix w is denoted by $w_{i,\ell}$, and w satisfies the constraint

$$\sum_{\ell=1}^{M} w_{i,\ell} = 1. \tag{14}$$

The eFCCM [15] is obtained by solving the following optimization problem:

$$\underset{u,w}{\text{maximize}} \sum_{i=1}^{C} \sum_{k=1}^{N} \sum_{\ell=1}^{M} u_{i,k} \log(w_{i,\ell}) R_{k,\ell} - \lambda^{-1} \sum_{i=1}^{C} \sum_{k=1}^{N} u_{i,k} \log(u_{i,k}) \tag{15}$$

subject to Eqs. (2) and (14), where $\lambda > 0$ is a fuzzification parameter. The tFCCM [13] are obtained by solving the following optimization problem

$$\underset{u,w}{\text{maximize}} \sum_{i=1}^{C} \sum_{k=1}^{N} \sum_{\ell=1}^{M} u_{i,k}^m \log(w_{i,\ell}) R_{k,\ell} - \frac{\lambda^{-1}}{m-1} \sum_{i=1}^{C} \sum_{k=1}^{N} [u_{i,k}^m - u_{i,k}] \tag{16}$$

subject to Eqs. (2) and (14), where $\lambda > 0$ and $m > 1$ are fuzzification parameters. Because the optimization problem described in Eq. (15) is similar to Eqs. (1) and (10), and because the optimization problem described in Eq. (16) is similar to Eqs. (3) and (11), it is possible to generalize FCCM in the same manner in which eFCM was modified into ePCM. This fact motivated this work.

3   Proposed Method

3.1   Modifying ePCM and tPCM

In this subsection, we modify ePCM and tPCM as a preparatory procedure to derive the proposed methods.

ePCM and tPCM objective functions are slightly generalized from Eqs. (4) and (5) to

$$\underset{u,v}{\text{minimize}} \sum_{k=1}^{N} u_{1,k} d_{1,k} + \lambda^{-1} \sum_{k=1}^{N} u_{1,k} \log(u_{1,k}) - \alpha \sum_{k=1}^{N} u_{1,k}, \qquad (17)$$

$$\underset{u,v}{\text{minimize}} \sum_{k=1}^{N} u_{1,k}^{m} d_{1,k} + \frac{\lambda^{-1}}{m-1} \sum_{k=1}^{N} (u_{1,k}^{m} - u_{1,k}) - \alpha \sum_{k=1}^{N} u_{1,k}, \qquad (18)$$

where the factor of the last term in the original problems in Eqs. (4) and (5), λ^{-1}, is replaced by another parameter $\alpha \in (-\infty, +\infty)$ for modified ePCM and $\alpha \in (-1/(\lambda(m-1)), +\infty)$ for modified tPCM. These optimal solutions for membership are described as

$$u_{1,k} = \beta \exp(-\lambda d_{1,k}), \qquad (19)$$

for modified ePCM, where $\beta = \exp(\lambda\alpha - 1) \in (0, +\infty)$, and

$$u_{1,k} = \beta(1 - \lambda(1-m)d_{1,k})^{1/(1-m)}, \qquad (20)$$

for modified tPCM, where $\beta = (1 + \alpha\lambda(m-1))/m)^{1/(m-1)} \in (0, +\infty)$, and the optimal solutions of cluster center are the same as the original forms in Eqs. (7) and (9). We can observe that $\alpha = \lambda^{-1}$ recovers the original problems. We note that the membership value is 1 at $d_{1,k} = 0$ for the original membership form; this is not the case in Eqs. (19) and (20), except for the case in which $\alpha = \lambda^{-1}$. However, this does not imply that the modified ePCM and the modified tPCM do not contain defects. First, in the possibilistic theory, the maximal membership value does not need to be 1. Second, such a modification does not affect cluster center updating, as explained in the following procedure. Denote the membership and the cluster center in the modified ePCM as $\tilde{u}$ and $\tilde{v}$ to distinguish them from those in the original ePCM. Then, we have

$$\tilde{v}_1 = \frac{\sum_{k=1}^{N} \tilde{u}_{1,k} x_k}{\sum_{k=1}^{N} \tilde{u}_{1,k}} = \frac{\sum_{k=1}^{N} \beta u_{1,k} x_k}{\sum_{k=1}^{N} \beta u_{1,k}} = \frac{\beta \sum_{k=1}^{N} u_{1,k} x_k}{\beta \sum_{k=1}^{N} u_{1,k}} = v_1, \qquad (21)$$

which means that such a modification does not affect the updating of the cluster centers, and simply changes the scale of membership. The case of tPCM also leads to the same result. Hereafter, the modified versions of ePCM and tPCM are used to derive the proposed methods.

3.2 Possibilistic Clustering for Spherical Data

In this subsection, we propose two possibilistic clustering methods for spherical data, ePCS and tPCS.

ePCS is obtained by solving the optimization problem

$$\underset{u,v}{\text{minimize}} \sum_{k=1}^{N} u_{1,k}(1 - x_k^{\mathsf{T}} v_1) + \lambda^{-1} \sum_{k=1}^{N} u_{1,k} \log(u_{1,k}) - \alpha \sum_{k=1}^{N} u_{1,k}, \qquad (22)$$

subject to Eq. (12). This optimization problem is derived by subtracting the cosine correlation between an object and a cluster center from 1 $(1 - x_k^\mathsf{T} v_1)$ to obtain the object-cluster dissimilarity, instead of using the squared Euclidean distance between an object and a cluster center ($\|x_k - v_1\|_2^2$) applied in ePCM, which was described in Eq. (17). The optimal solutions for the membership and cluster center are described as

$$u_{1,k} = \beta \exp(\lambda x_k^\mathsf{T} v_1), \tag{23}$$

$$v_1 = (\sum_{k=1}^{N} u_{1,k} x_k)/(\| \sum_{k=1}^{N} u_{1,k} x_k \|_2), \tag{24}$$

where $\beta = \exp(-\lambda - 1 + \lambda\alpha)$. ePCS is also derived from eFCS by subtracting the possibilistic constraint term $\alpha \sum_{k=1}^{N} u_{1,k}$ from the eFCS objective function described in Eq. (10), omitting the probabilistic constraint in Eq. (2), and considering the spherical constraint in Eq. (12). The ePCS membership in Eq. (23) is described for arbitrary object x as $u_1(x) = \beta \exp(\lambda x^\mathsf{T} v_1)$; this is the unnormalized von Mises-Fisher distribution. This membership function for a one-dimensional sphere is depicted in Fig. 1 for several parameter values of λ, where β is set such that $\max_x u_1(x) = 1$. The ePCS optimization problem is described as the following maximizing problem:

$$\text{Eq. (22)} \Leftrightarrow \underset{u,v}{\text{maximize}} \sum_{k=1}^{N} u_{1,k} x_k^\mathsf{T} v_1 - \lambda^{-1} \sum_{k=1}^{N} u_{1,k} \log(u_{1,k}) + (\alpha - 1) \sum_{k=1}^{N} u_{1,k}$$

$$\Leftrightarrow \underset{u,v}{\text{maximize}} \sum_{k=1}^{N} u_{1,k} x_k^\mathsf{T} v_1 - \lambda^{-1} \sum_{k=1}^{N} u_{1,k} \log(u_{1,k}) + \alpha' \sum_{k=1}^{N} u_{1,k}, \tag{25}$$

where $\alpha' = \alpha - 1$. This optimization problem is also obtained from the maximizing problem of eFCS in Eq. (13), by adding the possibilistic constraint term $\alpha' \sum_{k=1}^{N} u_{1,k}$ from the ePCS objective function described in Eq. (13), while omitting the probabilistic constraint in Eq. (2) and considering the spherical constraint in Eq. (12). This maximizing problem is used to derive ePCCM in the next subsection.

The tPCS method is obtained by solving the optimization problem

$$\underset{u,v}{\text{minimize}} \sum_{k=1}^{N} u_{1,k}^m (1 - x_k^\mathsf{T} v_1) + \frac{\lambda^{-1}}{m-1} \sum_{k=1}^{N} (u_{1,k}^m - u_{1,k}) - \alpha \sum_{k=1}^{N} u_{1,k}, \tag{26}$$

subject to Eq. (12). This optimization problem is derived by subtracting the cosine correlation between an object and a cluster center from 1 $(1 - x_k^\mathsf{T} v_1)$ to obtain the object-cluster dissimilarity. This value replaces the squared Euclidean distance between an object and an cluster center ($\|x_k - v_1\|_2^2$), which is used in the tPCM method described in Eq. (5). The optimal solutions for the membership and cluster center are described as

$$u_{1,k} = \beta(1 - \lambda(1 - m)(1 - x_k^\mathsf{T} v_1))^{\frac{1}{1-m}}, \tag{27}$$

$$v_1 = (\sum_{k=1}^{N} u_{1,k}^m x_k)/(\| \sum_{k=1}^{N} u_{1,k}^m x_k \|_2), \tag{28}$$

where $\beta = ((1 - \alpha\lambda(1 - m))/m)^{1/(m-1)}$. tPCS is also derived from tFCS by subtracting the possibilistic constraint term $\alpha \sum_{k=1}^{N} u_{1,k}$ from tFCS objective function described in Eq. (10), omitting the probabilistic constraint in Eq. (2), and considering the spherical constraint in Eq. (12). The membership is rewritten using $\lambda' = \lambda/(1 - \lambda(1 - m))$ and $\beta' = \beta(\lambda/\lambda')^{1/(1-m)}$ as

$$u_{1,k} = \beta'(1 + \lambda'(1 - m)x_k^\mathsf{T} v_1)^{1/(1-m)}. \tag{29}$$

This membership in Eq. (29) is described for arbitrary object x as $u_1(x) = \beta'(1 + \lambda'(1 - m)x^\mathsf{T} v_1)^{1/(1-m)}$; this is a deformation of the unnormalized von Mises-Fisher distribution when $\lambda = 1$, i.e., $u_1(x)$ recovers a von Mises-Fisher distribution with $m \to 1$, which is similar to the method used by the Tsallis distribution [20] to recover a Gaussian distribution. This membership function for a one-dimensional sphere is depicted in Figs. 2 and 3 for several parameter values of (λ, m), where β is set such that $\max_x u_1(x) = 1$. The tPCS optimization problem is described as the following maximizing problem:

$$\text{Eq. (26)} \Leftrightarrow \underset{u,v}{\text{maximize}} \sum_{k=1}^{N} u_{1,k}^m x_k^\mathsf{T} v_1 - \frac{1 + \lambda(m-1)}{\lambda(m-1)} \sum_{k=1}^{N} [u_{1,k}^m - u_{1,k}]$$

$$+ (\alpha - 1) \sum_{k=1}^{N} u_{1,k}$$

$$\Leftrightarrow \underset{u,v}{\text{maximize}} \sum_{k=1}^{N} u_{1,k}^m x_k^\mathsf{T} v_1 - \frac{\lambda' - 1}{m - 1} \sum_{k=1}^{N} [u_{1,k}^m - u_{1,k}] + \alpha' \sum_{k=1}^{N} u_{1,k}, \tag{30}$$

where $\alpha' = \alpha - 1$ and $\lambda' = \lambda/(1 - \lambda(1 - m))$. This maximizing problem is used to derive tPCCM in the next subsection.

3.3 Possibilistic Clustering for Categorical Multivariate Data

In this subsection, we propose two possibilistic clustering methods for categorical multivariate data, ePCCM and tPCCM.

First, we reconfigure the objective function of eFCCM described in Eq. (15) as

$$\sum_{i=1}^{C} \sum_{k=1}^{N} \sum_{\ell=1}^{M} u_{i,k}(\log(w_{i,\ell})R_{k,\ell} - \log(\Gamma(R_{k,\ell} + 1))) - \lambda^{-1} \sum_{i=1}^{C} \sum_{k=1}^{N} u_{i,k} \log(u_{i,k}), \tag{31}$$

by adding the term

$$-\sum_{i=1}^{C} \sum_{k=1}^{N} \sum_{\ell=1}^{M} u_{i,k} \log(\Gamma(R_{k,\ell} + 1)) \tag{32}$$

to the original objective function. This term originates from the third term of the following lower bound, for the log-likelihood of a multinomial mixture model

$$-\sum_{k=1}^{N}\sum_{i=1}^{C}u_{i,k}\log(u_{i,k})\sum_{k=1}^{N}\sum_{i=1}^{C}u_{i,k}+\log(\Gamma(\sum_{\ell=1}^{M}R_{k,\ell}+1))$$

$$-\sum_{k=1}^{N}\sum_{i=1}^{C}u_{i,k}\sum_{\ell=1}^{M}\log(\Gamma(R_{k,\ell}+1))+\sum_{k=1}^{N}\sum_{i=1}^{C}u_{i,k}\sum_{\ell=1}^{M}\log(w_{i,\ell})R_{k,\ell} \tag{33}$$

$$\leq\sum_{k=1}^{N}\sum_{i=1}^{C}u_{i,k}\log(\frac{1}{u_{i,k}}\frac{\Gamma(\sum_{\ell=1}^{M}R_{k,\ell}+1)}{\prod_{\ell=1}^{M}\Gamma(R_{k,\ell}+1)}\prod_{\ell=1}^{M}w_{i,\ell}^{R_{k,\ell}}) \tag{34}$$

$$=\sum_{k=1}^{N}\log(\sum_{i=1}^{C}\frac{\Gamma(\sum_{\ell=1}^{M}R_{k,\ell}+1)}{\prod_{\ell=1}^{M}\Gamma(R_{k,\ell}+1)}\prod_{\ell=1}^{M}w_{i,\ell}^{R_{k,\ell}}). \tag{35}$$

The added term described in Eq. (32) does not affect the optimal solution of eFCCM because of the constraint described in Eq. (2), whereas it plays a role in constituting the membership function in a possibilistic manner; this is discussed later.

Next, similar to the manner in which ePCS is derived from eFCS, the ePCCM optimization problem is constructed from eFCCM. The objective function of the eFCS maximizing problem described in Eq. (13) is quite similar to that of eFCCM in Eq. (31) if $s_{i,k}=x_k^{\mathsf{T}}v_i$ in eFCS (Eq. (13)) and $s_{i,k}=\sum_{\ell=1}^{M}\log(w_{i,\ell})R_{k,\ell}-\log(\Gamma(R_{k,\ell}+1))$ in eFCCM (Eq. (31)). Based on this information, an ePCCM optimization problem is proposed as

$$\underset{u,w}{\text{maximize}}\sum_{k=1}^{N}\sum_{\ell=1}^{M}u_{1,k}(\log(w_{1,\ell})R_{k,\ell}-\log(\Gamma(R_{k,\ell}+1)))-\lambda^{-1}\sum_{k=1}^{N}u_{1,k}\log(u_{1,k})$$

$$+\alpha\sum_{k=1}^{N}u_{1,k} \tag{36}$$

subject to the constraint in Eq. (14), which is obtained from the eFCCM objective function in Eq. (31) by setting $C=1$, omitting the constraint in Eq. (2), and adding the possibilistic constraint $\alpha\sum_{k=1}^{N}u_{1,k}$ to the eFCCM objective function. By solving this optimization problem, we obtain the optimal solutions for memberships (u,w) as

$$u_{1,k}=\beta\exp(\lambda\sum_{\ell=1}^{M}\log(w_{1,\ell})R_{k,\ell}-\log(\Gamma(R_{k,\ell}+1))), \tag{37}$$

$$w_{1,\ell}=(\sum_{k=1}^{N}u_{1,k}R_{k,\ell})/(\sum_{r=1}^{M}\sum_{k=1}^{N}u_{1,k}R_{k,r}), \tag{38}$$

where $\beta=\exp(-\lambda-1+\alpha\lambda)$. The ePCCM membership in Eq. (37) is described for arbitrary object $R=(R_1,\cdots,R_M)$ as $u_1(R)=\beta\exp(\lambda\sum_{\ell=1}^{M}\log(w_{1,\ell})-\log(\Gamma(R_\ell+1)))$ is the unnormalized multinomial distribution when $\lambda=1$. This membership function for $M=2$ and $w_1=(0.2,0.8)$ is depicted in Fig. 4 for several parameter values of λ, where β is set such that $\max_R u_1(R) = 1$. Here, we can observe the purpose of adding the term in Eq. (32). If this term is omitted, such a membership function is described as

$$u_1(R) = \beta \exp(\lambda \sum_{\ell=1}^{M} \log(w_{1,\ell}) R_\ell), \tag{39}$$

and is depicted in Fig. 5 for $\lambda = 1$ where β is set such that $\max_R u_1(R) = 1$. From this figure, we can observe that such membership functions cannot capture the mode of densities; when the mode with $w_1 < 0.5$ is at the minimal value of R_1, $R_1 = 0$, the mode with $w_1 > 0.5$ is at the maximal value of R_1, $R_1 = 20$, and the mode with $w_1 = 0.5$ is disappears. On the other hand, by adding the term in Eq. (32), we can observe in Fig. 4 that the membership functions can capture the mode of densities.

The tPCCM optimization problem is obtained from ePCCM in a similar manner to how tPCS is derived from ePCS, i.e., by replacing $u_{i,k}$ in the first term of Eq. (36) and Shannon entropy in the second term of Eq. (36) by $u_{i,k}^m$ and Tsallis entropy, respectively, as

$$\underset{u,w}{\text{maximize}} \sum_{k=1}^{N} \sum_{\ell=1}^{M} u_{1,k}^m (\log(w_{1,\ell}) R_{k,\ell} - \log(\Gamma(R_{k,\ell} + 1))) - \frac{\lambda^{-1}}{m-1} \sum_{k=1}^{N} [u_{1,k}^m - u_{1,k}]$$

$$+ \alpha \sum_{k=1}^{N} u_{1,k} \tag{40}$$

subject to the constraint in Eq. (14). By solving this optimization problem, we obtain the optimal solutions of memberships as

$$u_{1,k} = \beta(1 + (1-m)\lambda \sum_{\ell=1}^{M} \log(w_{1,\ell}) R_{k,\ell} - \log(\Gamma(R_{k,\ell} + 1)))^{1/(1-m)}, \tag{41}$$

$$w_{1,\ell} = (\sum_{k=1}^{N} u_{1,k}^m R_{k,\ell}) / (\sum_{r=1}^{M} \sum_{k=1}^{N} u_{1,k}^m R_{k,r}), \tag{42}$$

where $\beta = ((1 - \alpha\lambda(1 - m))/m)^{1/(m-1)}$. This membership in Eq. (41) is derived for arbitrary object $R = (R_1, \cdots, R_M)$, as $u_k = \beta(1 + (1 - m)\lambda \sum_{\ell=1}^{M} \log(w_{1,\ell}) R_\ell - \log(\Gamma(R_\ell + 1)))^{1/(1-m)}$, which is a deformation of the unnormalized multinomial distribution when $\lambda = 1$, i.e., $u_1(x)$ recovers multinomial distribution with $m \to 1$ by setting an adequate normalization factor β. This membership function for $M = 2$ and $w_1 = (0.2, 0.8)$ is depicted in Figs. 6 and 7 for several parameter values of (λ, m) where β is set such that $\max_R u_1(R) = 1$.

4   Numerical Example

This section provides numerical examples based on artificial and actual datasets. The first example illustrates the performance of ePCS and tPCS using a dataset

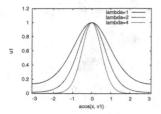

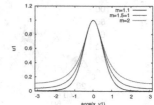

Fig. 1. ePCS membership functions

Fig. 2. tPCS membership functions with $m = 2$

Fig. 3. tPCS membership functions with $\lambda' = 1$

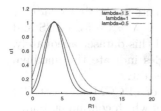

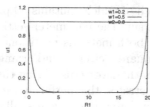

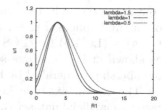

Fig. 4. ePCCM membership functions

Fig. 5. Incomplete ePCCM membership functions

Fig. 6. tPCCM membership functions with $m = 2$

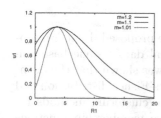

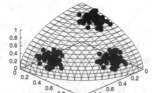

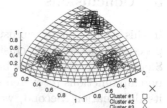

Fig. 7. tPCCM membership functions with $\lambda = 1$

Fig. 8. Artificial Dataset #1

Fig. 9. Result for Artificial Dataset #1

-0.8pc

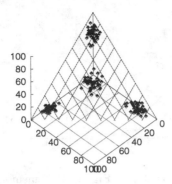
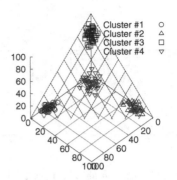

Fig. 10. Artificial dataset #2

Fig. 11. Results for Artificial dataset #2 obtained with ePCCM and tPCCM

containing three clusters, each of which contains 50 points in the first quadrant of the unit sphere (Fig. 8). Using the parameter settings $\lambda = 1.0$ for ePCS and $(\lambda, m) = (1.0, 1.5)$ for tPCS, both methods partitioned this dataset adequately, as shown in Fig. 9, where squares, circles, and triangles indicate the maximal memberships generated by both algorithms during the test.

The second example illustrates the performance of ePCCM and tPCCM using an artificial dataset containing four clusters, all of which contain 50 points obtained from a random sampling of multinomial distributions with parameters $(0.8, 0.1, 0.1)$, $(0.1, 0.8, 0.1)$, $(0.1, 0.1, 0.8)$, and $(1/3, 1/3, 1/3)$ (Fig. 10). With the parameter settings $\lambda = 1.0$ for ePCCM and $(\lambda, m) = (1.0, 1.5)$ for tPCCM, both methods partitioned this dataset adequately, as shown in Fig. 11. The maximal membership of the data is depicted by squares, circles, triangles, and reverse triangles.

5   Conclusions

In this study, four possibilistic clustering methods were proposed. First, we proposed two possibilistic clustering methods for spherical data — one based on Shannon entropy, and one based on Tsallis entropy. It was shown that the membership functions recovered the unnormalized von Mises-Fisher distribution and its deformation. Second, we proposed two possibilistic clustering methods for categorical multivariate data. It was shown that these membership functions recovered the unnormalized multinomial distribution and its deformation. The validity of the proposed methods was confirmed through numerical examples.

In future work, we will (1) apply the proposed methods to larger and more complex datasets, (2) investigate how fuzzification parameters affect clustering accuracy and propose a method to automatically set the best parameter values, (3) apply the fuzzified method used in [16], (4) compare the proposed methods with other clustering methods, (5) apply the sequential cluster extraction [24], which is another algorithm for possibilistic clustering, and (6) develop a possibilistic clustering approach for other data types.

References

1. Bezdek, J.: Pattern Recognition with Fuzzy Objective Function Algorithms. Plenum Press, New York (1981)
2. MacQueen, J.B.: Some methods of classification and analysis of multivariate observations. In: Proceedings of the 5th Berkeley Symposium on Mathematical Statistics and probability, pp. 281–297 (1967)
3. Miyamoto, S., Mukaidono, M.: Fuzzy c-means as a regularization and maximum entropy approach. In: Proceedings of 7th International Fuzzy Systems Association World Congress (IFSA 1997), vol. 2, pp. 86–92 (1997)
4. Ménard, M., Courboulay, V., Dardignac, P.: Possibilistic and probabilistic fuzzy clustering: unification within the framework of the non-extensive thermostatistics. Pattern Recogn. **36**, 1325–1342 (2003)
5. Krishnapuram, R., Keller, J.M.: A possibilistic approach to clustering. IEEE Trans. Fuzzy Syst. **1**, 98–110 (1993)
6. Krishnapuram, R., Keller, J.M.: The possibilistic c-means algorithm: insights and recommendations. IEEE Trans. Fuzzy Syst. **4**, 393–396 (1996)
7. Strehl, A., Ghosh, J., Mooney, R.: Impact of similarity measures on web-page clustering. In: Proceedings of AAAI 2000, pp. 58–64 (2000)
8. Dhillon, I.S., Modha, D.S.: Concept decompositions for large sparse text data using clustering. Mach. Learn. **42**, 143–175 (2001)
9. Miyamoto, S., Mizutani, K.: Fuzzy multiset model and methods of nonlinear document clustering for information retrieval. In: Torra, V., Narukawa, Y. (eds.) MDAI 2004. LNCS (LNAI), vol. 3131, pp. 273–283. Springer, Heidelberg (2004)
10. Mizutani, K., Inokuchi, R., Miyamoto, S.: Algorithms of nonlinear document clustering based on fuzzy set model. Int. J. Intell. Syst. **23**(2), 176–198 (2008)
11. Kanzawa, Y.: Maximizing model of bezdek-like spherical fuzzy c-means clustering. In: Proceedings of the FUZZ-IEEE 2014, pp. 2482–2488 (2014)
12. Kanzawa, Y.: On kernelization for a maximizing model of bezdek-like spherical fuzzy c-means clustering. In: Torra, V., Narukawa, Y., Endo, Y. (eds.) MDAI 2014. LNCS, vol. 8825, pp. 108–121. Springer, Heidelberg (2014)
13. Kanzawa, Y: Fuzzy clustering based on α-divergence for spherical data and for categorical multivariate data. In: Proceedings of the FUZZ-IEEE 2015, (to appear)
14. Oh, C., Honda, K., Ichihashi, H.: Fuzzy clustering for categorical multivariate data. In: Proceedings of the IFSA World Congress and 20th NAFIPS International Conference, pp. 2154–2159 (2001)
15. Honda, K., Oshio, S., Notsu, A.: FCM-type fuzzy co-clustering by K-L information regularization. In: Proceedings of 2014 IEEE International Conference on Fuzzy Systems, pp. 2505–2510 (2014)
16. Honda, K., Oshio, S., Notsu, A.: Item membership fuzzification in fuzzy co-clustering based on multinomial mixture concept. In: Proceedings of 2014 IEEE International Conference on Granular Computing, pp. 94–99 (2014)
17. Kummamuru, K., Dhawaie, A., Krishnapuram, R.: Fuzzy co-clustering of document and keywords. In: IEEE Proceedings of FUZZ 2003, vol. 2, pp. 772–777 (2003)

18. Miyamoto, S., Inokuchi, R., Kuroda, Y.: Possibilistic and fuzzy c-means clustering with weighted objects. In: Proceedings of 2006 IEEE International Conference on Fuzzy Systems, pp. 869–874 (2006)
19. Miyamoto, S., Ichihashi, H., Honda, K.: Algorithms for Fuzzy Clustering. Springer, Heidelberg (2008)
20. Tsallis, C.: Possible generalization of boltzmann-gibbs statistics. J. Stat. Phys. **52**, 479–487 (1988)
21. Buchta, C., Kober, M., Feinerer, I., Hornik, K.: Spherical k-means clustering. J. Stat. Softw. **50**(10), 1–22 (2012)
22. Barber, M.J.: Modurality and community detection in bipartite networks. Phys. Rev. E **76**, 066102 (2007)
23. Davis, A., Gardner, B.B., Gardner, M.R.: Deep South. University of Chicago Press, Chicago (1941)
24. Kanzawa, Y.: Sequential cluster extraction using power-regularized possibilistic c-means. JACIII **19**(1), 67–73 (2015)

A Unified Theory of Fuzzy c-Means Clustering Models with Improved Partition

László Szilágyi[1,2,3]($\boxtimes$)

[1] Department of Control Engineering and Information Technology,
Budapest University of Technology and Economics, Budapest, Hungary
[2] Faculty of Technical and Human Science of Tîrgu-Mureş, Sapientia - Hungarian
Science University of Transylvania, Tîrgu-Mureş, Romania
lalo@ms.sapientia.ro
[3] Canterbury University of Christchurch, Christchurch, New Zealand

Abstract. This paper attempts to unify the theory of a certain class of modified variants and another class of manipulated versions of the fuzzy c-means algorithm. Starting from the objective function of the so-called fuzzy c-means algorithm with generalized improved partition (GIFP-FCM), and defining its rewarding term in a more flexible way, we obtain a unified algorithm that can model all algorithm variants in question including the wide family of suppressed and generalized suppressed FCM. Numerical tests were carried out to provide a comparison of the modeled algorithms in terms of accuracy and cluster size insensitivity. The suppression of the probabilistic fuzzy partition obtained at high values of the fuzzy exponent m proved the most effective.

Keywords: Fuzzy c-means algorithm · Improved partition · Suppressed fuzzy c-means algorithm

1 Introduction

C-means clustering algorithms belong to unsupervised classification methods which group a set of input vectors into a previously defined number (c) of classes. Initially there was the hard c-means (HCM) algorithm [9], which employed the bivalent (crisp) logic to describe partitions. HCM usually converges quickly, but is considerably sensitive to initialization, and frequently gets stuck in local minima leading to mediocre partitions.

The introduction of fuzzy logic [12] into clustering problems led to the definition of the fuzzy partition, in which every input vector can belong to several classes with various degrees of membership. The first c-means clustering algorithm that employs fuzzy partitions is the so-called fuzzy c-means (FCM) introduced by Bezdek [1], which uses a probabilistic constraint to define the fuzzy membership functions. FCM reportedly creates finer partitions than HCM, has

Research supported by the Hungarian National Research Funds (OTKA), Project no. PD103921.

© Springer International Publishing Switzerland 2015
V. Torra and Y. Narakawa (Eds.): MDAI 2015, LNAI 9321, pp. 129–140, 2015.
DOI: 10.1007/978-3-319-23240-9_11

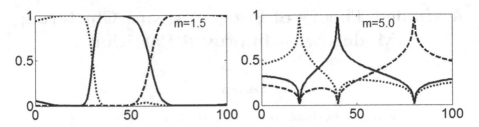

Fig. 1. Fuzzy membership functions produced by FCM in a single dimensional problem. The multimodality escalates as the fuzzy exponent m grows.

a reduced sensitivity to initial cluster prototypes, but it converges much slower. In spite of this drawback, FCM is one of the most popular clustering algorithms, not only in engineering studies, but also in a series of sciences from biology to sociology.

The fuzzy membership functions provided by FCM are multimodal, the phenomenon being exhibited in Fig. 1. Those input data, which are at approximately equal distance from two cluster prototypes, receive illogically high fuzzy memberships with respect to distant clusters. Improved partition were introduced in the theory of fuzzy c-means clustering to alleviate the effects of this multimodality. The so-called FCM with improved partition (IFP-FCM) added a term to the objective function of FCM that pushed fuzzy membership functions away from $1/2$ [4]. This method later received a generalization [14], which not just forces fuzzy membership functions towards 0 or 1, but also has a metric-based foundation.

The suppressed fuzzy c-means (s-FCM) algorithm, introduced by Fan et al. [2], proposed to make a step from FCM towards HCM, by manipulating with the fuzzy membership functions computed in each iteration of the FCM's alternating optimization (AO) scheme. The authors defined a previously set constant suppression rate $\alpha \in [0,1]$, which determined the behavior of the algorithm. In each iteration, after having determined the new fuzzy membership functions for input vector with index k, denoted by $u_{1k}, u_{2k}, \ldots, u_{ck}$, the algorithm looks for the largest (winner) membership value u_w, with $w = \arg\max\{u_{ik} | i = 1, 2, \ldots, c\}$, suppresses all non-winner memberships by multiplication with α and raises the winner membership value u_w in such a way that the probabilistic constraint is not affected. Several selection rules for the suppression rate have been introduced later (e.g. [3,5,8]), most of them proposing to compute α using the current cluster prototypes and median or average distance between input vectors. Further on, s-FCM also has several generalized versions that assign each input vector a dedicated, context-dependent suppression rate in each iteration [11].

This paper introduces an optimal clustering model based on the minimization of a quadratic objective function, which can act like any of the above mentioned c-means clustering algorithm. The flexibility is achieved using separate parameter selection rules for each algorithm. Besides giving a unified framework for many existing methods, we also show that suppressed fuzzy c-means clustering models are optimal, which is not an obvious fact based on their definition.

The rest of this paper is structured as follows. Section 2 takes into account the existing c-means clustering models with improved partition. Section 3 introduces a unified framework for the implementation of improved partition c-means clustering models. Section 4 gives a comparative evaluation of improved and suppressed partitions. Conclusions are given in the last section.

2   Background

2.1   The Fuzzy c-Means Algorithm

The conventional fuzzy c-means (FCM) algorithm partitions a set of object data $\mathbf{X} = \{\mathbf{x}_1, \mathbf{x}_2, \ldots, \mathbf{x}_n\}$ into a number of c clusters based on the minimization of a quadratic objective function, defined as:

$$J_{\text{FCM}} = \sum_{i=1}^{c} \sum_{k=1}^{n} u_{ik}^m \|\mathbf{x}_k - \mathbf{v}_i\|_{\mathbf{A}}^2 = \sum_{i=1}^{c} \sum_{k=1}^{n} u_{ik}^m d_{ik}^2, \tag{1}$$

where $\mathbf{v}_i$ represents the prototype or centroid of cluster i $(i = 1 \ldots c)$, $u_{ik} \in [0,1]$ is the fuzzy membership function showing the degree to which vector $\mathbf{x}_k$ belongs to cluster i, $m > 1$ is the fuzzyfication parameter, and d_{ik} represents the distance (any inner product norm defined by a symmetrical positive definite matrix $\mathbf{A}$) between $\mathbf{x}_k$ and $\mathbf{v}_i$. FCM uses a probabilistic partition, meaning that the fuzzy memberships assigned to any input vector $\mathbf{x}_k$ with respect to clusters satisfy the probability constraint $\sum_{i=1}^{c} u_{ik} = 1$. The minimization of the objective function J_{FCM} is achieved by alternately applying the optimization of J_{FCM} over $\{u_{ik}\}$ with $\mathbf{v}_i$ fixed, $i = 1 \ldots c$, and the optimization of J_{FCM} over $\{\mathbf{v}_i\}$ with u_{ik} fixed, $i = 1 \ldots c$, $k = 1 \ldots n$ [1]. In each loop, the optimal values are deduced from the zero gradient conditions and Lagrange multipliers, and obtained as follows:

$$u_{ik}^{\star} = \frac{d_{ik}^{-2/(m-1)}}{\sum_{j=1}^{c} d_{jk}^{-2/(m-1)}} \quad \forall i = 1 \ldots c, \forall k = 1 \ldots n, \tag{2}$$

$$\mathbf{v}_i^{\star} = \frac{\sum_{k=1}^{n} u_{ik}^m \mathbf{x}_k}{\sum_{k=1}^{n} u_{ik}^m} \quad \forall i = 1 \ldots c. \tag{3}$$

According to the alternating optimization (AO) scheme of the FCM algorithm, Eqs. (2) and (3) are alternately applied, until cluster prototypes stabilize.

Hard c-means [9] is a special case of FCM, which uses $m = 1$, and thus the memberships are obtained by the winner-takes-all rule. Each cluster prototype will be the average of the input vectors assigned to the given cluster.

2.2   Fuzzy c-Means with Improved Partition

Partitions provided by FCM have an undesired property: in the proximity of the boundary between two neighbor clusters, fuzzy memberships with respect to other clusters have local maxima, instead of being close to zero. To suppress this phenomenon without losing the fuzzy nature of the algorithm, Höppner and Klawonn [4] introduced the so-called FCM with improved partition (IFP-FCM), which is derived from an objective function that additionally contains a rewarding term:

$$J_{\text{IFP-FCM}} = \sum_{i=1}^{c} \sum_{k=1}^{n} u_{ik}^m d_{ik}^2 - \sum_{k=1}^{n} a_k \sum_{i=1}^{c} (u_{ik} - 1/2)^2, \qquad (4)$$

where parameters a_k are positive numbers. The second term has the effect of pushing the fuzzy membership values u_{ik} ($i = 1 \ldots c$, $k = 1 \ldots n$) towards 0 or 1, while maintaining the probabilistic constraint. Later, Zhu et al. [14] introduced a generalized version of this algorithm, derived from the objective function

$$J_{\text{GIFP-FCM}} = \sum_{i=1}^{c} \sum_{k=1}^{n} u_{ik}^m d_{ik}^2 + \sum_{k=1}^{n} a_k \sum_{i=1}^{c} u_{ik}(1 - u_{ik}^{m-1}), \qquad (5)$$

whose optimization leads to the partition update formula

$$u_{ik}^\star = \frac{(d_{ik}^2 - a_k)^{-1/(m-1)}}{\sum_{j=1}^{c} (d_{jk}^2 - a_k)^{-1/(m-1)}} \qquad \forall i = 1 \ldots c, \forall k = 1 \ldots n. \qquad (6)$$

Equation (6) explains us the behavior of GIFP-FCM: for any input vector $\mathbf{x}_k$, the square of its distances measured from all cluster prototypes are virtually reduced by a constant positive value a_k. The authors also proposed a formula for the choice of a_k: $a_k = \omega \min_i \{d_{ik}^2 | i = 1 \ldots c\}$, with $\omega \in [0.9, 0.99]$, thus keeping the square of all distorted distances positive. Using $\omega = 1$ would reduce GIFP-FCM to HCM.

Both versions of the improved clustering models keep FCM's prototype update formula given in Eq. (3).

2.3   Suppression of Fuzzy c-Means

The suppressed fuzzy c-means (s-FCM) algorithm [2] had the declared goal of reducing the execution time of FCM by improving its convergence speed, while preserving its fine partition quality. The s-FCM algorithm does not minimize J_{FCM}, and it was not introduced as an algorithm that minimizes any objective function. Instead of that, it manipulates with the AO scheme of FCM, by inserting an extra computational step in each iteration, placed between the partition update formula (2) and prototype update formula (3). This new step deforms the partition (fuzzy membership functions) according to the following rule:

$$\mu_{ik} = \begin{cases} 1 - \alpha + \alpha u_{ik} & \text{if } i = w_k \equiv \arg\max_j \{u_{jk}\} \\ \alpha u_{ik} & \text{otherwise} \end{cases}, \qquad (7)$$

where μ_{ik} ($i = 1 \ldots c$, $k = 1 \ldots n$) represents the fuzzy memberships obtained after suppression. During the iterations of s-FCM, these suppressed membership values μ_{ik} will replace u_{ik} in Eq. (3). Suppression rate $\alpha = 1$ makes s-FCM identical with FCM, while $\alpha = 0$ with HCM. Any other values of α lead to algorithms that differ from all above mentioned ones.

Although w_k depends on the index of the input vector x_k, in the following we will denote it by w for the sake of simplicity of formulas. Likewise, the largest fuzzy membership value assigned to vector x_k will be denoted by u_w instead of $u_{w_k k}$, which is not supposed to be confusing, as all formulas are referring to each single input vector separately.

Later, several papers [3,5,13] proposed various schemes to select the parameter of suppression, α, either as a constant, or a value updated once per iteration.

In an earlier paper [10], we have shown that for any $\alpha > 0$, the suppression given in Eq. (7) is equivalent with virtually reducing the distance between x_k and its winner cluster prototype v_w, from $d_{wk} = ||x_k - v_w||$ to:

$$d'_{wk} = \frac{d_{wk}}{\left(1 + \frac{1-\alpha}{\alpha u_w}\right)^{(m-1)/2}} = \frac{d_{wk}}{\sqrt{\left(1 + \frac{1-\alpha}{\alpha u_w}\right)^{m-1}}}, \tag{8}$$

where the denominator is a positive power of a number greater or equal to 1. All other distances d_{ik}, $i \in \{1, 2, \ldots, c\} \setminus \{w\}$ remain unmodified. The so-called quasi learning rate of the s-FCM algorithm was defined as $\eta = 1 - d'_{wk}/d_{wk}$, just as in case of competitive clustering algorithms [6], and obtained as

$$\eta = \begin{cases} 1 - \left(1 + \frac{1-\alpha}{\alpha u_w}\right)^{(1-m)/2} & \text{if } \alpha \neq 0 \\ 1 & \text{if } \alpha = 0 \end{cases}. \tag{9}$$

Although the suppression rate was introduced as a constant α, the quasi learning rate depends on three variables: fuzzy exponent m, winner (highest) fuzzy membership value for the given vector u_w, and the suppression rate α.

The formulas given in Eqs. (8) and (9) were later used to introduce so-called generalized suppressed FCM (gs-FCM) algorithms that apply context-dependent suppression rates α_k for each input vector x_k in each iteration, according to some suppression rules [11].

In the following, we will show that all algorithms mentioned in this section are related and can be described within a certain framework of fuzzy c-means algorithms with manipulated partition.

3   The Unified Framework

Let us introduce the unified clustering algorithm now. The objective function is quite similar to the one of GIFP-FCM, but there is a slight modification:

$$J = \sum_{i=1}^{c} \sum_{k=1}^{n} u_{ik}^m d_{ik}^2 + \sum_{k=1}^{n} s_{ik} \sum_{i=1}^{c} u_{ik}(1 - u_{ik}^{m-1}), \tag{10}$$

where the slight modification consists in the nature of the rewarding parameter denoted this time by s_{ik}: we made it dependent on both the cluster (i) and the input vector (k) it refers to. The objective function is minimized under the probabilistic constraint $\sum_{i=1}^{c} u_{ik} = 1$, $\forall k = 1 \ldots n$. For any $k = 1 \ldots n$, parameters s_{ik} ($i = 1 \ldots c$) are treated as constants during each iteration, being determined at the beginning of the loop based on the distances $d_{1k}, d_{2k}, \ldots d_{ck}$. There is one strict condition to be held: $\forall i = 1 \ldots c$, $\forall k = 1 \ldots n$, $s_{ik} \leq \min\{d_{jk}^2 | j = 1 \ldots c\}$. Without this condition some squared distances could become negative, causing certain fuzzy membership functions have complex (imaginary) values.

The alternating optimization formulas of this objective function, similarly to those of GIFP-FCM, are obtained as:

$$u_{ik}^{*} = \frac{(d_{ik}^2 - s_{ik})^{-1/(m-1)}}{\sum_{j=1}^{c}(d_{jk}^2 - s_{jk})^{-1/(m-1)}} \quad \forall i = 1 \ldots c, \forall k = 1 \ldots n, \qquad (11)$$

$$\mathbf{v}_i^{*} = \frac{\sum_{k=1}^{n} u_{ik}^m \mathbf{x}_k}{\sum_{k=1}^{n} u_{ik}^m} \quad \forall i = 1 \ldots c. \qquad (12)$$

The above optimization formulas should be applied until convergence occurs, just as in case of the FCM algorithm.

Now we will show this objective function and its optimization scheme is compatible with all algorithms mentioned in the previous section:

1. The most obvious choice is setting each $s_{ik} = \omega \min\{d_{jk}^2 | j = 1 \ldots c\} = a_k$, $\forall i = 1 \ldots c$, $\forall k = 1 \ldots n$, with previously set $\omega \in [0.9, 0.99]$. This way we obtain the GIFP-FCM algorithm, as introduced in [4]. Setting $\omega = 1$ would reduce the algorithm to HCM.

2. If we denote by w the index of the cluster whose prototype is closest to vector $\mathbf{x}_k$, that is, $w = \arg\min_{j}\{d_{jk}, j = 1, 2, \ldots, c\}$, and we set

$$s_{ik} = \begin{cases} d_{ik}^2 \left[1 - \left(\dfrac{\alpha d_{ik}^q}{\alpha d_{ik}^q + (1-\alpha) \sum\limits_{j=1}^{c} d_{jk}^q} \right)^{m-1} \right] & \text{if } i = w \\ 0 & \text{if } i \neq w \end{cases}, \qquad (13)$$

with fuzzy exponent $m > 1$, $q = -2/(m-1)$, and $\alpha \in (0,1]$, we obtain a clustering algorithm equivalent to the suppressed FCM at suppression rate α. The above formula reduces a lot in the very popular case of $m = 2$:

$$s_{ik} = \begin{cases} d_{ik}^2 \dfrac{(1-\alpha) \sum\limits_{j=1}^{c} d_{jk}^{-2}}{(1-\alpha) \sum\limits_{j=1}^{c} d_{jk}^{-2} + \alpha d_{ik}^{-2}} & \text{if } i = w \\ 0 & \text{if } i \neq w \end{cases}. \qquad (14)$$

It is easy to admit that $\alpha = 1$ reduces the above formula to $s_{ik} = 0$ and the algorithm to FCM. In order to get the HCM algorithm (equivalent to the $\alpha = 0$ case), one needs to set

$$s_{ik} = \begin{cases} d_{ik}^2 & \text{if } i = w \\ 0 & \text{if } i \neq w \end{cases}. \qquad (15)$$

3. Clustering models that use the same suppression rate α for each input vector, which is updated at the beginning if each loop, based on cluster prototype positions $v_1, v_2, \ldots, v_c$, and distances d_{ik} $(i = 1, \ldots, c;\ k = 1, \ldots, n)$, can be modeled as follows: in each iteration, first compute the suppression rate α according to the rule prescribed by the algorithm, and then apply Eq. (13) to get the necessary s_{ik} values.

4. If we wish to model the generalized s-FCM algorithm of type θ [11], namely the one with constant learning rate $\eta = \theta \in [0, 1]$, we need to set:

$$s_{ik} = \begin{cases} d_{ik}^2 \theta(2 - \theta) & \text{if } i = w \\ 0 & \text{if } i \neq w \end{cases}. \tag{16}$$

Obviously, $\theta = 0$ causes $s_{ik} = 0$, which corresponds to the FCM algorithm, while $\theta = 1$ reduces our algorithm to HCM.

5. We may also want to model the family of generalized s-FCM algorithms defined by $\eta = f(u_w^{(\text{FCM})})$ introduced in [11], where $u_w^{(\text{FCM})}$ stands for the greatest valued fuzzy membership assigned to input vector x_k by the FCM algorithm using the same fuzzy exponent m:

$$u_w^{(\text{FCM})} = \frac{(\min\{d_{jk} | j = 1, \ldots, c\})^{-2/(m-1)}}{\sum\limits_{j=1}^{c} d_{jk}^{-2/(m-1)}} = \frac{d_{wk}^{-2/(m-1)}}{\sum\limits_{j=1}^{c} d_{jk}^{-2/(m-1)}}. \tag{17}$$

In [11], we defined for example the gs-FCM algorithm of type ρ described by $\eta = 1 - \rho u_w^{(\text{FCM})}$, and gs-FCM of type β given by $\eta = 1 - (u_w^{(\text{FCM})})^{\beta/(1-\beta)}$. No matter which one the function $f(\cdot)$ is, to achieve the behavior of the algorithm, we need to set:

$$s_{ik} = \begin{cases} d_{ik}^2 \left[1 - \left(1 - f^2\left(u_w^{(\text{FCM})}\right)\right)\right] & \text{if } i = w \\ 0 & \text{if } i \neq w \end{cases}, \tag{18}$$

where $u_w^{(\text{FCM})}$ can be computed from distances d_{ik} $(i = 1, \ldots, c)$ using Eq. (17).

6. For the family of generalized s-FCM algorithms defined by direct formula between the largest fuzzy membership before (u_w) and after (μ_w) suppression, we may proceed as follows. Let the general suppression formula be $\mu_w = g(u_w)$. We need to define s_{ik} as:

$$s_{ik} = \begin{cases} d_{ik}^2 \left[1 - \left(\frac{u_w(1-g(u_w))}{g(u_w)(1-u_w)}\right)^{m-1}\right] & \text{if } i = w \\ 0 & \text{if } i \neq w \end{cases}.$$

For example, the gs-FCM algorithm of type τ, whose definition formula $\mu_w = (u_w + \tau)/(1 + u_w\tau)$, with $\tau \in [0, 1]$, was inspired by the relativistic addition of velocities, is achieved via defining s_{ik} as:

$$s_{ik} = \begin{cases} d_{ik}^2 \left[1 - \left(\frac{(1-\tau)d_{ik}^q}{d_{ik}^q + \tau \sum\limits_{j=1}^{c} d_{jk}^q}\right)^{m-1}\right] & \text{if } i = w \\ 0 & \text{if } i \neq w \end{cases},$$

Data: Input data $X = \{x_1, x_2, \ldots, x_n\}$
Data: Fuzzy exponent $m > 1$
Result: Cluster prototype vectors $v_1, v_2, \ldots v_c$
Result: Labels $\xi_k \in \{1, 2, \ldots, c\}$, $\forall k = 1 \ldots n$
repeat
    for $i \in \{1, 2, \ldots c\}$ **do**
       | $V_i^{(\mathrm{up})} \leftarrow \mathbf{0}$; $V_i^{(\mathrm{dn})} \leftarrow 0$
    end
    $q \leftarrow -2/(m-1)$
    Updating partition
    for $k \in \{1, 2, \ldots n\}$ **do**
       $w \leftarrow \arg\min_i \{d_{ik} | i = 1, \ldots, c\}$
       if $d_{wk} = 0$ **then**
          for $i \in \{1, 2, \ldots c\}$ **do**
             | $u_i \leftarrow 0$
          end
          $u_w \leftarrow 1$
       end
       else
          $u_w^{(\mathrm{FCM})} = d_{wk}^q / \sum_{j=1}^{c} d_{jk}^q$
          Use chosen formula to find s_{ik} $(i = 1 \ldots c)$ using $u_w^{(\mathrm{FCM})}$
          for $i \in \{1, 2, \ldots c\}$ **do**
             | $\delta_i = d_{ik}^2 - s_{ik}$
          end
          $\Sigma_u \leftarrow 0$
          for $i \in \{1, 2, \ldots c\}$ **do**
             $u_i \leftarrow \delta_i^{q/2}$
             $\Sigma_u \leftarrow \Sigma_u + u_i$
          end
          for $i \in \{1, 2, \ldots c\}$ **do**
             | $u_i \leftarrow u_i/\Sigma_u$
          end
       end
       for $i \in \{1, 2, \ldots c\}$ **do**
          $V_i^{(\mathrm{up})} \leftarrow V_i^{(\mathrm{up})} + u_i^m x_k$
          $V_i^{(\mathrm{dn})} \leftarrow V_i^{(\mathrm{dn})} + u_i^m$
       end
    end
    Updating cluster prototypes;
    for $i \in \{1, 2, \ldots c\}$ **do**
       | $v_i \leftarrow V_i^{(\mathrm{up})}/V_i^{(\mathrm{dn})}$
    end
until *convergence occurs*;
Labeling each input vector
for $k \in \{1, 2, \ldots n\}$ **do**
    | $\xi_k \leftarrow \arg\min_i \{||x_k - v_i||\}$
end

Algorithm 1: The unified algorithm, in a memory efficient implementation, similarly to [7]. Partition information u_{ik} and rewarding parameters s_{ik}, need not be stored for each input vector separately.

with $q = -2/(m-1)$, which in the popular case of fuzzy exponent $m = 2$ reduces to

$$s_{ik} = \begin{cases} d_{ik}^2 \dfrac{\sum\limits_{j=1}^{c} d_{jk}^{-2} - d_{ik}^{-2}}{\sum\limits_{j=1}^{c} d_{jk}^{-2} + d_{ik}^{-2}/\tau} & \text{if } i = w \\ 0 & \text{if } i \neq w \end{cases}.$$

A memory efficient implementation of the unified algorithm is summarized in Algorithm 1. As all variants of FCM modeled within the proposed framework take a step from the base fuzzy c-means algorithm towards hard c-means, let us refer to the unified framework as hardened fuzzy c-means clustering algorithm (HFCM).

4 Comparative Evaluation

The aim of this section is to show how the algorithms modeled by HFCM influence the fuzzy membership functions and evaluate the way they alleviate the effects of membership function multimodality.

First we consider an input data set consisting of all scalar integer values from 0 to 1000, and set as base algorithm the FCM with fuzzy exponent $m = 3$. Now let us apply three different ways of stepping towards HCM: (1) using a lower fuzzy exponent in FCM, e.g. $m = 2$; (2) apply suppression of this base FCM with a constant $0 < \alpha < 1$; (3) add a rewarding term to this base FCM to act like GIFP-FCM at constant $0 < \omega < 1$.

Figure 2 shows some examples of obtained fuzzy membership functions for all input values. The base algorithm roughly produces triangular membership functions. Reducing the value of the fuzzy exponent m within the FCM algorithm changes the aspect of membership functions toward the bell-like shape, meaning that input data in the proximity of cluster prototypes belong to that given cluster with membership close to 1. This kind of shape change does not occur in any nontrivial case of HFCM. Suppression at any fixed rate $0 < \alpha < 1$ causes a sudden step in the membership functions in the vicinity of those input data that are at equal distance from two (or more) cluster prototypes. Otherwise, suppression maintains the triangular fuzzy membership functions. On the other hand, GIFP-FCM produces fuzzy membership functions with the shape of a dome with sharp peak. All variants visibly suppress the strength of multimodality.

Now let us create a test scenario to investigate the alleviated multimodality. We will have a set of $n = n_1 + n_2 + n_3$ two-dimensional vectors, organized in three compact groups, with their centers in $(0, -3)$, $(0, 3)$, and $(10, 0)$, respectively, each group normally distributed with a unit variance. The first two groups will contain an equal number of vectors, $n_1 = n_2 = 2000$, while the third group will be much smaller. We will cluster the input data set into $c = 3$ clusters using s-FCM and GIFP-FCM with various parameter values. In any of the cases, if n_3 is too small, it will be neglected by the clustering algorithm, and all three cluster prototypes will be situated in the area covered by the vectors of the first two groups. We will investigate what is the minimum size of the third cluster

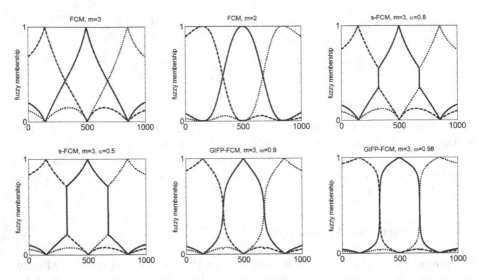

Fig. 2. Improved partitions in various tested scenarios. Stepping from FCM ($m = 3$) towards HCM: via keeping the algorithm but moving fuzzy exponent towards 1, via keeping fuzzy exponent but suppressing (s-FCM algorithm), and via providing generalized improved partitions (GIFP-FCM algorithm).

needed for a correct clustering. The lower this limit value is, the more efficient the suppression of the multimodality effect will be. What we need to know from the beginning, the FCM algorithm at $m = 3$ cannot distinguish the third group from the other two if $n_3 < 119$ even if we initialize the cluster prototypes with their ideal position, which actually means cheating in the favor of the algorithm. For $m = 5$, the limit condition is $n_3 < 241$.

We have employed s-FCM and GIFP-FCM in various circumstances to cluster the above presented data set. But in this case cluster prototypes were initialized in a fair way to optimally cover the area of input data. Table 1 exhibits the lowest values of n_3 for which the third cluster was successfully distinguished. As it was expected, the limit value of n_3 is lower as the applied algorithm approaches HCM (α decreases and ω increases). However, there is something strange in the outcome of s-FCM indicating, that better accuracy is achieved in case of choosing a higher fuzzy exponent m and suppressing that partition.

In order to provide a fair comparison between suppressed and improved partitions, we need to match the α and ω values somehow. For this purpose we employed the means of s_{ik} rewarding parameters and asserted that for any ω that value of α should be chosen, for which their corresponding s_{wk} values are equal. In case of $m = 3$, the formula of corresponding suppression rate is

$$\alpha = \frac{u_w \sqrt{1 - \omega} - (1 - u_w)(1 - \omega)}{u_w^2 - (1 - \omega)(1 - u_w)^2}. \tag{19}$$

Supposing a fixed winner fuzzy membership of $u_w = 0.9$, we obtain the comparison indicated in Table 2: s-FCM seems to be significantly more effective.

Table 1. Limit values of n_3, for which the clusters are identified correctly in case of tested algorithms is various circumstances

Algorithm		$m = 2$	$m = 2.5$	$m = 3$	$m = 3.5$	$m = 4$	$m = 5$
FCM		96	149	188	258	266	310
	m	$\alpha = 0.4$	$\alpha = 0.5$	$\alpha = 0.6$	$\alpha = 0.7$	$\alpha = 0.8$	$\alpha = 0.9$
s-FCM	3	10	15	24	38	60	106
s-FCM	5	3	3	8	18	38	98
	m	$\omega = 0.99$	$\omega = 0.95$	$\omega = 0.9$	$\omega = 0.8$	$\omega = 0.7$	$\omega = 0.5$
GIFP-FCM	3	3	12	23	42	66	98
GIFP-FCM	5	6	22	40	75	130	167

All algorithms modeled within the unified framework belong to the same class of optimality. All of them minimize the same objective function. All of them make exclusively optimal steps when updating the partition and cluster prototypes using Eqs. (11) and (12). However, changing rewarding term values at the beginning of each iteration adjusts the problem to be optimized.

Table 2. Limit values of n_3 in case of matched parametrization

Fixed ω for GIFP-FCM	0.99	0.95	0.9	0.8	0.7	0.5	0.3
Corresponding α for s-FCM	0.110	0.242	0.339	0.473	0.574	0.728	0.838
GIFP-FCM at $m = 3$	3	12	23	42	66	98	139
s-FCM at $m = 3$	2	2	6	12	22	42	73

5 Conclusions

In this paper we proposed a unified framework for several kinds of FCM algorithms with improved partition. Starting from the objective function of the GIFP-FCM algorithm, and generalizing its rewarding term, were able to define parameter settings which make the behavior of the unified algorithm rigorously correspond to the ones of suppressed FCM, and various kinds of generalized suppressed FCM, e.g. with constant learning rate, with learning rate defined as function of winner fuzzy membership, or with direct formula between winner fuzzy membership before and after suppression. The unified algorithm has also shown that all s-FCM variants fall in the same class of optimal algorithms as GIFP-FCM. Numerical tests have shown the superiority of partition suppression over partition improvement, in terms of cluster size insensitivity.

References

1. Bezdek, J.C.: Pattern Recognition with Fuzzy Objective Function Algorithms. Plenum, New York (1981)
2. Fan, J.L., Zhen, W.Z., Xie, W.X.: Suppressed fuzzy c-means clustering algorithm. Pattern Recogn. Lett. **24**, 1607–1612 (2003)
3. Fan, J.L., Li, J.: A fixed suppressed rate selection method for suppressed fuzzy c-means clustering algorithm. Appl. Math. **5**, 1275–1283 (2014)
4. Höppner, F., Klawonn, F.: Improved fuzzy partition for fuzzy regression models. Int. J. Approx. Reason. **5**, 599–613 (2003)
5. Hung, W.L., Yang, M.S., Chen, D.H.: Parameter selection for suppressed fuzzy c-means with an application to MRI segmentation. Pattern Recogn. Lett. **27**, 424–438 (2006)
6. Kohonen, T.: The self-organizing map. Proc. IEEE **78**, 1474–1480 (1990)
7. Kolen, J.F., Hutcheson, T.: Reducing the time complexity of the fuzzy c-means algorithm. IEEE Trans. Fuzzy Syst. **10**, 263–267 (2002)
8. Li, J., Fan, J.L.: Parameter selection for suppressed fuzzy c-means clustering algorithm based on fuzzy partition entropy. In: 11th International Conference on Fuzzy Systems and Knowledge Discovery, pp. 82–87. Xiamen (2014)
9. Steinhaus, H.: Sur la division des corp materiels en parties. Bull. Acad. Pol. Sci. C1 III **IV**, 801–804 (1956)
10. Szilágyi, L., Szilágyi, S.M., Benyó, Z.: Analytical and numerical evaluation of the suppressed fuzzy c-means algorithm: a study on the competition in c-means clustering models. Soft. Comput. **14**, 495–505 (2010)
11. Szilágyi, L., Szilágyi, S.M.: Generalization rules for the suppressed fuzzy c-means clustering algorithm. Neurocomput. **139**, 298–309 (2014)
12. Zadeh, L.A.: Fuzzy sets. Inf. Control **8**, 338–353 (1965)
13. Zhao, F., Fan, J.L., Liu, H.Q.: Optimal-selection-based suppressed fuzzy c-means clustering algorithm with self-tuning non local spatial information for image segmentation. Expert Syst. Appl. **41**, 4083–4093 (2014)
14. Zhu, L., Chung, F.L., Wang, S.: Generalized fuzzy c-means clustering algorithm with improved fuzzy partition. IEEE Trans. Syst. Man Cybern. B. **39**, 578–591 (2009)

Data Mining and Data Privacy

Effective MVU via Central Prototypes and Kernel Ridge Regression

Carlotta Orsenigo[✉]

Dept. of Management, Economics and Industrial Engineering,
Politecnico di Milano, via Lambruschini 4b, 20156 Milano, Italy
carlotta.orsenigo@polimi.it

Abstract. Maximum variance unfolding (MVU) is one of the most prominent manifold learning techniques for nonlinear dimensionality reduction. Despite its effectiveness it has proven to be considerably slow on large data sets, for which fast extensions have been developed. In this paper we present a novel algorithm which combines classical MVU and multi-output kernel ridge regression (KRR). The proposed method, called Selective MVU, is based on a three-step procedure. First, a subset of distinguished points indicated as central prototypes is selected. Then, MVU is applied to find the prototypes embedding in the low-dimensional space. Finally, KRR is used to reconstruct the projections of the remaining samples. Preliminary results on benchmark data sets highlight the usefulness of Selective MVU which exhibits promising performances in terms of quality of the data embedding compared to renowned MVU variants and other state-of-the-art nonlinear methods.

Keywords: Nonlinear dimensionality reduction · Manifold learning · Maximum variance unfolding · Prototype selection

1 Introduction

Dimensionality reduction is the process of converting high dimensional data into meaningful representations of reduced dimensionality. As a preliminary step it plays a fundamental role in several machine learning tasks by favoring data visualization, clustering and classification. Dimensionality reduction techniques are usually divided into linear and nonlinear approaches. Within the family of nonlinear methods manifold learning algorithms have drawn great interest by attempting to recover the low dimensional manifold along which data are supposed to lie. These include, among others, Isometric feature mapping [1], Locally linear embedding [2], Laplacian eigenmaps [3], Local tangent space alignment [4] and Maximum variance unfolding [5].

Maximum variance unfolding (MVU), also known as Semidefinite embedding, relies on the notion of isometry which can be defined as a smooth invertible mapping that behaves locally like a rotation plus a translation. The final low-dimensional embedding is therefore locally-distance preserving, since it is derived

© Springer International Publishing Switzerland 2015
V. Torra and Y. Narakawa (Eds.): MDAI 2015, LNAI 9321, pp. 143–154, 2015.
DOI: 10.1007/978-3-319-23240-9_12

by keeping unchanged the distances and angles between neighboring points. The unfolding process requires the solution of a semidefinite program (SDP) which maximizes the variance of the points in the feature space, represented by the trace of the corresponding Gram matrix, under linear equality constraints which impose the local-isometry conditions.

Despite its effectiveness MVU turns out to be considerably slow when the number of points increases since solving large semidefinite programs is time-consuming. To overcome this drawback different approaches can be followed. One may resort to greedy optimization procedures, as those described in [6] and [7], able to efficiently achieve the global optimum of semidefinite programs on large data sets, or to iterative algorithms which transform graph embeddings into MVU feasible solutions [8]. A further strategy to speed up the algorithm is to reduce the original SDP to a smaller problem by means of the Gram matrix factorization. Based on this last approach two fast variants have been developed. In the first the Gram matrix is reconstructed from a smaller submatrix of inner products between randomly chosen landmarks [9]. In the second variant matrix factorization is obtained by expanding the solution of the initial SDP in terms of the bottom eigenvectors of the graph Laplacian [10].

In this paper we present a novel method for nonlinear dimensionality reduction which combines MVU and kernel ridge regression [11] and [12]. The proposed algorithm, called Selective MVU, is based on a three-step procedure. A subset of distinguished points, indicated as prototypes, is first selected from the original data set. Classical MVU is then applied on the collection of prototypes to find their embedding in the low d-dimensional space. The projections of the remaining points are finally derived by learning the nonlinear mapping through multi-output kernel ridge regression (KRR), which has been successfully used as out-of-sample extension for manifold learning [13]. The proposed algorithm enables significant computational savings compared to classical MVU that is in this case applied only to the set of representative points. It draws inspiration from both landmark methods for fast embedding [14] and [9], which place a point in the feature space according to its distance from the projected landmarks, and the spectral regression paradigm [15], in which the subspace learning problem is cast into a regression framework. To designate the collection of prototypes we also propose a novel method based on K-means algorithm that behaves more effectively than random selection. Experiments on eight benchmark data sets highlight the usefulness of Selective MVU which provides promising results compared to well-known MVU fast variants and other prominent nonlinear dimensionality reduction techniques.

The remainder of the paper is organized as follows. Section 2 briefly recalls maximum variance unfolding. Section 3 presents the novel Selective MVU algorithm and the prototype selection method. Computational experiments and results are described in Sect. 4. Conclusions and future developments are discussed in Sect. 5.

2   Maximum Variance Unfolding

Maximum variance unfolding imposes local-isometry constraints aimed at preserving both distances and angles between points and their neighbors, in order to find low-dimensional projections which faithfully represent the input data.

Let $S_m = \{\mathbf{x}_i, i \in \mathcal{M} = \{1, 2, \ldots, m\}\} \subset \Re^n$ be a set of m points approximately confined to a nonlinear manifold of intrinsic dimension d ($d \ll n$). The unfolding process starts with the construction of the neighborhood graph in which nodes represent data points and edges neighborhood relations. Then, it requires the solution of a quadratic optimization problem which maximizes the variance of the embedding subject to the local-isometry conditions. In practice, the problem is reformulated as the following semidefinite program over the Gram matrix $\mathbf{G}_m = [g_{ij}]$ of the points in the feature space, with $g_{ij} = \langle \mathbf{z}_i, \mathbf{z}_j \rangle, \forall i, j \in \mathcal{M}$,

$$\max_{\mathbf{G}} \ \mathrm{tr}\,(\mathbf{G}) \tag{SD}$$

$$\text{s.to} \quad g_{ii} + g_{jj} - 2g_{ij} = d_{ij}^2 \quad \forall i, j \in \mathcal{M}, \ \eta_{ij} = 1, \tag{1}$$

$$\sum_{i,j \in \mathcal{M}} g_{ij} = 0, \tag{2}$$

$$\mathbf{G} \succeq 0, \tag{3}$$

where $d_{ij} = \|\mathbf{x}_i - \mathbf{x}_j\|$ and the coefficient $\eta_{ij} \in \{0, 1\}$ takes the value 1 if $\mathbf{x}_j$ is among the k-nearest neighbors of $\mathbf{x}_i$ or $\mathbf{x}_i$ and $\mathbf{x}_j$ are common neighbors of another point in the data set. The first constraints of problem SD preserve the distances between neighboring points. The second yields a unique solution by centering the projections on the origin and the third forces the Gram matrix to be positive semidefinite. The objective function, finally, maximizes the trace of $\mathbf{G}$ which is tantamount to maximizing the total variance of the points in the low-dimensional space.

Once the matrix $\mathbf{G}$ is learned via semidefinite programming the final embedding is obtained by computing its d largest eigenvalues and setting the projections to $\mathbf{Z} = \mathbf{V}\mathbf{\Lambda}^{1/2}$, where $\mathbf{Z}_{m \times d}$ is the matrix of embedded vectors $\mathbf{z}_i$, $\mathbf{\Lambda}_d$ is the square diagonal matrix of leading eigenvalues and $\mathbf{V}_{m \times d}$ is the matrix of corresponding eigenvectors.

Although efficient solvers for semidefinite programming exist, problem SD hardly scales to large data sets. The computational effort increases with the number of constraints and the size of $\mathbf{G}$. It is possible to show, however, that for well-sampled manifolds the Gram matrix can be reasonably approximated as the product of smaller matrices $\mathbf{G} \approx \mathbf{Q}\mathbf{Y}\mathbf{Q}'$, where $\mathbf{Q}_{m \times l}$ ($l \ll m$) must be properly determined. This results in a semidefinite program over the square matrix $\mathbf{Y}$ of size l, which has to be optimized under the local distance constraints. The low-rank expansion of $\mathbf{G}$ represents the key point of two fast MVU extensions given by Landmark MVU (L-MVU) [9] and Graph Laplacian Regularized MVU (GL-MVU) [10].

3   Selective MVU

The distinctive traits of MVU are the maximization of the variance of the embedding and the preservation of the distances between neighboring points. L-MVU and GL-MVU algorithms represent a substantial improvement over classical MVU from the computational viewpoint. However, they both diverge from the original paradigm due to the Gram matrix factorization.

In this paper we present a novel MVU extension, called Selective MVU (S-MVU), in which the required computing effort is reduced according to a different framework. Instead of resorting to modified MVU formulations applied to the entire set of data, the proposed method uses classical MVU to find the embedding of a collection of distinguished points indicated as prototypes. The low-dimensional coordinates of the remaining samples are then reconstructed via multi-output kernel ridge regression (KRR) [13]. The aim of this study, therefore, is to empirically investigate whether the solution of the original MVU model over a subset of representative points combined with an accurate regression method for learning the nonlinear mapping may provide higher quality low-dimensional projections compared to both MVU fast variants. The proposed Selective MVU algorithm can be summarized as follows.

Procedure. Selective MVU (S-MVU)

1. Define a collection $P \subseteq S_m$ of prototypes, where $\text{card}(P) = p$. Let $\mathcal{P} \subseteq \mathcal{M}$ be the set of their indices.
2. Find the embedding of P by solving problem SD over the corresponding Gram matrix $\mathbf{G}_p = [g_{ij}]$, where $g_{ij} = \langle \mathbf{z}_i, \mathbf{z}_j \rangle, \forall i, j \in \mathcal{P}$. Then, set $\mathbf{Z} = \mathbf{V}\boldsymbol{\Lambda}^{1/2}$, where $\mathbf{Z}_{p \times d}$ contains the projections of the prototypes in the feature space, $\boldsymbol{\Lambda}_d$ collects the d leading eigenvalues of $\mathbf{G}$ and $\mathbf{V}_{p \times d}$ the corresponding eigenvectors.
3. Learn the mapping via multi-output kernel ridge regression. To this aim, define a Mercer kernel $\rho : \Re^n \times \Re^n \mapsto \Re$ inducing a nonlinear projection $\phi : \Re^n \mapsto \mathcal{H}$ from the original input space $\Re^n$ to a Hilbert space $\mathcal{H}$. Formulate the kernel ridge regression model as $\min \|\mathbf{Z} - \langle \boldsymbol{\Phi}, \mathbf{W} \rangle\|_F^2 + \lambda \|\mathbf{W}\|_{\mathcal{H}}^2$, where $\| \cdot \|_F$ is the Frobenius norm of a matrix, the vector $\boldsymbol{\Phi}$ collects the images $\phi(\mathbf{x}_i)$, $i \in \mathcal{P}$, in $\mathcal{H}$ and the parameter λ controls the trade-off between the error and the penalty term. The regression coefficients can be computed in close form as $\mathbf{W} = \boldsymbol{\Phi}' \left(\mathbf{U} + \lambda \mathbf{I}_p \right)^{-1} \mathbf{Z}$, where $\mathbf{I}_p$ is the identity matrix of size p and $\mathbf{U}_p = [u_{ij}]$ is the kernel matrix associated to ρ, with $u_{ij} = \langle \phi(\mathbf{x}_i), \phi(\mathbf{x}_j) \rangle$, $\forall i, j \in \mathcal{P}$.
4. Embed the other points $\mathbf{x}_k, k \in \mathcal{M} \backslash \{\mathcal{P}\}$, by setting $\mathbf{z}_k = \langle \mathbf{W}', \phi(\mathbf{x}_k) \rangle = \mathbf{Z}' \left(\mathbf{U} + \lambda \mathbf{I}_p \right)^{-1} \mathbf{T}(\mathbf{x}_k)$, where the generic element of the p-dimensional vector $\mathbf{T}$ is given by $t_j = \langle \phi(\mathbf{x}_j), \phi(\mathbf{x}_k) \rangle, j \in \mathcal{P}$.

Notice that, the distance-preserving constraints of problem SD are in this case imposed only to the collection of prototypes, which are the pivotal elements for data embedding.

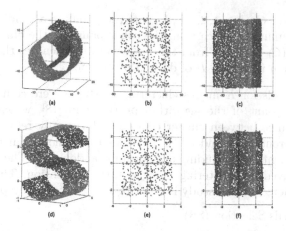

Fig. 1. Embedding of two artificial data sets. Panels (a) and (d) illustrate the Swiss roll and the S-curve data sets in the original three-dimensional space. Panels (b) and (e) show the two-dimensional projection of the randomly selected prototypes by means of MVU. Panels (c) and (f) depict the final mapping obtained by applying KRR on the embedded prototypes.

To illustrate the projection based on Selective MVU we applied the proposed algorithm to two artificial data sets obtained by sampling 6000 points from a Swiss roll and a S-curve surface, respectively. In particular, we computed the two-dimensional embedding from the three-dimensional space by setting $k = 6$, using the radial basis function (RBF) kernel for KRR and randomly choosing 10 % of the available points as prototypes. The projections obtained by S-MVU are depicted in Fig. 1. As we may observe, although based on a very small number of representative points the final mapping of both data sets faithfully correspond to the structure of the manifold in the native three-dimensional space.

3.1 Central Prototypes Selection

The most straightforward way to designate the set of prototypes is to select them randomly. Random selection is usually applied to identify landmark points in landmarks-based manifold learning algorithms [14] and [9]. However, it may generate misleading data projections [16] and [17], especially when data are affected by noise.

To find the collection of representative points we resorted to a simple but effective procedure based on clustering. In particular, we first applied K-means algorithm to partition the points into K distinct clusters. From each cluster we then selected a predefined number of central prototypes, defined as the points for which the maximum distance from the other points in the cluster is minimized. The algorithm can be summarized as follows:

Procedure. Central Prototypes Selection (CPS)

1. Let $\gamma = p/m$ be the fraction of points in S_m to use as prototypes.
2. Identify a set of K points as initial seeds and partition S_m into K clusters by applying K-means algorithm.

3. For each cluster $C_h, h = 1, 2, \ldots, K$, sort the points in ascending order based on their maximum distance from the other points in the cluster. Select the first $\lfloor \gamma \cdot \text{card}(C_h) \rfloor$ points from the list, where $\lfloor \cdot \rfloor$ denotes the integer part, and insert them in the set P of desired prototypes.

The initial seeds for K-means clustering at step 2 were computed through a multivariate variant of the algorithm proposed in [18], which introduces a measure of distance between cluster centers and virtually reduces to zero the variance of different runs. Indeed, it provides the same initial clusters across multiple experiments by excluding any form of randomness. The algorithm was originally conceived for clustering along a single dimension. The multivariate extension considered in this study is described by the following procedure.

Procedure. Seeds Selection (SS)

1. Sort the points in S_m in terms of increasing magnitude, given by their norms $\|\mathbf{x}_i\|$, $i \in \mathcal{M}$. Let F be the set of sorted points.
2. Compute the distances $D_j = \|\mathbf{x}^{j+1} - \mathbf{x}^j\|$, $j = 1, 2, \ldots, m - 1$, between all pairs of consecutive points, where $\mathbf{x}^j$ denotes point $\mathbf{x}$ at position j in F.
3. Identify the indices $\{j_1, j_2, \ldots, j_{(K-1)}\}$ corresponding to the $K - 1$ highest distance values and sort them in ascending order. Define the sets of indices $\mathcal{U} = \{j_1, j_2, \ldots, j_{(K-1)}, j_K\}$ and $\mathcal{V} = \{j_0, j_1 + 1, \ldots, j_{(K-1)} + 1\}$ of the points serving as upper and lower bounds, respectively, where $j_K = m$ and $j_0 = 1$.
4. Compute the K initial seeds as the mean vectors between the upper and lower bound points defined above.

To highlight the usefulness of central prototypes selection we considered a data set composed by 6000 points randomly sampled from a Swiss roll manifold with 1% of uniform distributed outliers, and computed the low-dimensional embedding by means of alternative techniques. In particular, we analyzed the effect of randomness in the worst case scenario when outliers are used as landmarks in L-MVU and prototypes in S-MVU. The different embeddings, obtained by setting $k = 6$ for all methods, are illustrated in Fig. 2. As one may notice, the presence of noise interferes with the unfolding process and induces a major distortion when the projections are based on the outliers (Panels d and e). The use of central prototypes in S-MVU, however, mitigates this effect preserving the structure of the underlying manifold (Panel f). As shown in Fig. 2, a major robustness of S-MVU compared to GL-MVU and L-MVU was also observed when injecting 10% of outliers (Panels g, h and i).

3.2   Complexity of Selective MVU

Classical MVU runs in $O\left(m^3 + c^3\right)$ over a set of m points, where c is the number of constraints in the semidefinite program [5]. The time-complexity of KRR is $O\left(p^3\right)$ [13] whereas the prototype selection procedure runs in $O\left(m^2 + mKIn\right)$, where I is the number of iterations in the K-means algorithm and the quadratic

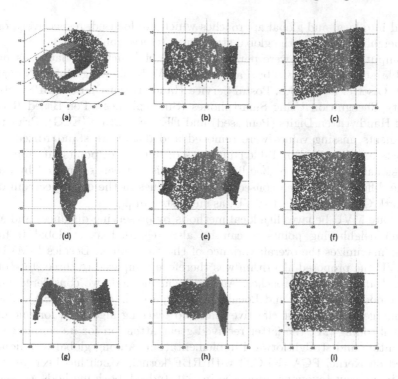

Fig. 2. Embedding of a Swiss roll with noise. Panel (a) represents the Swiss roll composed by 6000 points with 1% of uniform distributed outliers (60 outliers). Panels (b) and (c) describe the projections obtained by GL-MVU (12 Laplacian eigenvectors used) and L-MVU (30 landmarks used). Panels (d) and (e) illustrate the mapping of L-MVU (30 outliers used as landmarks) and S-MVU (600 prototypes composed by the 60 outliers and 540 randomly chosen points). Panel (f) shows the embedding of S-MVU based on 600 central prototypes. Finally, panels (g), (h) and (i) display the unfolding of GL-MVU, L-MVU and S-MVU, respectively, on the Swiss roll data set with 10% of outliers.

term refers to the intra-clusters distances computation. The overall complexity of S-MVU is, therefore, $O\left(m^2 + mKIn + p^3 + c^3\right)$. Major computational advantages are obtained when $p \ll m$, where p can be naturally expressed as a fraction $\gamma \in (0, 1]$ of the available points, $p = \lfloor \gamma m \rfloor$. Experiments on artificial manifold data sets and on medium-size data sets from the UCI Repository [19] empirically showed that γ can be fixed to a very small value ($\gamma \approx 0.1$) to obtain a fast low-dimensional unfolding at the expense of a reduced loss in the quality of the embedding.

4   Experiments and Results

To evaluate the usefulness of Selective MVU we resorted to a variety of criteria based on the concept of loss of quality, which is supposed to be strongly related to the preservation of the data geometry [20]. Most of these criteria can be divided into local and global approaches which are focused, respectively, on the local neighborhood and the global structure preservation.

Computational tests were performed on eight benchmark data sets mostly available at [19] and given by Page Blocks, Wall-Following Robot Navigation (Wall), Gisette, Isolet, U.S. Postal Service Handwritten Digits (USPS), Human Activity Recognition Using Smartphones (Smartphones), Pen-Based Recognition of Handwritten Digits (Penbased) and EEG Eye State (EEG). Prior to the experiments missing values were removed and data were standardized. These data sets are described in Table 1 in terms of number of points, attributes and dimensionality d^* of the embedding space. This last was estimated by analyzing the difference between consecutive eigenvalues in the eigenspectrum of the landmark Gram matrix in L-MVU, as suggested in [9].

Despite MVU behaves like local methods by preserving distances and angles between neighboring points it can be also regarded as a global technique since it maximizes the overall variance of the embedding. Besides L-MVU and GL-MVU the proposed algorithm was therefore compared to three state-of-the-art local and global approaches. Among local methods we considered Locally linear embedding (LLE) and Local tangent space alignment (LTSA). The former emerged as the most effective manifold learning algorithm for microarray data embedding [21]; the latter received great attention for its simple geometric intuition and straightforward implementation. Among global techniques we focused on Kernel PCA (KPCA) with RBF kernel, which has been related to MVU in a recent taxonomy proposed in [22]. Indeed, both methods are spectral techniques which convert the dimensionality reduction problem into the eigendecomposition of a kernel matrix.

In the following experiments some parameters were fixed. The number of Laplacian eigenvectors in GL-MVU was set to 12 whereas 30 landmarks were used in L-MVU to limit the computing time. The percentage of points taken as prototypes in S-MVU was fixed to 0.1 for all data sets except for EEG, for which it was set to 0.05. The number of clusters in procedure CPS was found through

Table 1. Description of the data sets. The last column indicates the estimated dimensionality of the embedding space.

ID	Data set	Points	Attributes	d^*
1	*Pageblocks*	5406	10	5
2	*Wall*	5456	24	6
3	*Gisette*	7000	5000	10
4	*Isolet*	7797	617	6
5	*USPS*	9298	256	6
6	*Smartphones*	10299	561	5
7	*Penbased*	10992	16	6
8	*EEG*	14980	14	6

a grid search so to minimize the neighborhood size for which the connection of the neighborhood graph was achieved. Finally, the RBF kernel was applied in the KRR model by setting $\lambda = 0.1$ and fixing the RBF parameter to 10^j for a given j in the interval $[-3, -1]$. The same RBF parameter's values were tested for KPCA. All methods were implemented in MATLAB. Computations were run on a 3.40 GHz quad-core processor with 16 GB RAM.

To analyse the local neighborhood preservation we resorted to two local criteria which measure the degree of overlap between the neighboring sets of a point and of its embedding. The first is represented by the Local-Continuity Meta-Criterion (Q_L) [23], which is defined as the average size of the overlap of the neighboring sets. The second is given by Trustworthiness and Continuity (Q_{TC}) [24], which is based on the exchange of indices of neighboring samples in the input and the feature space according to the pairwise Euclidean

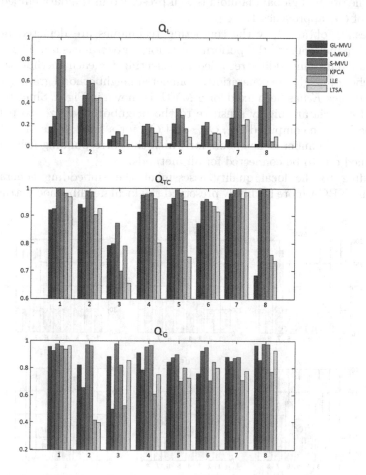

Fig. 3. Local and global quality assessment. Each panel indicates the performance of the competing algorithms on the eight data sets.

distances. Q_{TC} is defined as a linear combination of two measures which evaluate, respectively, the degree of trustworthiness that points farther away enter the neighborhood of a sample in the embedding and the degree of continuity that points originally included in the neighborhood are pushed farther away. The coefficient of the linear combination was here fixed to 0.5. The above criteria have proven to be good estimates of the embedding quality. In particular, the greater their values are in the interval $[0, 1]$, the better is the projection.

The global structure holding performance was, instead, analyzed by means of a global metric recently proposed in [25]. This metric, here denoted as Q_G, evaluates the difference of the transforming scales of the embedding set compared to the original data manifold along various directions. This is achieved by computing a shortest path tree of the neighborhood graph and using the Spearman's rank order correlation coefficient defined on the rankings of the main branches lengths. The original global manifold is well preserved in the data embedding as the value of Q_G approaches 1.

The results obtained by the competing techniques are depicted in Fig. 3, where each panel collects the performances for a given measure. The computing time for data embedding recorded by selecting for each method the minimum number of neighbors generating a connected neighborhood graph (provided $k \geq 4$), and once K has been fixed for S-MVU, is shown in Fig. 4. Since the aforementioned criteria are highly sensitive to the neighborhood size, to perform a fair comparison we computed Q_L, Q_{TC} and Q_G for a fixed value of k that was set equal to the number of neighbors for which the corresponding neighborhood graph turned out to be connected for all methods.

According to the local quality assessment the embedding generated by S-MVU and KPCA more faithfully preserved the local neighborhood structure of

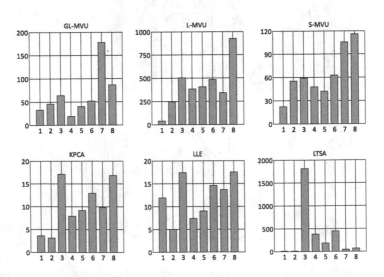

Fig. 4. Computing time (secs) for embedding each of the eight data sets.

the original manifold. In particular, S-MVU dominated both MVU extensions, LLE and LTSA, and provided better results on the majority of the data sets (6 out of 8) compared to KPCA. The proposed algorithm exhibited also notable performances in terms of global structure preservation, as indicated by Q_G. Therefore, the embedding set of S-MVU encountered a smaller distortion of the global shape of the manifold on most data sets. It is worth to notice that, whereas L-MVU and LLE performed generally better than GL-MVU and LTSA according to the local measures, they were often dominated by the latter based on the global metric. The proposed S-MVU algorithm, instead, behaved well both in terms of local and global quality assessment.

5   Conclusions and Future Extensions

In this paper we described a novel method for nonlinear dimensionality reduction indicated as Selective MVU (S-MVU). In the proposed algorithm the unfolding process is guided by a subset of distinguished points called central prototypes, whose embedding is computed by means of classical MVU. The projections of the remaining samples are thereafter reconstructed via multi-output kernel ridge regression. S-MVU was empirically compared to two well-known fast MVU extensions and to three prominent nonlinear dimensionality reduction methods. On several benchmark data sets it achieved noteworthy performances and emerged as a valid alternative to state-of-the-art techniques in terms of quality of the data projection.

The present study can be extended in several directions. First, novel procedures for selecting the representative points or embedding the set of non-prototypes samples could be developed. It would be also worthwhile to investigate the effectiveness of greedy optimization algorithms for solving problem SD in Selective MVU to speed up the unfolding process. Finally, further computational tests could be performed by comparing the accuracy that alternative classification algorithms achieve on the different data projections.

References

1. Tenenbaum, J.B., de Silva, V., Langford, J.C.: A global geometric framework for nonlinear dimensionality reduction. Science **290**, 2319–2323 (2000)
2. Roweis, S.T., Saul, L.K.: Nonlinear dimensionality reduction by locally linear embedding. Science **290**, 2323–2326 (2000)
3. Belkin, M., Niyogi, P.: Laplacian eigenmaps for dimensionality reduction and data representation. Neural Comput. **15**, 1373–1396 (2003)
4. Zhang, Z., Zha, H.: Principal manifolds and nonlinear dimension reduction via local tangent space alignment. SIAM J. Sci. Comput. **26**, 313–338 (2004)
5. Weinberger, K.Q., Saul, L.K.: Unsupervised learning of image manifolds by semidefinite programming. In: IEEE International Conference on Computer Vision and Pattern Recognition, pp. 988–995 (2004)
6. Kleiner, A., Rahimi, A., Jordan, M.I.: Random conic pursuit for semidefinite programming. In: Advances in Neural Information Processing Systems, pp. 1135–1143 (2010)

7. Hao, Z., Yuan, G., Ghanem, B.: Bilgo: Bilateral greedy optimization for large scale semidefinite programming. Neurocomputing **127**, 247–257 (2014)
8. Chen, W., Weinberger, K.Q., Chen, Y.: Maximum variance correction with application to A* search. In: Proceedings of the 30th International Conference on Machine Learning, pp. 302–310 (2013)
9. Weinberger, K.Q., Packer, B.D., Saul, L.K.: Nonlinear dimensionality reduction by semidefinite programming and kernel matrix factorization. In: Proceedings of the 10th International Workshop on Artificial Intelligence and Statistics, pp. 381–388 (2005)
10. Weinberger, K.Q., Sha, F., Zhu, Q., Saul, L.K.: Graph laplacian regularization for large-scale semidefinite programming. In: Advances in Neural Information Processing Systems, vol. 19, p. 1489 (2007)
11. Hastie, T., Tibshirani, R., Friedman, J.: The Elements of Statistical Learning. Springer, New York (2001)
12. Shawe-Taylor, J., Cristianini, N.: Kernel Methods for Pattern Analysis. Cambridge University Press, Cambridge (2004)
13. Orsenigo, C., Vercellis, C.: Kernel ridge regression for out-of-sample mapping in supervised manifold learning. Expert Syst. Appl. **39**, 7757–7762 (2012)
14. de Silva, V., Tenenbaum, J.B.: Global versus local methods in nonlinear dimensionality reduction. In: Advances in Neural Information Processing Systems, vol. 15, pp. 705–712 (2003)
15. Cai, D., He, X., Han, J.: Spectral regression for efficient regularized subspace learning. In: IEEE 11th International Conference on Computer Vision, pp. 1–8 (2007)
16. Chen, Y., Crawford, M. M., Ghosh, J.: Improved nonlinear manifold learning for land cover classification via intelligent landmark selection. In: IEEE International Geoscience & Remote Sensing Symposium, pp. 545–548 (2006)
17. Gu, R.J., Xu, W.B.: An improved manifold learning algorithm for data visualization. In: Proceedings of the 2006 International Conference on Machine Learning and Cybernetics, pp. 1170–1173 (2006)
18. Khan, F.: An initial seed selection algorithm for k-means clustering of georeferenced data to improve replicability of cluster assignments for mapping application. Appl. Soft Comput. **11**, 3698–3700 (2012)
19. Bache, K., Lichman, M.: UCI Machine Learning Repository. University of California, School of Information and Computer Science, Irvine (2013). http://archive.ics.uci.edu/ml
20. Gracia, A., González, S., Robles, V., Menasalvas, E.: A methodology to compare dimensionality reduction algorithms in terms of loss of quality. Inf. Sci. **270**, 1–27 (2014)
21. Orsenigo, C., Vercellis, C.: A comparative study of nonlinear manifold learning methods for cancer microarray data classification. Expert Syst. Appl. **40**, 2189–2197 (2013)
22. van der Maaten, L., Postma, E., van den Herik, H.: Dimensionality reduction: A comparative review (2007)
23. Chen, L., Buja, A.: Local multidimensional scaling for nonlinear dimension reduction, and proximity analysis. J. Am. Stat. Assoc. **104**, 209–219 (2009)
24. Venna, J., Kaski, S.: Local multidimensional scaling. Neural Networks **19**, 889–899 (2006)
25. Meng, D., Leung, Y., Xu, Z.: A new quality assessment criterion for nonlinear dimensionality reduction. Neurocomputing **74**, 941–94 (2011)

Cooperative Multi-agent Learning in a Large Dynamic Environment

Wiem Zemzem$^{(\boxtimes)}$ and Moncef Tagina

National School of Computer Science, University of Manouba, Manouba, Tunisia
{wiem.zemzem,moncef.Tagina}@ensi-uma.tn

Abstract. In this work, we are addressing the problem of cooperative multi-agent learning for distributed decision making in non stationary environments. Our principal focus is to improve learning by exchanging information between local neighbors (agents) and to ensure the adaption to the new environmental form without ignoring knowledge already acquired. First, a distributed dynamic correlation matrix based on multi-Q learning method, presented in [1], is evaluated. To overcome the shortcomings of this method, a new multi-agent reinforcement learning approach and a new cooperative action selection strategy are developed. Several simulation tests are conducted using a cooperative foraging task with a single moving target and show the efficiency of the proposed methods in the case of large, unknown and temporary dynamic environments.

Keywords: Cooperative multi-agent learning · Distributed decision making · Reinforcement learning · Cooperative foraging task · Exchanging information · Adaptation to environmental changes · Large · Unknown and dynamic environments

1 Introduction

Multi-robot systems (MRSs) have drawn considerable attentions to both industry and academia in the last two decades. A team of robots, though may be not powerful individually, can effectively compensate and compromise their limitations by cooperation [2]. The multi-robot system could accomplish a task in a faster, cheaper and more efficient way than a single robot. However, the complexity of many required tasks makes them difficult to solve with preprogrammed agent behaviors. The robots must instead discover a solution on their own, using learning. A significant part of the research on multi-agent learning concerns reinforcement learning techniques. Robots uses multi-agent reinforcement learning (MARL) to acquire a wide spectrum of skills including navigation [1–8], object transportation [9] and playing soccer [10].

MARL is, however, faced with several problems including, the necessity of a huge number of learning trials especially if the state space is high and the need of a particular parameter settings. The tradeoff between exploration and exploitation is also a popular challenge of RL. Each agent faces the following

© Springer International Publishing Switzerland 2015
V. Torra and Y. Narakawa (Eds.): MDAI 2015, LNAI 9321, pp. 155–166, 2015.
DOI: 10.1007/978-3-319-23240-9_13

dilemma: either it chooses a random action in order to explore new area and enlarge its knowledge about the environment (exploration), or it chooses an action which looks optimal given the past observations and rewards (exploitation) [3,4,8]. Algorithms such as ε-greedy, Boltzmann distribution, Simulated annealing, Probability Matching and Optimistic Initial Values are proposed to solve this dilemma [3]. However, they are designed for simply assuming the environment is stationary.

Learning in unknown and non stationary environments is yet more challenging. In this case, each agent needs to constantly explore the environment in order to integrate the most recent changes into its knowledge of the world but such exploration should not be done excessively that performance is greatly degraded. Many previous works which deal with reinforcement learning in multi-robot systems [11,12] keep a significant exploration rate to ensure the convergence of the learning algorithm. However, this exploration is randomly and doesn't directly benefit from other robots' knowledge. This will lead to unnecessary exploration intervals and therefore the slowdown of the system's convergence. Other distributed methods where agents help each other by exchanging information between them like those using joint-Qtables (as examples: the policy averaging (PA) method [13] and the experience counting (EC) method [6,7]) or the DDCM method [14] are promising in stationary environments but fail to solve non stationary problems. As the environment changes, some previous knowledge become incorrect. Even if an agent detects the change, it still uses incorrect information received from other collaborators. A successful adaptation to the new environmental form is then difficult, even impossible, especially when there is a limited communication range between agents. An example of a distributed RL method using local coordination is the D-DCM-Multi-Q method (a distributed dynamic correlation matrix based multi-Q method) [1]. The cooperation is ensured by considering the correlated Q-values of neighboring robots in addition to each individual robot's Q-value when updating this individual information.

The present work is motivated by the same problem as in [1]: a group of autonomous mobile agents should learn and collaborate together in order to efficiently perform a cooperative foraging task in a large, unknown and temporary dynamic environment. Our previous work, presented in [15], deals with non stationary environments but is restricted to single-agent systems. The present paper extends it to multi-agent systems. Improving multi-agent learning in dynamic environment is more difficult due to the problem of coordination and new ideas are proposed to deal with it, especially accelerating learning by exchanging information between local neighboring agents. Moreover, agents should be able to adapt to the new environmental form without ignoring knowledge already acquired. Firstly, we will demonstrate the shortcomings of the D-DCM-MultiQ approach in both stationary and non-stationary environments. To overcome these limitations, a new model-free approach, called CMRL-MRMT (The Cooperative Multi-agent learning approach based on the most Recently Modified Transitions), is developed. This method is inspired from the D-DCM-MultiQ approach as regards the distribution of learning over several cooperative agents but uses the most recently modified Qvalues in order to facilitate dealing with the dynamics of the environment. A new cooperative action selection strategy, called CG-MPA (the Cooperative Greedy policy favoring the most promising actions), is also established. The most promising action is whose Qvalue is the highest and the most recently modified and which leads to a more promising State. A more promising next state is the least recently visited state whose hightest Qvalue is greater than that of the actual state. Using this EEP, the exploration is done only if needed wich ensures a quick adaptation to changes as well as the convergence to a constant solution when the environment remains stationnary.

The rest of the paper is organized as follows. Problem statement is described in Sect. 2. In Sect. 3, the D-DCM-Multi-Q approach is presented and briefly reviewed. Section 4 is dedicated to present our suggestions for improvement. Several experiences are conducted in Sect. 5 showing the efficiency of our proposals. Some concluding remarks and future works are discussed in Sect. 6.

2   Problem Statement

Our objective is to provide learning capabilities to a group of agents in order to perform a common task in a distributed manner. The work assigned to the agents is a multi-agent foraging task with a single moving target in a large and discrete 2D environment. Agents make decisions only based on the interaction with the environment as well as the interaction with local neighbors. The general state transition process of the system is a tuple of $(n, \mathbf{S}, \mathbf{A}, \mathbf{T}, \mathbf{R})$, where n is the number of agent's neighbors. $\mathbf{S}$ is a set of states which is defined as $\mathbf{S} = [s_1, .., s_m]$, where m is the number of states. $\mathbf{A}$ is a set of actions available to an agent,which is defined as $\mathbf{A} = [a_1, .., a_p]$. $\mathbf{R} : \mathbf{S} \times \mathbf{A} \longrightarrow r$ is the reward function for an agent and $\mathbf{T}$ is a state transition function ($\mathbf{T}$ is a probability distribution over $\mathbf{S}$) [16]. The action set for each agent are defined as "up", "right", "down" and "left". It is assumed that all agents are initially in the nest (the starting position), and each agent can locate itself using its on-board sensors such as encoders and can detect the target or obstacles using sonar sensors. Each robot has limited onboard communication range and can share its state information with its neighbors that are within its communication range. The foraging task may be abstractly viewed as a sequence of two alternating tasks for each agent:

- A hunting phase: The agents Start from the nest and try to catch the moving prey(target) together. Agents cannot pass through any obstacles but more than one agent can occupy the same cell. Unlike [1], the prey will no longer move at each time step (intensive displacements) but performs several successive displacements after N iterations. Once one agent captures the prey (occupy the same cell as the prey), the prey stops moving and this agent will wait for other agents to reach the target (Waiting Phase).
- Ferrying phase: Once all agents capture the prey, they start a collective transport phase from their actual position to the nest. As agents must follow the same path, they select the shortest path among all agents' hunting paths.

3   A Review of the D-DCM-MultiQ Method

D-DCM-MultiQ is a distributed reinforcement learning method for multi-agent cooperation proposed by Hongliang Guo and Yan Meng [1]. In this method, the Q-value of the ith agent at the state s_i is updated with its own state value as well as the state values of its neighboring agents at the next state s'_i. The updated equation of the ith agent's Q-value is as follows:

$$Q_{k+1,i}(s_i, a_i) = (1 - \alpha_k) Q_{k,i}(s_i, a_i) + \alpha_k *$$

$$\left(R_{k,i}(s_i, a_i) + \gamma \sum_{j=1}^{N} f(i, j) \; max_{a_j} Q_{k,j} \left(s'_i, a_j \right) \right) \quad (1)$$

In what follows, we explain the failure of this method to solve the hunting task in both stationary and non stationary environments.

3.1   Failure to Complete the Hunting Task (Blockage of Learning)

By exploiting the knowledge of its neighbors, an agent can spread promising information (related to the prey) to intermediate states even before detecting the prey. However, the Boltzmann action selection strategy, which is applied in [1], isn't cooperative. When choosing the next action to perform, the agent can only exploit its own information. Therefore, the selected action may be not the best; i.e. it doesn't lead to another state having higher Qvalues than the actual state. The agent will loop repeatedly between two intermediate states. The opportunity to exit from the loop is very low since, using the same parameter settings as in [1] (*Reward* = −90 if the agent hits obstacles, 180 if it reaches the goal and −1 otherwise), the probability assigned to the transition causing the blockage is much higher than other probabilities. The blockage can occur before the convergence of the multi-agent system at any stage of learning: near the target since first episodes and away from the target when the Boltzmann Temperature decreases.

3.2   Inappropriate Dynamic Scenario

When dealing with a non stationary environment, Hongliang Guo and Yan Meng [1] represent the state space of each agent by the relative position between its current position and the target's position. Such a representation is employed in order to reduce the number of state-action pairs compared with the number using absolute positions for state definitions. For example, for the agent at $(3, 4)$ and the target at $(4, 3)$, the state, defined as $(1, -1)$, is the same as the one where the agent at $(6, 5)$ and the target at $(7, 4)$. Conducted experiments in [1] demonstrate that the D-DCM-MultiQ algorithm converges in the dynamic case with small oscillations. This oscillation is caused by the random movement of the target which may lead to probabilistic rewards into the agents' Qvalues. However:

– The representation of the state space by relative positions is possible only if the environment is without obstacles given that obstacles may limit the observability of the agent. That's why the efficiency of this representation doesn't suit the majority of robotic applications: presence of obstacles, interference between robots, etc.

– The successful adaptation of the robotic system to the environmental changes isn't ensured by the D-DCM-MultiQ method and the Boltzmann policy, as outlined in [1], but by the random movements of the target. More clearly, the scenario adopted in [1,14] assumes that, at each time step, the target has the equal probability to move to one of its neighboring grids or stay at the original grid and that it still moves even after being captured by one or more robots; the episode finishes only if the target is grabbed by all agents. As the environment is small (a 10×10 grid environment without obstacles) and the target is able to move at any time step, even if the agent fails to catch the target, the target ends up by finding itself in the same cell as the blocked agent and then makes it move again.

Consequently, by expanding the environment's size and decreasing the movement of the target over time, the D-DCM-MultiQ method is no longer able to deal with non stationary problems. On another hand, by using the Boltzmann policy and the D-DCM-MultiQ method, blocking situations which are observed in static environments are increased in the dynamic case: when an agent detects the disappearance of the target, it still uses its previous knowledge related to the ancient position of this target. As a result, incorrect information is spreading. Moreover, by adopting the Boltzmann policy, actions having the highest Qvalue are all the more privileged as the learning progresses and the Boltzmann temperature decreases.

4   Our Proposed Learning Approach for Cooperative Multi-agents Systems in Dynamic Environments

We aim to find a solution that solves the blockage caused by the non-continuous propagation of the goal's reward and improves the cooperation between neighboring agents to ensure an efficient adaptation to environment's changes without ignoring knowledge already acquired.

4.1   First Proposal: The Action Selection Strategy

As we are interested in cooperative multi-agent systems, we aim to ensure that each agent gets the most out of its neighboring knowledge in order to limit the risk of blockage. One possible solution is to introduce the information of neighboring agents in the exploration/exploitation policy (EEP) in addition to when updating the agent's state-action pair. More clearly, at each learning step, the agent chooses the next action to perform based on its own information as well as the information of its neighbors which increases the efficiency of the selected action since the propagation of promising information from the target will be correctly followed.

In another hand,when the environment changes, some previous information becomes invalid. Thus, it's useful to give a top priority to recent learning steps.

To this end, we propose a new EEP, called CG-MPA (the Cooperative Boltzmann policy favoring the most promising actions). This EEP, summurized in Algorithm 1 (Fig. 1), promotes the action whose Qvalue is the highest and the most recently modified and which leads to a more promising State.

Algorithm 1: The CG_MPA policy

Initialize L and L' to be empty

Initialize $neighbor_{a_1}, ..., neighbor_{a_M}$ to be zero

1: $\forall a \in A$ $(A = \{a_1, ..., a_M\})$: $neighbor_a(s) = argmax_j \left(LastChanged_j(s, a) \right)$,
where $j \in neighborhood$ of the $actual$ $agent$ $including$ the $agent$ $itself$

2: $maxQ(s) = max_a Q_{neighbor_a}(s, a)$
3: $\forall a \in A, if$ $Q_{neighbor_a}(s, a) = maxQ(s)$ $then$ $L \leftarrow a$
4: If $|L| = 1$ and the action a, stored in L, leads to the goal state then $action_{chosen} = a$

5: Else $(|L| > 1)$

6:     $\forall a \in L$
7:         Determine the eventual state s' resulting from executing action a in the actual state s

8:         Calculate maxQ(s') as described in steps 1 and 2

9:         If $(maxQ(s) < maxQ(s'))$ then $L' \leftarrow a$
10:        If $(|L'| >= 1)$
11:            select an action from L'
12:        else (L' is empty)
13:            select an action from all possible actions in the state s

---

Fig. 1. The CG-MPA policy

As shown in Fig. 1, the list L stores the most promising actions in the current state s while the list L' stores the actions of L leading to a more promising next state s'.

Each learner stores a table $Q : S \times A \rightarrow R$. In RL problems, the action-value function is defined as the expected infinite discounted sum of rewards that the agent will gain if it chooses the action a in the state s and follows the optimal policy. In addition to that, a second table $lastChange : S \times A \rightarrow R$ is used. This table memorizes, for each pair (s, a), the moment of the last modification of the corresponding Qvalue. The update of $LastChange$ is as follows:

- Initially, LastChange(s,a)=0, for all (s,a) SxA.
- At each time step t, after updating the last experimented transiton (s, a),

$$if(Q_t(s, a) \neq Q_{t-1}(s, a)) : LastChange(s, a) = t \qquad (2)$$

For each action a in the actual state s, the agent has to determine the neighbor having the most recently modified Qvalue $Q_{neighbor_a}(s)$, as mentioned in step 1 of the algorithm 1. As a result, M Qvalues are obtained relatively to the M possible actions in the state s. Then, it selects the highest Qvalue of them, as shown in step2. The obtained value, $maxQ(s)$, presents the highest and the most recently modified Qvalue of the state s.

As mentioned in step 4, in the case that the list L contains a single action a leading to the goal state, this action will be immediately chosen without using the second list L'. Consequently, the position of the goal must be known without really executing a in s. One possible solution is to store the position of the goal s_{target} and update it as follows:

- Initially s_{target}=null
- After reaching the goal, update s_{target}; $s_{target} \leftarrow actualPosition$
- If the agent visits the ancient goal position without hitting the prey, update s_{target}; $s_{target} \leftarrow null$

According to step 7, the agent has to identify the next state s' resulting from executing action a in the actual state s. Action a isn't really experienced and according to reinforcement learning properties, an agent can determine the characteristics of any state only after visiting it. However, in our case study, agents move in a discrete and a two-dimensional space, defined by the coordinates (x, y), by performing the following actions: "up", "right", "down" and "left". So, to calculate the coordinates of the next state, agents don't need to move neither to know the nature of that state (an obstacle, the prey or an intermediate state). In the contrary, knowing the next action to perform, the coordinates of the next state is automatically deduced from those of the actual state (For example: Next-action= $up \longrightarrow$ Next-state=$(x - 1, y)$).

As noted in step 9, each action in the list L will be stored in the second list L' only if the agent trying the action a in the actual state s will move to a more promising state s', i.e., a state s' having a higher Qvalue than the current state s. The chosen action will be then one of those stored in L' (step 11). In the worse case (L' is empty), the agent will resort to an exploration.

In order to promote the exploration of new areas, we can alternate between a random selection and a selection favoring the least recently tested states. To do that, each agent memorizes a table $Anc : S \rightarrow R$, storing the seniority of each visited state. $Anc(s) = t$ means that the last visit of state s was occurred at time t. As mentioned in step 15, when the agent chooses the next action to perform according to the least visited states, it should avoid known obstacles. This is in the aim to avoid stuck near corners (for example: L obstacle or the extremities of the environment, etc.). The random exploration (step 17) is without avoiding known obstacles. This is in order to detect moving obstacles and to find new areas that become recently reachable.

To avoid obstacles, a specific distribution of reward is needed:

- $R < 0$ if the agent hits an obstacle,
- $R > 0$ if the agent captures the prey,
- $R = 0$ otherwise.

By this definition, actions related to negative Qvalues are actions leading to collision with obstacles.

4.2 Second Proposal: The Proposed Learning Method Strategy

To ensure an efficient adaptation to the environmental changes, we propose a new reinforcement learning approach, entitled CMRL-MRMT (The Cooperative Multi-agent learning approach based on the most Recently Modified Transitions). This method is inspired from the D-DCM-MultiQ approach as regards the distribution of learning over local neighboring agents but uses the most recently modified Qvalues in order to facilitate dealing with non-stationarity. The Q-value of the ith agent is updated through the following equation:

$$Q_{t,i}(s,a) = (1 - \alpha) Q_{t-1,i}(s,a) + \alpha (R_{t,i}(s,a) + \gamma max_b Q_{t,k}(s',b)) \quad (3)$$

where $Q_{t,i}(s,a)$ represents the Q-value of the ith agent at the time step t relatively to the state s and the action a. $R_{t,i}(s,a)$ is the immediate reward of the ith agent at time step t after executing the action a at the state s and s' is its actual state. γ is the decaying factor and α is the learning rate. $max_b Q_{t,k}(s',b)$ represents the most promising Qvalue at the state s'; $k = argmax_j (LastChanged_j(s',b))$, where $b \in A$, j refers to each agent in the neighborhood including the ith agent and k refers to the agent having the most recently modified Qvalue of the transition (s',b). There's $M=|A|$most recently modified Qvalues relative to M possible transitions at the state s'. The most promising Qvalue at the state s' is therefore the highest of these M values.

The learning algorithm is described in Fig. 2.

5   Experiments and Analysis

In this section, we will evaluate the impact of our proposed learning strategies on a distributed multi-agent system. As described earlier, the testing scenario is a cooperative foraging task with a single moving target (prey). We aim to evaluate our learning algorithm in a wide unknown environment. For that, we extend our testing environment from 10×10 (as in [1]) to 30×30 grid world.

The testing scenario is simulated using Simbad, a java3d robot simulator [17]. As shown in Fig. 3, the cell, called the nest, is the starting position. A ball, situated at the bottom right corner, refers to the prey. The maze is surrounded by walls and contains obstacles.

In the following experiments, the decline factor γ and the learning rate α are defined as 0.9 for all the agents. The neighborhood range for each agent is defined as 9 ($< \frac{size}{3}$). For the rewards, it is defined as 0 for each regular action without hitting any obstacle or the target. If an agent hits a wall, it gets a penalty of -90 and if it captures the prey, it gets a reward of 180. All entries of the tables Q and $LastChange$ are initialized to zero.

For each test, 40 runs are conducted. Every run consists of 200 episodes and starts initially with an environment in the form of the Fig. 3-a. After 100 episodes, the environment changes to the form of Fig. 3-b.

```
Initialization
{
  For (agent i=1,..., N)
  {
    Initialize agent's position;
    For (state s=1,..., M)
    {
      Initialize Q(s,a), LastChange(s,a) and Anc(s) to be zero;
    }
  }
}
While (TRUE)
{
  For (agent i=1,...,N)
  {
    Select action according to Algorithm 1 (Fig. 1);
    Execute action;
    Reward(i)=getReward()
  }
  For (agent i=1,...,N)
  {
    Update Q(s,a) according to eq.3
  }
}
```

Fig. 2. The pseudo code of the CMRL-MRMT algorithm

5.1 Impact of Least Recently Visited States on the Selection of the Next State

In this section, the impact of the selection of actions leading to least recently visited states on the learning performance is studied. Two systems of three agents are compared. The CG-MPA policy is adopted with varying the value of the parameter p. As described earlier, p defines the probability to choose a random action instead of that leading to the least recently visited states:

- 1st system with p=1 (a fully random choice)
- 2nd system with p=0.5 (50 % a random choice and 50 % the selection of the least recently visited state with avoiding known obstacles)

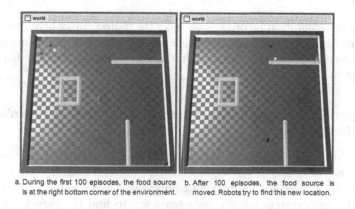

a. During the first 100 episodes, the food source   b. After 100 episodes, the food source is
is at the right bottom corner of the environment.   moved. Robots try to find this new location.

Fig. 3. The testing environment with four agents

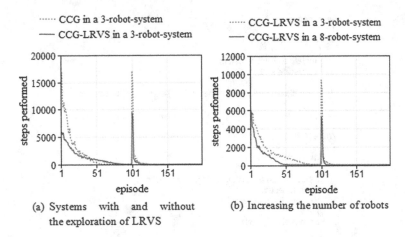

(a) Systems with and without the exploration of LRVS

(b) Increasing the number of robots

Fig. 4. The number of time steps needed on each epsisode over time (average of forty experiences)

From Fig. 4-a, we can see that, in both systems, the exploration is important only when it's needed, i.e., when an unordered backward-propagation of the goal's reward is detected. This can happen during first episodes ([0-50] episodes) given that the environment is unknown and after the displacement of the prey ([101-117] episodes) to ensure the adaptation to the new form of the enviornment. Using the CMRL-MRMT algorithm, both systems converges to a near-optimal solution during the first 100 episodes (a static environment) as well as after the environment's change. The constructed path is also constant:

– Before the displacement of the prey: about 117 iterations for the 1st system and 118 iterations for the 2nd system (the length of the optimal path is 116 iterations)
– After the displacement of the prey: about 90 iterations for the 1st system and 89 iterations for the 2nd system (the length of the optimal path is 82 iterations)

However, the exploration favoring least recently visited states accelerates considerably learning. Using the CG-LRVS policy, a reduction of time steps needed for the convergence is ensured in the stationary case as well as after the displacement of the prey.

5.2 Impact of Increasing the Number of Agents on Learning Performance

We aim to study the impact of increasing the agents' number on the learning performance. For that, two systems of three and eight agents are considered. Figure 4-b shows that, during first four episodes, the 8 agents' system is slower than the 3 agents'system since more agents try to find the target and because learning is still at the beginning. However, learning is considerably accelerated by increasing the agents' number during the first 100 episodes. The number of episodes and time steps needed for the adaptation to the new form of the environment is also decreased.

These relevant results remained valid when changing the size of the environment. Several tests have been conducted while varying the environment's size from 10×10 cells to 40×40 cells and have demonstrate the effeciency of our method.

6   Conclusion

In this paper, the problem of cooperative multi-agent learning for distributed decision making is studied. We have shown the shortcomings of the D-DCM-Multi-Q learning approach in both stationary and dynamic environments.

As a solution, we have formalized a new learning approach where the update of Qvalues depends principally on most recently modified information. A new EEP is also established by which the exploration is priviliged only if an unordered backward-propagation of the goal's reward is detected. Combining these two proposals is necessary for the successful completion of the learning task. It allows the learners to avoid blocking situations, to adapt to the new changes and to converge to a constant and near-optimal path when the environment remains stationary. On another hand, an exploration favoring least recently visited states is considered. This one is not essential for the system convergence but is extremely useful in terms of accelerating learning in presence of large state spaces and therefore increasing the system efficiency. Our method was validated through a cooperative foraging task containing a temporary moving target. Several satisfactory results are shown demonstrating the effectiveness of this approach. These relevant results remain valid when changing the location of obstacles. Agents succeed to construct a new path if the old one is blocked by obstacles. However, they can't detect the existance of a shortcut since the exploration of new areas is only possible if the actual solution is no longer adequate. In our futur works, we expect to further improve our method to overcome this limitation. Testing more complex scenarios, as continuous state spaces, is also interesting.

References

1. Hongliang, G., Yan, M.: Distributed reinforcement learning for coordinate multi-robot foraging. J. Intell. Rob. Syst. **60**(3–4), 531–551 (2010)
2. Yifan, C.: Intelligent Multi-robot Cooperation for Target Searching and Foraging Tasks in completely Unknown Environments. Ph.D. dissertation, University of Guelph, Guelph (2013)
3. Yogeswaran, M., Ponnambalam, S.G.: Reinforcement learning: explorationexploitation dilemma in multi-agent foraging task. OPSEARCH **49**(3), 223–236 (2012)
4. Chen, K., Lin, F., Tan, Q., Shi, Z.: Adaptive action selection using utility-based reinforcement learning. In: IEEE International Conference on Granular Computing GRC 2009, pp. 67–72. IEEE, August 2009

5. Jaradat, M.A.K., Al-Rousan, M., Quadan, L.: Reinforcement based mobile robot navigation in dynamic environment. Robot. Comput. Integr. Manuf. **27**(1), 135–149 (2011)
6. Cunningham, B., Cao, Y.: Non-reciprocating sharing methods in cooperative Q-learning environments. In: Proceedings of the the 2012 IEEE/WIC/ACM International Joint Conferences on Web Intelligence and Intelligent Agent Technology, Vol. 02, pp. 212–219. IEEE Computer Society, December 2012
7. Torrey, L., Taylor, M.E.: Help an agent out: student/teacher learning in sequential decision tasks. In: Proceedings of the Adaptive and Learning Agents workshop (at AAMAS-120) (2012)
8. Coggan, M.: Exploration and exploitation in reinforcement learning. In: Proceedings of the Fourth International Conference on Computational Intelligence and Multimedia Applications (ICCIMA 2001), Shonan International Village Yokosuka City (2001)
9. Mataric, M.J.: Learning in multi-robot systems. In: Weiss, G., Sen, S. (eds.) IJCAI-WS 1995. LNCS, vol. 1042, pp. 152–163. Springer, Heidelberg (1996)
10. Stone, P., Sutton, R.S., Kuhlmann, G.: Reinforcement learning for robocup soccer keepaway. Adaptive Behav. **13**(3), 165–188 (2005)
11. Panait, L., Luke., S.: A pheromone-based utility model for collaborative foraging. In: Proceedings of the Third International Joint Conference on Autonomous Agents and Multiagent Systems, pp. 36–43, Washington, DC (2004)
12. Hrolenok, B., Luke, S., Sullivan, K., Vo, C.: Collaborative Foraging using Beacons. In: van der Hoek., et al. (ed.) Proceedings of the Ninth International Conference on Autonomous Agents and Multiagent Systems (AAMAS 2010), pp. 1197–1204 (2010)
13. Tan, M.: Multi-agent reinforcement learning: independent vs. cooperative agents. In: Proceedings of the Tenth International Conference on Machine Learning, pp. 330–337 (1993)
14. Guo, H., Meng, Y.: Dynamic correlation matrix based multi-q learning for a multi-robot system. In: 2008 IEEE/RSJ International Conference on Intelligent Robots and Systems IROS 2008, pp. 840–845. IEEE, September 2008
15. Zemzem, W., Tagina, M.: A Novel exploration/exploitation policy accelerating learning in both stationary and non-stationary environment navigation tasks. Int. J. Comput. Electr. Eng. **7**(3), 149–158 (2015)
16. Sutton, S., Barto, G.: Reinforcement Learning: An Introduction. The MIT Press, Cambridge (1998)
17. http://simbad.sourceforge.net/

Optimized and Parallel Query Processing in Similarity-Based Databases

Petr Krajča[(✉)]

Department of Computer Science, Palacky University,
Olomouc 17. Listopadu 12, 77146 Olomouc, Czech Republic
petr.krajca@upol.cz

Abstract. We present a novel method of query execution in similarity-based databases which adopts techniques commonly used in traditional programming language compilers. Our method is based on decomposition of relational algebra operators into a small set of simple operations which are subject of further optimizations. It shows up that with a small set of optimizations rules our system itself is able to infer efficient algorithms for data processing. Furthermore, operations we propose are compatible with the map/reduce approach to data processing, and thus, allows for implicitly parallel or distributed data processing.

Keywords: Domain similarities · Relational model of data · Query execution · Query optimization · Fuzzy logic · Parallel data processing

1 Introduction

In this paper we deal with database systems based on generalization of the Codd data model [4,6] which allows queries with imprecise matches. This generalization results by considering complete residuated lattices as a structure representing degrees to which particular tuple (row in data table) satisfies given condition. We may consider, for instance, a database query "show all cars which cost $10,000". Traditional database system will return a list of all cars costing *exactly* $10,000, if there is a car for $9.900, it will not be included into the result, even though that from the viewpoint of the user it might be a reasonable result. Generalization we use allows approximate queries like "show all cars which cost *about* $10,000" and result of such query is a set of tuples with ranks indicating proximity of a tuple to the given query. If the car costs $10,000 the rank of the tuple will be 1, if not, the rank will be lower. In essence, the ranks have comparative meaning—higher ranks represent better matches. Let us note that our model is not an extension built on top of the classic model like most of query systems going in this direction. Unlike other approaches [7,11,15,16,20], the model is not aimed solely on ranking and querying. Indeed, the formalization of similarities and use of the general scales of ranks allows us to properly

Supported by grant no. 202/12/P167 of the Czech Science Foundation and IGA UP 2014, reg. č. PrF_2014_034.

© Springer International Publishing Switzerland 2015
V. Torra and Y. Narakawa (Eds.): MDAI 2015, LNAI 9321, pp. 167–179, 2015.
DOI: 10.1007/978-3-319-23240-9_14

formalize various similarity-based dependencies in data, e.g., functional dependencies and keys [4,6].

The aim of this paper is to propose a general framework for database query processing which allows implicit query optimizations and efficient data retrieval. In order to achieve this goal we adopt techniques used by traditional compilers.

The paper is organized as follows. First, we provide a short survey on existing approaches and introduction to the generalized relational model of data we use. Afterwards, we introduce operations for data processing and their relationship to the relational algebra. Subsequently, we describe optimizations that are used to improve data processing. The paper is concluded with the section on parallelization and experimental evaluation of our approach.

Related Works. In previous works, the data model was studied in terms of relational algebra and calculus [4,6], query language was proposed [18] as well as algorithms for some particular types of queries [19]. In [3] is presented an approach which maps operators of the generalized relational model [4,6] to operation of the RDBMS Oracle. This approach exploits robustness of the RDBMS, however is not efficient, since the RDBMS does not have any information about the generalized relational model and is unable to perform optimizations. In [19] are presented basic optimizations techniques for some common types of queries, especially those with the *top-k* operator (returning top k best matching results) where it is possible to use a variant of the Fagin's algorithm [11]. Several similar database systems involving ranking in relational databases, including the probably the best known RankSQL [20] are using a rank-relational model (a model built on top of the Codd model) in which tuples in relations are annotated by numerical scores computed from so-called predicate scores by monotone scoring functions. Since our model is more general it is not always possible to use algorithms from these database management systems along with our model.

2   Generalized Relational Model of Data

We outline here the foundations of our model and introduce basic notions necessary for understanding of the basic type of queries considered in this paper, more details can be found in [4,6,19].

Our model can be seen as a generalization of the classic RM which results by substituting the two-element Boolean algebra which is the implicit structure of yes/no matches in the classic RM by a more general structure, namely a (complete) residuated lattice [12].

Model we use departs from the yes/no matches and allows general "degrees of matches" upon which we build the generalized relational model. In the classic RM, the concept of a relation on a relation scheme R (a finite set of attributes), which is considered as a finite subset of a cartesian product $\prod_{y \in R} D_y$ of domains D_y of attributes $y \in R$ can be identified with an indicator function

$$\mathcal{D} \colon \prod_{y \in R} D_y \to \{0, 1\} \tag{1}$$

so that for only finitely many tuples $r \in \prod_{y \in R} D_y$ we have $\mathcal{D}(r) = 1$. If $\mathcal{D}$ is viewed as a result of query Q, then $\mathcal{D}(r) = 1$ is interpreted so that "the tuple r matches the query Q". In our model, we replace $\{0, 1\}$ by a set L of degrees which is assumed to be equipped with a partial order $\leq$ so that $\langle L, \leq \rangle$ is a complete lattice, i.e., an arbitrary subset of L has its infimum (greatest lower bound) and supremum (least upper bound) in L. We adhere to the *comparative meaning* of degrees from L (higher degrees represent better matches) as it is usual in fuzzy logics in the narrow sense (FLns), see [10,13,14]. Under this assumption, we may replace (1) by

$$\mathcal{D} \colon \prod_{y \in R} D_y \to L \tag{2}$$

so that for only finitely many tuples $r \in \prod_{y \in R} D_y$ we have $\mathcal{D}(r) \neq 0$. Clearly, (2) is a map which assigns to each r a value $\mathcal{D}(r)$ from L, we call the value *the rank of r in $\mathcal{D}$* and if $\mathcal{D}$ is interpreted as a result of a query Q, then $\mathcal{D}(r)$ is the *degree to which r matches the query Q*. The notion of a relation of a relation scheme which appears in the ordinary RM can be then seen as a particular case for $L = \{0, 1\}$ with its natural ordering (i.e., $0 < 1$).

Furthermore, the lattice of degrees should be equipped with operations to aggregate degrees. Thus, reasonable choice for a structure of degrees which replaces the two-element Boolean algebra in our model is a complete residuated lattice $\mathbf{L} = \langle L, \wedge, \vee, \otimes, \to, 0, 1 \rangle$ where $\langle L, \wedge, \vee, 0, 1 \rangle$ is a complete lattice, i.e., partially ordered set in which arbitrary infima and suprema exists, 0 and 1 denote the least and greatest element, respectively. Operations $\otimes$ and $\to$ are binary operations on L such that $\langle L, \otimes, 1 \rangle$ is a commutative monoid, i.e., is commutative, associative, neutral with respect to 1 (full match), and operations $\otimes$, $\to$ satisfy the following *adjointness property*:

$$a \otimes b \leq c \text{ iff } a \leq b \to c \tag{3}$$

for all $a, b, c \in L$ ($\leq$ denotes lattice ordering). Recall that the adjointness of $\otimes$ and $\to$ is a crucial property of structures of degrees used in FLns, see [2,12,14]. Typical choice for L is a real unit interval with Łukasiewicz operations (i.e., $a \otimes b = \max(a + b - 1, 0)$ and $a \to b = \min(1 - a + b, 1)$), Gödel operations (i.e., $a \otimes b = \min(a, b)$ and $a \to b = 1$ if $a \leq b$, otherwise $a \to b = b$), or product operations (i.e., $a \otimes b = a \cdot b$ and $a \to b = 1$ if $a \leq b$, otherwise $a \to b = \frac{b}{a}$). Nonetheless, class of residuated lattices is larger and includes other popular t-norm based structures [17], finite structures, and various nonlinear structures.

These operations and in particular (truth functions of) general conjunctions, i.e., operations $\otimes$, appear in our model as we consider counterparts to relational operations like the natural join. Indeed, in the ordinary RM, for relations $\mathcal{D}_1$ and $\mathcal{D}_2$ on relation schemes $R \cup S$ and $S \cup T$ such that R, S, T are pairwise disjoint, we consider the natural join of $\mathcal{D}_1$ and $\mathcal{D}_2$ as a relation on $R \cup S \cup T$, denoted by $\mathcal{D}_1 \bowtie \mathcal{D}_2$ which consists of concatenation of all joinable tuples from $\mathcal{D}_1$ and $\mathcal{D}_2$. Identifying the relations with their indicator functions as in (1) and using the usual notation for tuple concatenation (i.e., rs stands for the set-theoretic union of maps r and s, see [21]), we have $(\mathcal{D}_1 \bowtie \mathcal{D}_2)(rst) = 1$ iff $\mathcal{D}_1(rs) = 1$ and $\mathcal{D}_2(st) = 1$. Therefore, we may rewrite the natural join as follows

$$(\mathcal{D}_1 \bowtie \mathcal{D}_2)(rst) = \mathcal{D}_1(rs) \otimes \mathcal{D}_2(st), \tag{4}$$

where $\otimes$ is a binary operation $\otimes \colon \{0,1\}^2 \to \{0,1\}$ which coincides with the truth function of the logical connective "conjunction" in the usual sense (i.e., $1 \otimes 1 = 1$ and $1 \otimes 0 = 0 \otimes 1 = 0 \otimes 0 = 0$).

Thus, for considering analogues of natural joins in our model, we use a generalization of $\otimes$. Note that $\otimes$ is distributive over arbitrary suprema, i.e.,

$$a \otimes \bigvee_{i \in I} b_i = \bigvee_{i \in I} (a \otimes b_i) \tag{5}$$

holds true for any $a \in L$ and all $b_i \in L$ ($i \in I$). As a consequence, $\otimes$ is *monotone* which is a desirable property since then better results of subqueries (e.g., $\mathcal{D}_1$ and $\mathcal{D}_2$) yield better results of composed queries whose results are computed by $\otimes$ as in case of (4).

The need to have a reasonable generalization of a natural join in our model justifies the presence of $\otimes$. Analogously, one can say that $\to$ is crucial for expressing a "graded containment" which is essential, e.g., for expressing universal queries of the form "all As are Bs".

There are wide benefits of using complete residuated lattices as structures for degrees of matches. First, the structures are reasonably strong (the adjointness ensures that $\mathbf{L}$ and $\otimes$ and $\to$ have reasonable properties). Second, with residuated lattices we get reasonable logical background for our model. As a consequence, database instances can be seen as safe interpretations of (many-sorted) predicate languages [8,14], predicate formulas (with free variables) can be seen as prescribing queries in our model, and evaluation of the formulas in structures can be seen as a way of query evaluation, see [6] for details.

Following the previous arguments, the basic notion which appears in our model and which replaces the ordinary notion of a relation on a relation scheme is introduced as follows.

Definition 1 (Ranked Data Tables). Let $\mathbf{L}$ be a complete residuated lattice, $R \subseteq Y$ be a finite set of attributes (a relation scheme). Then, any map $\mathcal{D}$ of the form (2) such for only finitely many tuples $r \in \prod_{y \in R} D_y$ we have $\mathcal{D}(r) \neq 0$ is called a *ranked data table* (an RDT). The number of tuples $\mathcal{D}(r) \neq 0$ is called a size of RDT.

In order to be able to express similarity-based queries, each domain D_y is equipped with a *similarity* $\mathbf{L}$-relation [2], i.e., a map $\approx_y \colon D_y \times D_y \to L$ which assigns to each value $d_1, d_2 \in D_y$ a degree $d_1 \approx_y d_2$ to which d_1 is similar to d_2. We assume that each $\approx_y$ is at least *reflexive* ($d \approx_y d = 1$ for all $d \in D_y$) and *symmetric* ($d_1 \approx_y d_2 = d_2 \approx_y d_1$ for all $d_1, d_2 \in D_y$).

3   Query Processing

Queries in the generalized relational model may be represented as so called *relational expressions*. The same way as the ordinary expressions consist of variables and operators, relational expressions consist of *relational variables* containing RDTs as their values and *relational operators* taking an RDT as its input and evaluating again to an RDT. In this paper we outline only the most important relational expressions we consider in our model.

The simplest relational expression one can consider is a *relational variable*, in other words, it is a variable which has an RDT as its value. Relational variable typically represents data stored in the database and shall be denoted $\mathcal{D}_{table}$.

Restriction is represented by an operator $\sigma_\theta(\mathfrak{R})$ having two arguments θ and $\mathfrak{R}$ representing a condition and a relational expression, respectively. Restriction operator takes each tuple r from the RDT $\mathcal{D}$ which is a value of the relational expression $\mathfrak{R}$ and returns the same tuple with a rank which is given by $\mathcal{D}(r) \otimes \|\theta\|_r$. In other words, it assigns to each tuple the degree to which r satisfies condition θ, e.g. $\sigma_{price \approx 10,000}(\mathcal{D}_{cars})$.

Projection is expressed by an operator $\pi_S(\mathfrak{R})$ with two parameters $\mathfrak{R}$ and S representing relational expression and a subset of the relation scheme of $\mathfrak{R}$, respectively. For each tuple r from $\mathfrak{R}$ is returned a new tuple s containing all attributes from S and their values from tuple r. Notice, that it may happen that some tuple appears in the result set multiple times. In order to preserve all demanded features of RDTs we take supremum of their ranks and return the tuple only once.

Set-theoretic nature of RDTs allows to introduce a wide range of operations, like union, intersection, strong intersection, etc. These operators have two arguments, relational expressions $\mathfrak{R}_1$, $\mathfrak{R}_2$ evaluating to RDTs $\mathcal{D}_1$ and $\mathcal{D}_2$ with the same relation scheme. For each tuple r is computed new rank as $\mathcal{D}_1(r) \odot \mathcal{D}_2(r)$, where $\odot$ is a binary operation defined on L. *Union* operator is defined with $\vee$, *intersection* is defined with $\wedge$, strong intersection with $\otimes$, etc.

The last important operator is *natural join* but its meaning was already discussed in Sect. 2, therefore we omit its detailed description in this section.

3.1 Compilation

Query processing in database management systems typically consists of several steps. (i) Query is transformed from a query language (e.g., SQL, RESIQL [18]) into a relational algebra expression. (ii) Afterwards, rules of relational algebra are used to create more efficient expression. (iii) The relational expression is transformed into an execution plan which consists of so called physical operators directly working with the data. (iv) Finally, data are retrieved using physical operators. In this paper we primarily focus on step (iii) and by extension on step (iv).

Typically, database management system has a set of physical operators for particular use cases, e.g., for natural joins, access to the sorted data table, etc. and the database system creates an execution plan by picking the most suitable combination of these operators. In case of similarity-based databases only a relatively small number of algorithms for physical operators is known, and furthermore, these algorithms often solves only a particular problem, for instance, return *top-k* results. Moreover, often further conditions have to be fulfilled, for instance, input data have to be in some particular order, random access to tuples is required, etc. Further, in case of similarity-based databases we may encounter new combinations of operators which were not yet thoroughly studied, for instance, similarity-based joins or similarity-based semijoins.

Our approach to query processing is different than the usual one. It is based on the idea that relational algebra operators are transformed into a small set of simple relational operations which can be subject of further optimizations. The basic building blocks are operators `for` and `emit`. The `for` operator has two parameters (i) the input relational expression and (ii) a *body*. For each tuple from the input relational expression, it performs operations in the body. It may be, for instance, another `for` or the `emit` operator. The `emit` operator simply emits a new tuple which is described by its argument and all emitted tuples are collected and returned by the `for` operator. Within the `emit` operator it is possible to apply on each tuple a transformation function f which is a map $f : \Gamma_{y \in R} D_y \to \Gamma_{y \in S} D_y$ where $\Gamma_{y \in R} D_y$ denotes set of all RDTs of size 1 with a relation scheme R. In other words, map f transforms one tuple to another.

Various combinations of these two operators are able to express the basic operators of the relational algebra. Restriction $\sigma_\theta(\mathfrak{R})$ where θ is a condition $r_i \approx c$ and $\mathfrak{R}$ is a relational expression evaluating to an RDT $\mathcal{D}$ with the relation scheme $\{r_1, \ldots, r_n\}$ can be expressed with operators `for` and `emit` as follows:

```
for u in ℜ
    emit (rank : 𝒟(u) ⊗ (u(rᵢ) ≈ c), r₁ : u(r₁), ..., rₙ : u(rₙ))
```

Basically, `for` takes every tuple from $\mathfrak{R}$ and evaluates its body, in this case, it evaluates operator `emit` which emits a new tuple where new rank is assigned to the tuple, according to the degree to which the tuple satisfies the given condition. If the rank of the tuple is 0, it is not emitted, since it means that the tuple does not satisfies condition at all.

Notice that these two operators have very similar meaning like the `map` function in Lisp, or other programming languages. Later we are to use this fact for automatic query parallelization. To make the notation simple and comprehensible, we depict transformation function as a sequence of attribute-value pairs $(attr_1 : value_1, \ldots, attr_n : value_n)$. Further, we use the same notation for ranks and for attributes, even though the *rank* is not an attribute.

We can nest multiple `for` operators together. This may be used for instance to represent joins—natural joins and cross joins. The natural join $\mathfrak{R}_1 \bowtie \mathfrak{R}_2$ of two relational expression evaluating to RDTs $\mathcal{D}_1$ and $\mathcal{D}_2$ with relation schemes $\{r_1 \ldots, r_n, t_1, \ldots, t_n\}$ and $\{s_1, \ldots, s_n, t_1, \ldots, t_n\}$, respectively, with common attributes $t_1, \ldots, t_n$, can be expressed as an expression:

```
for u in ℜ₁
  for v in ℜ₂
      emit (rank : 𝒟₁(u) ⊗ 𝒟₂(v) ⊗ u(t₁) = v(t₁) ⊗ ⋯ ⊗ u(tₙ) = v(tₙ),
            r₁ : u(r₁), ..., rₙ : u(rₙ),
            t₁ : u(t₁), ..., tₙ : u(tₙ),
            s₁ : v(s₁), ..., sₙ : v(sₙ))
```

Interpretation of this expression is straightforward, both `for` operators iterate over all combination of tuples from $\mathfrak{R}_1$ and $\mathfrak{R}_2$ and emit tuples which are their concatenations and are joinable, in other words, values in their common attributes are equal. Similarly, cross join is a special case of the natural join of two relational expression over disjoint schemes and is compiled analogously.

With appropriate transformation function we may represent other relational operators which are not discussed in this paper, e.g. extension, renaming, etc.

For projection the `for` and `emit` operators are not fully sufficient. It may happen that some tuple is emitted multiple times, which is in conflict with the definition of an RDT. Therefore, if some tuple is emitted multiple times, we consider tuple with rank equal to suprema of ranks of all duplicate tuples. For this purpose we introduce a new operator `reduce` having two arguments—relational expression and an aggregation function $f : 2^L \to L$. First, the relational expression is evaluated. Afterwards, tuples having the same values of all attributes are grouped and for each group new rank is computed with an aggregation function. Next, for each group is emitted a new tuple with an aggregated rank.

Using the `reduce` operator along with `for` and `emit` it is possible to represent the projection $\pi_{\{r_1,\dots,r_m\}}(\mathfrak{R})$ as follows:

```
reduce ⋁
  for u in 𝔑
    emit (rank : 𝒟(u), r₁ : u(r₁), ..., rₘ : u(rₘ))
```

It iterates over all tuples from RDT $\mathcal{D}$ which is a value of $\mathfrak{R}$, emits new tuples and for same tuples it aggregates their ranks, i.e., in this case $\bigvee$ computes a suprema of ranks. Note, that if $\{r_1,\dots,r_m\}$ contains a candidate key, the `reduce` operator may be omitted, since it is ensured that there are no duplicities.

The `reduce` operator has a wide range of applications and by choosing proper aggregation function we may represent set-theoretic operators, for instance, union $\mathfrak{R}_1 \cup \mathfrak{R}_2$ of two relational expression $\mathfrak{R}_1$ and $\mathfrak{R}_2$ with the same relation scheme may be represented as:

```
reduce ⋁
  for u in 𝔑₁
    emit u
  for v in 𝔑₂
    emit v
```

Intersection and other set-theoretic operations may be represented analogously, however, we have to pay attention to an aggregation function. It may happen, that some tuple is present only in one of the relations which implies it has a zero rank in the second relation, and thus, the resulting rank has to be zero as well. To solve this issue, we introduce the following aggregation function: $\bigwedge^n : 2^L \to L$ such that

$$\bigwedge{}^n(X) = \begin{array}{ll} 0, & \text{if } |X| < n \\ \bigwedge_{x \in X} x, & \text{otherwise.} \end{array}$$

In other words, this aggregation function has the same meaning as the usual one, it computes infima of all values, however more than n values have to be aggregated, otherwise its value is zero. Hence, intersection of two relational expressions may be represented analogously as an union but with reduce $\bigwedge^2$.

3.2   Optimizations

The optimization scheme we propose is based on three basic rules that transform combinations of operators to a more optimal one. We shall call these rules—composition, filtering, and index selection.

Composition. The first rule combines two for operators into a single one. Let us assume a relational expression $\mathfrak{R}_1$ that is represented by the for operator with a body that contains an emit operator with a transformation function f_1. Note that the emit operator may be nested inside another operator. Further, let us assume a relational expression $\mathfrak{R}_2$ that is represented by the for operator with an argument $\mathfrak{R}_1$ and with a body containing solely the emit operator with a transformation function f_2, like in the following case:

$\mathfrak{R}_1$: for u in $\mathfrak{R}$
     body
        emit $f_1(u)$
$\mathfrak{R}_2$: for v in $\mathfrak{R}_1$
     emit $f_2(v)$

This means, relational expression $\mathfrak{R}_2$ applies transformation function f_2 on every tuple from $\mathfrak{R}_1$. Hence, relational expression $\mathfrak{R}_2$ can be replaced with the relational expression $\mathfrak{R}_1$ where the emit transformation function is replaced with a composition of functions $f_2 \circ f_1$:

$\mathfrak{R}_2$: for u in $\mathfrak{R}$
     body
        emit $f_2(f_1(u))$

For instance, let us consider two restrictions $\sigma_{r_i \approx c_i}(\sigma_{r_j \approx c_j}(\mathcal{D}))$ on RDT $\mathcal{D}$ with relation scheme $\{r_1, \ldots r_n\}$. This expression is represented by two for-loops

$\mathfrak{R}_1$: for u in $\mathcal{D}$
     emit $(rank : \mathcal{D}(u) \otimes (u(r_i) \approx c_i), r_1 : u(r_1), \ldots, r_n : u(r_n))$
$\mathfrak{R}_2$: for v in $\mathfrak{R}_1$
     emit $(rank : \mathcal{D}(v) \otimes (v(r_j) \approx c_j), r_1 : v(r_1), \ldots, r_n : v(r_n))$

The composition rule transforms these two expressions into the following one:

$\mathfrak{R}_2$: for u in $\mathcal{D}$
     emit $(rank : \mathcal{D}(u) \otimes (u(r_i) \approx c_i) \otimes (u(r_j) \approx c_j),$
        $r_1 : u(r_1), \ldots, r_n : u(r_n))$

Notice that the simplified expression corresponds to the restriction $\sigma_{(r_i \approx c_i) \otimes (r_j \approx c_j)}(\mathcal{D})$ which is equivalent to $\sigma_{r_i \approx c_i}(\sigma_{r_j \approx c_j}(\mathcal{D}))$ from the point view of the relational algebra, see [5]. Nonetheless, this is only an example and the composition rule is able to correctly deal with other combinations of operators, e.g., joins and restrictions.

Filtering. The second rule takes care of abandoning useless computation. In case of natural joins (or as a consequence of the composition rule) expressions usually contains multiple nested loops. However the decision whether the tuple will be emitted or not is made within the `emit` operator which always lies in the innermost `for`-cycle, apparently, this is a source of inefficiency.

It is desirable to skip the computation in the inner loops if it is clear that the final rank of the tuple will be zero no matter what is in the inner loops. To solve this problem we utilize the fact that the `emit` operator is the only place where the rank is assigned, furthermore, if the expression which determines the rank is a composition of subexpressions which are aggregated by a monotone function (e.g., $\otimes$ or $\wedge$) we can determine from subexpressions if the final rank will be zero, or not. This follows from the monotonicity of the aggregation function, if one subexpression is zero, the final rank have to be also zero.

We introduce a new operator `filter` having two arguments—an expression representing *condition* which has to be satisfied and a *body*. This operator performs operations in its body if the expression is not 0, i.e., it works as an `if` command in many programming languages. To reduce useless computation as much as possible, the `filter` operator is always placed into the outer most `for`-loop which contains all variables necessary for condition evaluation. For placing the filter into an appropriate place it is possible to use known algorithms for *loop-invariant code motion* which is a common technique used by conventional compilers, for more details see, for instance, [1].

For example, expression $\sigma_{r_i=c}(\mathcal{D}_1 \bowtie \mathcal{D}_2)$ where $\mathcal{D}_1$ and $\mathcal{D}_2$ are RDTs with disjoint schemes may be compiled into the following form:

```
for u in D₁
  filter (u(rᵢ) ≈ c)
    for v in D₂
        emit (rank : D₁(u) ⊗ D₂(v) ⊗ (u(rᵢ) ≈ c),
              r₁ : u(r₁), ..., rₙ : u(rₙ),
              s₁ : v(s₁), ..., sₙ : v(sₙ))
```

With the `filter` operator is the expression more efficient since it considers only reasonable tuples which possibly may have non-zero ranks and skips irrelevant inner loops.

Index Selection. Database systems traditionally uses indexes to efficiently retrieve tuples from the physical storage. For similarity-based databases it is possible to utilize such methods as well [19]. Thus, we introduce the third rule and a new operator `index` having two arguments—relational variable (physical data table) and a condition. This operator returns tuples satisfying the given condition. In this paper we do not address how the data are obtained, we only assume that there is an efficient method of retrieving tuples satisfying the given condition.

In our representation it is easy to identify opportunities for index application. If the operator `for` has a relational variable as its argument and if its body contains a `filter` operator with a condition that can utilize existing index, the relational variable in the `for`-loop may be replaced with an **index** operator and `filter` operator may be optionally removed. For example, similarity-based join $\sigma_{r_i \approx s_i}(\mathcal{D}_1 \bowtie \mathcal{D}_2)$ involving two RDTs $\mathcal{D}_1$ and $\mathcal{D}_2$ with relation schemes $\{r_1 \ldots, r_n\}$ and $\{s_1, \ldots, s_n\}$, respectively, and an index on attribute r_i will be transformed as follows:

```
for u in 𝒟₁
   for v in index(𝒟₂, u(rᵢ) ≈ v(sᵢ))
      filter (u(rᵢ) ≈ v(sᵢ))
         emit (rank : 𝒟₁(u) ⊗ 𝒟₂(v) ⊗ (u(rᵢ) ≈ v(sᵢ)),
               r₁ : u(r₁), ..., rₙ : u(rₙ),
               s₁ : v(s₁), ..., sₙ : v(sₙ))
```

This example shows that with the set of simple rules that were described in this section we were able to automatically infer a variant of the nested-loop join algorithm for similarity-based join which efficiently uses indexes. More complex expressions would be optimized in a very similar way as well. Note that in this example we preserved the `filter` operator in the expression for greater clarity, but it could be removed.

Rule Application Order. The rules are applied in the following order. (1) The *composition* is applied on all suitable `for`-loops and all chained loops are transformed into the nested ones. (2) *Filtering* is applied on all suitable `for`-loops, this assures that each `for`-loop contains appropriate `filter` operator in its body. (3) *Index selection* is applied on suitable filters and arguments of `for`-loops.

4   Parallel Query Processing and Experimental Evaluation

As the multicore processors become commonly available it is important to take the parallelization of query processing into account. The set of operations we introduce is well-suited for parallelization because there is no shared mutable state that could be source of obstacles. Apparently, each `for` operator may partitionate its input data into multiple groups and each group can be processed separately in its own thread. The aggregation procedure **reduce** can be processed by multiple threads as well, for instance, in the following steps. First, tuples are split into groups and assigned to separate threads where tuples are pre-sorted and partially aggregated, afterwards, partially aggregated tuples from each thread are aggregated into a final dataset. This approach is in fact very similar to the Map/Reduce system [9] which is commonly used to process large datasets. We have successfully adopted this approach to process database queries on a shared memory computer.

Each `for` operator is a represented as a *map*-task reading input data and (i) if the body contains `emit`, new tuple is emitted, (ii) if the body contains `for`, the tuple is passed to another *map*-task, and (iii) if the body contains `filter` the condition is evaluated and if the value is not 0, the computation continues with the option (i) or (ii). Finally, data are passed to a *reduce*-task which collects (and optionally aggregates) tuples which are emitted by *map*-task. The data flow is outlined in the next figure depicting two nested loops and a projection. Note that each task may be performed by multiple threads.

The discussed optimization framework was implemented as a part of the RESIQL database and this allowed us to evaluate its efficiency using the real data. The following table presents results of experiments involving complex queries (each query contained at least one projection, at least one restriction, and at least one natural join) and real-world datasets from the UCI Machine Learning Repository and our own dataset (cars).

	dataset size	unoptimized		optimized		
		tuples	time	tuples	time	time (8 procs.)
adult	48,842	829,821	5.6 s	292,938	3.9 s	1.9 s
bank	45,211	2.04×10^9	151 min	65,705	2.3 s	1.2 s
cars	4,707	65,898	458 ms	43,964	323 ms	178 ms
wine quality	6,497	28,302,400	143.6 s	352,841	7.1 s	2.9 s

All experiments were performed twice—with and without the optimization framework. Along with the time required to get the results we also focused on number of tuples that were fetched from the physical storage, since this is a very time demanding operation. Results clearly confirm benefits of our approach. On the other hand, parallel execution on a computer with 8 CPU cores brings only speedup of factor 2. But we have to take into account that the results in this case are skewed since parts of the computation (e.g., query preparation) have to be always performed in a single thread.

We have tested our approach on a shared memory computer, however, it can be used in a straightforward way to compile queries for other computational models and systems, for instance, for already mentioned Map/Reduce systems [9] like Apache Hadoop, or for systems based on Resilient Distributed Datasets [22] like Apache Spark. Indeed, it suffices to skip the last step of query processing which interprets the query and emit Java/Scala code instead and leave the query processing on a framework.

References

1. Aho, A.V., Lam, M.S., Sethi, R., Ullman, J.D.: Compilers: Principles, Techniques, and Tools, 2nd edn. Addison-Wesley, Boston (2006)
2. Belohlavek, R.: Fuzzy Relational Systems: Foundations and Principles. Kluwer Academic Publishers, Norwell (2002)
3. Belohlavek, R., Opichal, S., Vychodil, V.: Relational algebra for ranked tables with similarities: properties and implementation. In: Berthold, M., Shawe-Taylor, J., Lavrač, N. (eds.) IDA 2007. LNCS, vol. 4723, pp. 140–151. Springer, Heidelberg (2007)
4. Bělohlávek, R., Vychodil, V.: Data tables with similarity relations: functional dependencies, complete rules and non-redundant bases. In: Li Lee, M., Tan, K.-L., Wuwongse, V. (eds.) DASFAA 2006. LNCS, vol. 3882, pp. 644–658. Springer, Heidelberg (2006)
5. Belohlavek, R., Vychodil, V.: Logical foundations for similarity-based databases. In: Chen, L., Liu, C., Liu, Q., Deng, K. (eds.) DASFAA 2009. LNCS, vol. 5667, pp. 137–151. Springer, Heidelberg (2009)
6. Belohlavek, R., Vychodil, V.: Query systems in similarity-based databases: logical foundations, expressive power, and completeness. In: ACM Symposium on Applied Computing (SAC), pp. 1648–1655. ACM (2010)
7. Buckles, B.P., Petry, F.E.: A fuzzy representation of data for relational databases. Fuzzy Sets Syst. 7(3), 213–226 (1982)
8. Cintula, P., Hájek, P.: Triangular norm based predicate fuzzy logics. Fuzzy Sets Syst. 161, 311–346 (2010)
9. Dean, J., Ghemawat, S.: Mapreduce: simplified data processing on large clusters. In: Brewer, E., Chen, P. (eds.) OSDI, pp. 137–150, USENIX Association (2004)
10. Esteva, F., Godo, L.: Monoidal t-norm based logic: towards a logic for left-continuous t-norms. Fuzzy Sets Syst. 124(3), 271–288 (2001)
11. Fagin, R.: Combining fuzzy information from multiple systems. J. Comput. Syst. Sci. 58(1), 83–99 (1999)
12. Goguen, J.A.: The logic of inexact concepts. Synthese 19, 325–373 (1979)
13. Gottwald, S.: Mathematical fuzzy logics. Bull. Symbolic Logic 14(2), 210–239 (2008)
14. Hájek, P.: Metamathematics of Fuzzy Logic. Kluwer Academic Publishers, Dordrecht (1998)
15. Ilyas, I.F., Beskales, G., Soliman, M.A.: A survey of top-k query processing techniques in relational database systems. ACM Comput. Surv. 40(4), 11 (2008)
16. Kieling, W., Köstler, G.: Preference SQL–design, implementation, experiences (2002)
17. Klement, E.P., Mesiar, R., Pap, E.: Triangular Norms, 1st edn. Springer, Heidelberg (2000)
18. Krajca, P., Vychodil, V.: Foundations of relational similarity-based query language resiql. In: FOCI, pp. 15–23. IEEE (2013)
19. Krajca, P., Vychodil, V.: Query optimization strategies in similarity-based databases. In: Torra, V., Narukawa, Y., Navarro-Arribas, G., Megías, D. (eds.) MDAI 2013. LNCS, vol. 8234, pp. 179–191. Springer, Heidelberg (2013)
20. Li, C., Chang, K.C.C., Ilyas, I.F., Song, S.: Ranksql: query algebra and optimization for relational top-k queries. In: Proceedings of the 2005 ACM SIGMOD, pp. 131–142 (2005)

21. Maier, D.: The Theory of Relational Databases. Computer Science Press, Cambridge (1983)
22. Zaharia, M., Chowdhury, M., Das, T., Dave, A., Ma, J., McCauley, M., Franklin, M.-J., Shenker, S., Stoica, S.: Resilient distributed datasets: a fault-tolerant abstraction for in-memory cluster computing. In: Proceedings of the 9th USENIX Conference on Networked Systems Design and Implementation (NSDI 2012), USENIX Association, Berkeley (2012)

An Evaluation of Edge Modification Techniques for Privacy-Preserving on Graphs

Jordi Casas-Roma[✉]

Universitat Oberta de Catalunya, Barcelona, Spain
jcasasr@uoc.edu

Abstract. Noise is added by privacy-preserving methods or anonymization processes to prevent adversaries from re-identifying users in anonymous networks. The noise introduced by the anonymization steps may also affect the data, reducing its utility for subsequent data mining processes. Graph modification approaches are one of the most used and well-known methods to protect the privacy of the data. These methods converts the data by edges or vertices modifications before releasing the perturbed data. In this paper we want to analyse the edge modification techniques found in the literature covering this topic, and then empirically evaluate the information loss introduced by each of these methods. We want to point out how these methods affect the main properties and characteristics of the network, since it will help us to choose the best one to achieve a desired privacy level while preserving data utility.

Keywords: Privacy · Social networks · Graphs · Data utility · Graph mining

1 Introduction

In recent years, a huge amount of social and human interaction networks have been made publicly available. Embedded within this data there is user's private information which must be preserved before releasing this data to third parties and researchers. The study of Ferri et al. [13] reveals that up to 90 % of user groups are concerned by data owners sharing data about them. Backstrom et al. [2] point out that the simple technique of anonymizing graphs by removing the identities of the vertices before publishing the actual graph does not always guarantee privacy. They show that an adversary can infer the identity of the vertices by solving a set of restricted graph isomorphism problems.

Therefore, anonymization processes become an important concern in this scenario. These methods add noise into the original data to hinder re-identification processes. Nevertheless, the noise introduced by the anonymization steps may also affect the data, reducing its utility for subsequent data mining processes. Usually, the larger the data modification, the harder the re-identification but also the less the data utility. Thus, it is necessary to preserve the integrity of the data to ensure that the data mining step is not altered by the anonymization step. The analysis

© Springer International Publishing Switzerland 2015
V. Torra and Y. Narakawa (Eds.): MDAI 2015, LNAI 9321, pp. 180–191, 2015.
DOI: 10.1007/978-3-319-23240-9_15

performed on the obfuscated data should produce results as close as possible to the ones the original data would have led to.

Several methods appeared recently to preserve the privacy of the data contained in a graph. One of the most used and well-known approaches is based on "graph modification". These methods first transform the data by edges or vertices modifications (adding and/or deleting) and then release the perturbed data. The data is thus made available for unconstrained analysis. There are two main approaches in the privacy-preserving literature [23]: (a) *random perturbation* of the graph structure by randomly adding/removing/switching edges and often referred to as edge randomization [4,6,16,17,24,25]; and (b) *constrained perturbation* of the graph structure via sequential edge modifications in order to fulfil some desired constraints – for example k-anonymity-based approaches that modify the graph so that every node is in the end indistinguishable from $k - 1$ other nodes (in terms of node degree for instance) [7,10,18–20,28,29].

All aforementioned algorithms use edge modification techniques, i.e. add, remove and/or switch edges to achieve a desired privacy level. Nevertheless, it is inevitable to introduce noise in the data, producing a certain amount of information loss, and consequently, deteriorating the utility of the anonymous data. Some authors claim that only adding edges better preserves the data utility, since none true relationship is removed. On the contrary, some other authors claim that removing an edge affects the structure of the graph to a smaller degree than adding an edge [5].

In this paper we want to analyse the edge modification techniques found in the literature covering this topic, and then empirically evaluate the information loss introduced by each of these methods during the perturbation process. We want to understand how these edge modifications affect the main properties and characteristics of the network. This will help us to choose the best edge modification technique to achieve a desired privacy level while keeping data utility and reducing information loss.

1.1   Our Contributions

In this paper we present an empirical evaluation of the basic edge modification techniques, which can help us to increase data utility in anonymous networks. We focus on simple, undirected and unlabelled graphs. Since these graphs have no attributes or labels in the edges, information is only in the structure of the graph itself and, due to this, evaluating edge modification techniques is an critical way to reduce information loss. We offer the following results:

- We analyse the most used and well-known edge modification techniques found in the graph privacy literature.
- We conduct an empirical evaluation of these techniques on several synthetic and real graphs, comparing information loss based on different graph's properties.
- We demonstrate that graph's structure must be considered in order to select the best edge modification technique, and it conducts the process to reduce the information loss and increase the data utility.

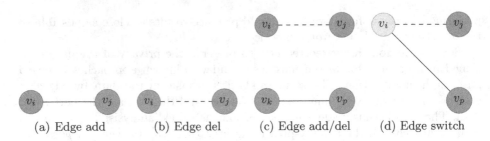

(a) Edge add   (b) Edge del   (c) Edge add/del   (d) Edge switch

Fig. 1. Basic operations for edge modification.

1.2   Notation

Let $G = (V, E)$ be a simple, undirected and unlabelled graph, where V is the set of vertices and E the set of edges in G. We define $n = |V|$ to denote the number of vertices and $m = |E|$ to denote the number of edges. We use $\{i, j\}$ to define an undirected edge from vertex v_i to v_j and $deg(v_i)$ to denote the degree of vertex v_i. Finally, we designate $G = (V, E)$ and $\widetilde{G} = (\widetilde{V}, \widetilde{E})$ to refer the original and the perturbed graphs, respectively. Note that in this work we use the terms graph and network indistinguishably.

1.3   Roadmap

This paper is organized as follows. In Sect. 2, we review the basic edge modification techniques for privacy-preserving on graphs. Section 3 introduces our experimental framework to analyse and compare the edge modification techniques on both synthetic and real networks. Then, in Sect. 4, we discuss the results in terms of information loss and data utility. Lastly, in Sect. 5, we present the conclusions of this work and some future remarks.

2   Edge Modification Techniques

We define four basic *edge modification* processes to change the network's structure by adding, removing or switching edges. These methods are the most basic ones, and they can be combined in order to create complex combinations. We are interested in them since they allow us to model, in a general and conceptual way, most of the privacy-preserving methods based on edge-modification processes. In the following lines we will introduce these basic methods, also called *perturbation methods*, due to the fact that they can model the perturbation introduced in anonymous data during the anonymization process.

There are four basic edge modifications illustrated in Fig. 1. Dashed lines represent existing edges which will be deleted and solid lines constitute the edges which will be added. Node color indicates whether a node changes its degree (dark grey) or not (light grey) after the edge modification has been carried out. These are:

- *Edge add* simply consists on adding a new edge $\{v_i, v_j\} \notin E$. Figure 1a illustrates this operation. The number of the edges will increase ($\tilde{m} > m$) when anonymization percentage increases. True relationships will be preserved in perturbed data.
- *Edge del* removes an existing edge $\{v_i, v_j\} \in E$, as depicted in Fig. 1b. Contrary to the previous method, the number of edges decreases $\tilde{m} < m$ and no fake relationships are included in the anonymous data, but several true relations are deleted from original data.
- *Edge add/del* is a combination of the previous pair methods. It simply consists of deleting an existing edge $\{v_i, v_j\} \in E$ and adding a new one $\{v_k, v_p\} \notin E$. Figure 1c illustrates this operation. In this case some true relations are deleted and some fake ones are created, but the total number of edges is preserved ($\tilde{m} = m$). All vertices involved in this operation change their degree.
- *Edge switch* occurs between three nodes $v_i, v_j, v_p \in V$ such that $\{v_i, v_j\} \in E$ and $\{v_i, v_p\} \notin E$. It is defined as deleting edge $\{v_i, v_j\}$ and creating a new edge $\{v_i, v_p\}$ as shown in Fig. 1d. As in the previous case, some true relations are removed, some fake ones are created and the number of edges is also preserved ($\tilde{m} = m$). However, two vertices change their degree (v_j and v_p) while the third one (v_i) does not.

For all perturbation methods, the number of vertices remains the same but the degree distribution changes. As previously stated, most of the anonymization methods rely on one (or more) of these basic edge modification operations. We believe that this covers the basic behavior of edge-modification-based methods for graph anonymization, even though some of them do not apply edge modification to the entire edge set.

As aforementioned, some algorithms are based on Edge add [10,19,21,28], since their authors usually claim that this edge modification technique better retain data utility. A similar situation occurs with Edge del [4,5]. Several random-based anonymization methods are based on the concept of Edge add/del [17,24,25] and most k-anonymity methods can be also modelled through the this concept [18,22,28,29]. Lastly, Edge switch is also used in many algorithms, such as [7,16,20,25]. Other methods consider to alter the vertex set to achieve anonymity. This concept is known as noise node addition [9,11,26]. We do not consider this algorithms in this paper due to space constraints and we propose it for future work.

3   Evaluating Framework

In this section we will post the experimental framework we have used to analyse and compare the information loss induced by our four edge modification techniques. Our experimental framework considers 10 independent executions of the edge modification methods with a perturbation parameter p in range between 0 % (original graph) and 25 % of total number of edges, i.e., $0 \leq p \leq 25$.

The process is the following: Firstly, we generate 10 independent sets of perturbed networks (from 0 % to 25 %) using each one of our edge modification

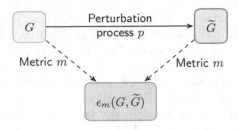

Fig. 2. Framework for evaluating information loss induced by edge modification techniques.

techniques (also called perturbation methods). Secondly, we compute the error between the original (G) and each perturbed network ($\widetilde{G}$) using several measures (defined in Sect. 3.2). Thirdly, we compute the average error over the 10 independent sets. We repeat the same process for all our tested networks (detailed in Sect. 3.1). This framework is depicted in Fig. 2.

3.1 Tested Networks

We use both synthetic and real networks in our experiments. We use software *igraph*[1] to generate two kinds of random graphs.

- Erdös-Rényi Model [12] is a classical random graph model. It defines a random graph as n vertices connected by m edges that are chosen randomly from the $n(n-1)/2$ possible edges. In our experiments, we set n=1,000 and m=5,000. This dataset is denoted as "ER-1000".
- Barabási-Albert Model [3], also called scale-free model, is a network whose degree distribution follows a power-law. That is, for degree d, its probability density function is $P(k) = d^{-\gamma}$. In our experiments, we set the number of vertices to be 1,000 and γ=1, i.e. linear preferential attachment. This dataset is denoted as "BA-1000".

Additionally, four different real networks are used in our experiments. Although all these sets are unlabelled, we have selected these datasets because they have different graph's properties. Table 1 shows a summary of their main features. They are the following ones:

- Zachary's Karate Club [27] is a network widely used in literature. The graph shows the relationships among 34 members of a karate club.
- Jazz musicians [14] is a collaboration graph of jazz musicians and their relationship.
- URV email [15] is the email communication network at the University Rovira i Virgili in Tarragona (Spain). Nodes are users and each edge represents that at least one email has been sent.
- Political blogosphere data (*polblogs*) [1] compiles the data on the links among US political blogs.

---

[1] Available at: http://igraph.org/.

Table 1. Network' properties. For each dataset we present the number of vertices (n), number of edges (m), average degree $(\overline{deg})$, average distance $(\overline{dist})$ and diameter (D).

Dataset	n	m	$\overline{deg}$	$\overline{dist}$	D
ER-1000	1,000	4,969	9.938	3.263	5
BA-1000	1,000	4,985	9.970	2.481	4
Zachary's Karate Club	34	78	4.588	2.408	5
Jazz musicians	198	2,742	27.697	2.235	6
URV email	1,133	5,451	9.622	3.606	8
Polblogs	1,222	16,714	27.31	2.737	8

3.2 Information Loss Measures

According to the authors in [8], we use some structural and spectral measures which are strongly or moderately correlated to clustering specific processes. We claim that choosing these measures our results will be applicable not only to graph's properties, but also to clustering and community detection processes. The first graph structural measure is *average distance* $(\overline{dist})$, which is defined as the average of the distances between each pair of vertices in the graph. *Transitivity* (T) is one type of clustering coefficient, which measures and characterizes the presence of local loops near a vertex. It measures the paths' percentage of length 2 which are also triangles. The above measures evaluate the entire graph as a unique score. We compute the error on these graph metrics as follows:

$$\epsilon_m(G, \widetilde{G}) = |m(G) - m(\widetilde{G}_p)| \tag{1}$$

where m is one of the graph characteristic metrics, G is the original graph and $\widetilde{G}_p$ is the p-percent perturbed graph, where $0 \leq p \leq 25$.

The following metrics evaluate specific structural properties for each vertex of the graph: the first one is *betweenness centrality* (C_B), which measures the fraction of the shortest paths that go through each vertex. This measure indicates the centrality of a vertex based on the flow between other vertices in the graph. A vertex with a high value indicates that this vertex is part of many of the shortest paths in the graph, which will be a key vertex in the graph structure. The second one is *closeness centrality* (C_C) and is defined as the inverse of the average distance to all accessible vertices. Finally, the third one is *degree centrality* (C_D), which evaluates the centrality of each vertex associated with its degree, i.e. the fraction of vertices connected to it. We compute the error on vertex metrics by:

$$\epsilon_m(G, \widetilde{G}) = \sqrt{\frac{1}{n}((g_1 - \widetilde{g}_1)^2 + \ldots + (g_n - \widetilde{g}_n)^2)} \tag{2}$$

where g_i is the value of the metric m for the vertex v_i of G and $\widetilde{g}_i$ is the value of the metric m for the vertex v_i of $\widetilde{G}$.

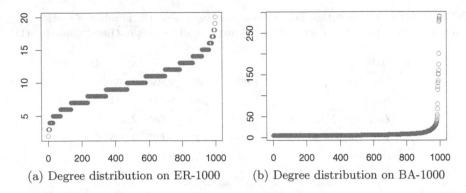

(a) Degree distribution on ER-1000   (b) Degree distribution on BA-1000

Fig. 3. Degree distribution on our synthetic networks. Horizontal axis represent the whole vertex set and vertical axis their degree values.

Moreover, one spectral measure which is closely related to many graph characteristics [25] is used. *The largest eigenvalue of the adjacency matrix A (λ_1) where λ_i are the eigenvalues of A and $\lambda_1 \geq \lambda_2 \geq \ldots \geq \lambda_n$.* The eigenvalues of A encode information about the cycles of a graph as well as its diameter.

4   Experimental Results

In this section we will discuss the results of our four edge modification techniques. Results are presented in Table 2. Each cell indicates the error value for the corresponding measure and method computed by Eqs. 1 and 2. Values are averaged from 10 independent executions. The lower the value, the better the method. Although deviation is undesirable, it is inevitable due to the graph's alteration process.

The first two tested networks are the synthetic ones. As we have commented previously, ER-1000 has been created using Erdös-Rényi model. Its degree histogram does not follow de power-law, as it can be seen in Fig. 3a. Most of the vertices have degree values between 7 and 13, while few have degree values lower than 7 or higher than 13. Edge add/del and Edge switch present the best values on almost all metrics used on our experiments, as we can see in Table 2. Last column corresponds to the cumulative normalized error (ε), which points out that Edge switch achieves the lowest information loss, closely followed by Edge add/del. Both Edge add and Edge del get worse results. On the contrary, the second network, BA-1000, has been constructed by applying scale-free model and its degree distribution follows a power-law. Figure 3b points out clearly a large number of vertices with small degree value and few vertices with very high degree value. It is important to underline the scale difference between this figure and the previous one. In this case, Edge add and Edge switch reach results with the lowest information loss. As in the previous case, the difference between these two methods and the other ones (Edge del and Edge add/del) is considerable.

Table 2. Results for *Edge add* (Add), *Edge Del* (Del), *Edge add/del* (Add/del) and *Edge Switch* (Switch) methods. For each dataset and method, we compare the results obtained on $\overline{dist}$, T, C_B, C_C, C_D and λ_1. Last column corresponds to the cumulative normalized error (ε) for each method on all evaluated metrics.

Network	Method	$\overline{dist}$	T	C_B	C_C	C_D	λ_1	ε
ER-1000	Add	0.1402	0.0012	0.0005	0.0149	0.0016	1.2454	4.407
	Del	0.1833	0.0013	0.0006	0.0197	0.0016	1.2262	5.984
	Add/del	0.0005	0.0002	0.0007	0.0073	0.0015	0.0122	1.077
	Switch	0.0003	0.0001	0.0005	0.0055	0.0010	0.0048	0.020
BA-1000	Add	0.0118	0.0025	0.0005	0.0030	0.0016	0.6507	0.667
	Del	0.1111	0.0038	0.0007	0.0315	0.0034	3.5769	6.000
	Add/del	0.0902	0.0014	0.0016	0.0230	0.0034	2.9250	4.279
	Switch	0.0488	0.0011	0.0005	0.0162	0.0019	1.4601	1.114
Karate	Add	0.1799	0.0060	0.0268	0.0428	0.0270	0.4312	2.772
	Del	0.1393	0.0223	0.0204	0.0696	0.0296	0.6171	4.104
	Add/del	0.0393	0.0166	0.0311	0.0404	0.0331	0.2352	2.730
	Switch	0.0935	0.0291	0.0297	0.0424	0.0233	0.1056	2.365
Jazz	Add	0.2290	0.0486	0.0073	0.0532	0.0199	1.9575	2.814
	Del	0.0653	0.0658	0.0021	0.0940	0.0223	4.7641	3.265
	Add/del	0.1888	0.1115	0.0077	0.0497	0.0179	2.9508	3.817
	Switch	0.1859	0.1129	0.0068	0.0451	0.0111	2.1005	2.622
URV email	Add	0.2142	0.0179	0.0011	0.0193	0.0014	0.5120	1.000
	Del	0.1238	0.0208	0.0007	0.2177	0.0017	2.3656	3.309
	Add/del	0.1028	0.0387	0.0013	0.1587	0.0016	1.9539	3.321
	Switch	0.1319	0.0429	0.0011	0.1481	0.0010	1.3955	2.385
Polblogs	Add	0.1738	0.0114	0.0013	0.1649	0.0031	1.0974	2.000
	Del	0.0569	0.0280	0.0005	0.1502	0.0050	9.0615	3.258
	AddDel	0.1158	0.0389	0.0015	0.1177	0.0045	7.8086	2.934
	Switch	0.1620	0.0459	0.0014	0.0991	0.0025	6.1445	2.531

The first tested real network is Zachary's Karate Club. Although Edge switch achieves the best values, Edge add and Edge add/del get values close to theirs. For instance, we can deepen on behaviour of λ_1 error in Fig. 4a. The $p = 0$ value (x-axis) represents the value of this metric on the original graph. Thus, values close to this point indicate low noise on perturbed data. As we can see, Edge switch remains closer to the original value than the other methods.

Jazz musicians is our second tested real network. The differences among our four methods are smaller using this dataset than the aforementioned experiments. Edge del reaches better results than previous cases and the method which introduces the most information loss is Edge add/del. However, Edge add and Edge switch get slightly lower information loss. For example, we analyse average distance in depth, which usually increases when applying Edge del and

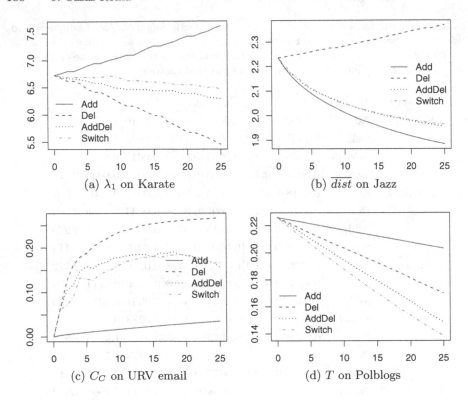

(a) λ_1 on Karate   (b) $\overline{dist}$ on Jazz

(c) C_C on URV email   (d) T on Polblogs

Fig. 4. Examples of the error evolution computed on our experimental framework. Perturbation parameter p varies along the horizontal axis from 0 % (original graph) to 25 %.

decreases when applying Edge add. It is obvious, since removing edges increases paths between vertices and adding new edges decreases paths. Nevertheless, it is interesting to see that perturbation introduced by removing edges is lower than others in this case, as can be seen in Fig. 4b.

Lastly, URV email and Polblogs represent the largest real networks in our experiments. Their structure is similar to BA-1000, since they are both scale-free networks but with parameter $\gamma \approx 0.5$. Results on URV email are similar to ones on BA-1000; Edge add achieves the best results, followed by Edge switch, and again Edge del and Edge add/del get the worst results. We can observe this behaviour in Fig. 4c, where Edge add obtains the lowest error on closeness centrality. The difference is quite important compared to Edge add/del and Edge switch, but even larger compared to Edge del. Similar behaviour can be observed on Polblogs dataset. Edge add achieves the best values, but Edge del and Edge switch also get also good values, close to the ones obtained by Edge add.

Figure 4d depicts transitivity, where all edge modification methods decrease values obtained on original network. As shown, Edge add gets values closer to the original ones on all perturbation percentage. Edge del and Edge switch obtain similar cumulative normalized error on this dataset, suggesting that both introduce similar noise on tested metrics.

As conclusions, we note that Edge switch gets lower information loss when it is applied to networks which do not fulfil the scale-free model, i.e. ER-1000 and Jazz musicians. On the other side, Edge add obtains the lowest information loss when dealing with scale-free networks, such as BA-1000, URV email and Polblogs. Edge switch also achieves good results on scale-free networks. That is not surprising, since Edge switch preserves the degree distribution keeping some related measures close to the original values. On the contrary, Edge del and Edge add/del introduce more perturbation on almost all analysed networks, except Polblogs where Edge del scores the second position and ER-1000 where Edge add/del also succeed to obtain the second position.

5 Conclusions

In this paper we have evaluated the basic edge modification techniques, which are commonly used on privacy-preserving algorithms. We have presented four basic types of edge modification methods, and a framework to assess the behaviour of some graph's properties during perturbation processes induced by these four edge modification methods. Our framework includes some experimental results both on synthetic and real-world networks.

As we have demonstrated, Edge switch better preserves graph's properties on networks with a degree distribution which does not follow the power-law. On the contrary, Edge add is the best method to keep graph's properties when perturbing scale-free networks. Edge del and Edge add/del introduce more noise during perturbation processes on both type of networks.

Many interesting directions for future research have been uncovered by this work. It would be interesting to also consider methods based on noise node addition [11] and information loss measures based on real graph-mining processes, such as clustering or community detection. It would be also very interesting to extend this analysis to other graph's types (directed or labelled graphs, for instance).

Acknowledgements. This work was partly funded by the Spanish MCYT and the FEDER funds under grants TIN2011-27076-C03 "CO-PRIVACY" and TIN2014-57364-C2-2-R "SMARTGLACIS".

References

1. Adamic, L.A., Glance, N.: The political blogosphere and the 2004 U.S. election. In: International Workshop on Link Discovery, pp. 36–43. ACM Press, New York (2005)
2. Backstrom, L., Dwork, C., Kleinberg, J.: Wherefore art thou r3579x? anonymized social networks, hidden patterns, and structural steganography. In: International Conference on World Wide Web, pp. 181–190. ACM, New York (2007)
3. Barabási, A.-L., Albert, R.: Emergence of scaling in random networks. Science **286**(5439), 509–512 (1999)

4. Bonchi, F., Gionis, A., Tassa, T.: Identity obfuscation in graphs through the information theoretic lens. In: International Conference on Data Engineering, pp. 924–935. IEEE, Washington (2011)
5. Bonchi, F., Gionis, A., Tassa, T.: Identity obfuscation in graphs through the information theoretic lens. Inf. Sci. **275**, 232–256 (2014)
6. Casas-Roma, J.: Privacy-preserving on graphs using randomization and edge-relevance. In: Torra, V., Narukawa, Y., Endo, Y. (eds.) MDAI 2014. LNCS, vol. 8825, pp. 204–216. Springer, Heidelberg (2014)
7. Casas-Roma, J., Herrera-Joancomartí, J., Torra, V.: An algorithm for k-Degree anonymity on large networks. In: International Conference on Advances on Social Networks Analysis and Mining, pp. 671–675. IEEE, Niagara Falls (2013)
8. Casas-Roma, J., Herrera-Joancomartí, J., Torra, V.: Anonymizing graphs: measuring quality for clustering. Knowl. Inf. Syst. (2014). (In press)
9. Chester, S., Kapron, B.M., Ramesh, G., Srivastava, G., Thomo, A., Venkatesh, S.: k-anonymization of social networks by vertex addition. In: ADBIS 2011 Research Communications, pp. 107–116, Vienna, Austria (2011). CEUR-WS.org
10. Chester, S., Gaertner, J., Stege, U., Venkatesh, S.: Anonymizing subsets of social networks with degree constrained subgraphs. In: IEEE International Conference on Advances on Social Networks Analysis and Mining, pp. 418–422. IEEE, Washington, USA (2012)
11. Chester, S., Kapron, B.M., Ramesh, G., Srivastava, G., Thomo, A., Venkatesh, S.: Why waldo befriended the dummy? k-anonymization of social networks with pseudo-nodes. Soc. Netw. Anal. Min. **3**(3), 381–399 (2013)
12. Erdös, P., Rényi, A.: On random graphs I. Publicationes Mathematicae **6**, 290–297 (1959)
13. Ferri, F., Grifoni, P., Guzzo, T.: New forms of social and professional digital relationships: the case of Facebook. Soc. Netw. Anal. Min. **2**(2), 121–137 (2011)
14. Gleiser, P.M., Danon, L.: Community structure in Jazz. Adv. Complex Syst. **6**(04), 565–573 (2003)
15. Guimerà, R., Danon, L., Díaz-Guilera, A., Giralt, F., Arenas, A.: Self-similar community structure in a network of human interactions. Phys. Rev. E **68**(065103), 1–4 (2003)
16. Hanhijärvi, S., Garriga, G.C., Puolamäki, K.: Randomization techniques for graphs. In: International Conference on Data Mining, pp. 780–791. SIAM, Sparks (2009)
17. Hay, M., Miklau, G., Jensen, D., Weis, P., Srivastava, S.: Anonymizing social networks, Technical report 07–19, UMass Amherst (2007)
18. Hay, M., Miklau, G., Jensen, D., Towsley, D., Weis, P.: Resisting structural re-identification in anonymized social networks. Proc. VLDB Endowment **1**(1), 102–114 (2008)
19. Kapron, B.M., Srivastava, G., Venkatesh, S.: Social network anonymization via edge addition. In: IEEE International Conference on Advances on Social Networks Analysis and Mining, pp. 155–162. IEEE, Kaohsiung (2011)
20. Liu, K., Terzi, E.: Towards identity anonymization on graphs. In: International Conference on Management of Data, pp. 93–106. ACM, New York (2008)
21. Lu, X., Song, Y., Bressan, S.: Fast identity anonymization on graphs. In: Liddle, S.W., Schewe, K.-D., Tjoa, A.M., Zhou, X. (eds.) DEXA 2012, Part I. LNCS, vol. 7446, pp. 281–295. Springer, Heidelberg (2012)
22. Stokes, K., Torra, V.: Reidentification and k-anonymity: a model for disclosure risk in graphs. Soft Comput. **16**(10), 1657–1670 (2012)

23. Wu, X., Ying, X., Liu, K., Chen, L.: A survey of privacy-preservation of graphs and social networks. In: Aggarwal, C.C., Wang, H. (eds.) Managing and mining graph data, pp. 421–453. Springer, New York (2010)
24. Ying, X., Pan, K., Wu, X., Guo, L.: Comparisons of randomization and k-degree anonymization schemes for privacy preserving social network publishing. In: Workshop on Social Network Mining and Analysis, pp. 10:1–10:10. ACM, New York (2009)
25. Ying, X., Wu, X.: Randomizing social networks: a spectrum preserving approach. In: International Conference on Data Mining, pp. 739–750. SIAM, Atlanta (2008)
26. Yuan, M., Chen, L., Yu, P.S., Yu, T.: Protecting sensitive labels in social network data anonymization. IEEE Trans. Knowl. Data Eng. **25**(3), 633–647 (2013)
27. Zachary, W.W.: An information flow model for conflict and fission in small groups. J. Anthropol. Res. **33**(4), 452–473 (1977)
28. Zhou, B., Pei, J.: Preserving privacy in social networks against neighborhood attacks. In: International Conference on Data Engineering, pp. 506–515. IEEE, Washington (2008)
29. Zou, L., Chen, L., Özsu, M.T.: K-automorphism: a general framework for privacy preserving network publication. Proc. VLDB Endowment **2**(1), 946–957 (2009)

Co-utile Collaborative Anonymization
of Microdata

Jordi Soria-Comas[✉] and Josep Domingo-Ferrer

Dept. of Computer Engineering and Mathematics, UNESCO Chair
in Data Privacy, Universitat Rovira i Virgili, Av. Països Catalans 26,
43007 Tarragona, Catalonia, Spain
{jordi.soria,josep.domingo}@urv.cat

Abstract. In surveys collecting individual data (microdata), each
respondent is usually required to report values for a set of attributes.
If some of these attributes contain sensitive information, the respondent
must trust the collector not to make any inappropriate use of the data
and, in case any data are to be publicly released, to properly anonymize
them to avoid disclosing sensitive information. If the respondent does not
trust the data collector, she may report inaccurately or report nothing
at all. The reduce the need for trust, local anonymization is an alternative whereby each respondent anonymizes her data prior to sending them
to the data collector. However, local anonymization by each respondent
without seeing other respondents' data makes it hard to find a good
trade-off minimizing information loss and disclosure risk. We propose a
distributed anonymization approach where users collaborate to attain an
appropriate level of disclosure protection (and, thus, of information loss).
Under our scheme, the final anonymized data are only as accurate as the
information released by each respondent; hence, no trust needs to be
assumed towards the data collector or any other respondent. Further, if
respondents are interested in forming an accurate data set, the proposed
collaborative anonymization protocols are self-enforcing and co-utile.

Keywords: Information security and privacy · Utility and decision
theory · Co-utility

1   Introduction

A microdata file contains data collected from individual respondents. Because
of the level of detail in the data, they can be useful for a variety of secondary
analyses by third parties other than the data collector. However, releasing the
original data is not feasible because it would lead to a violation of the privacy
of respondents. Statistical disclosure control (SDC), a.k.a. statistical disclosure
limitation, for microdata seeks to produce an anonymized version of the microdata file such that it enables valid statistical analyses but thwarts inference of
confidential information about any specific individual.

© Springer International Publishing Switzerland 2015
V. Torra and Y. Narakawa (Eds.): MDAI 2015, LNAI 9321, pp. 192–206, 2015.
DOI: 10.1007/978-3-319-23240-9_16

The mainstream literature on SDC for microdata (*e.g.* see [7]) focuses on centralized anonymization, which features a trusted data collector. The data collector (*e.g.* National Statistical Institute) gathers original data from the respondents and takes care of anonymizing them. While avoiding the computational burden of anonymization is confortable to respondents, it has the downside that they need to trust the data collector.

Local anonymization is an alternative disclosure limitation paradigm suitable for scenarios where the respondents do not trust (or trust only partially) the data collector. Each respondent anonymizes her own data before handing them to the data collector. In comparison to centralized anonymization, local anonymization usually results in greater information loss. The reason is that each respondent needs to protect her data without seeing the other respondents' data, which makes it difficult for her to find a good trade-off between the disclosure risk limitation achieved and the information loss incurred.

1.1 Contribution and Plan of this Paper

To overcome the limitations of the centralized and the local anonymization paradigms, we propose the notion of *collaborative anonymization*, which is in line with the novel notion of co-utility [4]. Co-utility models the interaction among a set of peers, each one with a selfish goal, in which peers help each other rationally. For collaboration to arise rationally, the best strategy for a peer to reach her goal must be to help another peer in reaching his goal. The advantage of co-utility is that it leads to a system that works smoothly without the need of external enforcement.

The rest of the paper is organized in the following manner. Section 2 provides a brief review of related work. Section 3 lists the requirements of collaborative anonymization and justifies why it is rationally preferable to centralized and local anonymization. Section 4 describes a collaborative anonymization technique that hides each respondent within a group of respondents. Section 5 describes a collaborative anonymization technique that masks the value of the confidential data. Conclusions and future research issues are summarized in Sect. 6. The Appendix gives background on several concepts this papers builds on: k-anonymity (Appendix A), reverse mapping (Appendix B) and co-utility (Appendix C).

2   Related Work

This work seeks to empower each respondent to anonymize her own data while preserving utility as in the centralized paradigm.

Related works exist that consider privacy-conscious data set owners, rather than privacy-conscious respondents. When dealing with privacy-conscious data set owners, one faces a data integration problem where the data owners do not want to share data that are more specific than those in the final anonymized data set to be jointly obtained. In [17] a top-down generalization approach for two owners of vertically partitioned data sets is proposed. Both owners start with the maximum level of generalization, and they iteratively and collaboratively refine the generalization. In [8,9] the same problem is tackled by using cryptographic techniques. In [10] the anonymization of horizontally partitioned data sets is considered. The main difference between the above proposals and our work is that the number of respondents is usually much greater than the number of data set owners (the latter are a small number in most realistic data integration settings). In our case, there is a different respondent for each data record being collected, which makes proposals oriented to a few data set owners unusable.

Among the related works specifically addressing respondent privacy, the local anonymization paradigm is closest to our approach in terms of trust requirements. Several local anonymization methods have been proposed. Many basic SDC techniques such as global recoding, top and bottom coding, and noise addition can be applied locally (check [7] for details on such techniques). There are, however, some techniques specifically designed for local anonymization that, in addition to helping a respondent to hide her response, allow the data collector to get an accurate estimation of the distribution of responses for groups of respondents. In randomized response [18], the respondent flips a coin before answering a sensitive dichotomous question (like "Have you taken drugs this month?"); if the coin comes up tails, the responder answers "yes", otherwise she answers truthfully. This protects the privacy of respondents, because the survey collector cannot determine whether a particular respondent's "yes" is random or truthful; but he knows that the "no" answers are truthful, so that he can estimate the real proportion of "no" as twice as much as the observed proportion of "no" (from which the real proportion of "yes" follows). FRAPP [1] can be seen as a generalization of random response. In FRAPP, the respondent reports the real value with some probability and, otherwise, it returns a random value from a known distribution. In AROMA [15] each respondent hides her confidential data within a set of possible confidential values drawn from some known distribution. In any case, to obtain an accurate result, the output of a query performed on the anonymized data must be adjusted according to the known distribution used to mask the actual data. While some kind of adjustment of the query results may also be needed in the centralized paradigm (e.g. when the generalization used for quasi-identifiers in a k-anonymous data set does not match the query; see Appendix A about quasi-identifiers and k-anonymity), the randomness introduced by local anonymization makes the estimate less accurate than in centralized anonymization.

An advantage of local anonymization, though, is that the respondent is given some capability to decide the amount of anonymization required, which is likely to increase her disposition to provide truthful data (rather than fake data). Yet, most privacy models/techniques give uniform disclosure limitation guarantees to all respondents, which may not suit the different perceptions of disclosure risk of the various respondents. To address this concern, [20] proposed a privacy model in which each individual determines the amount of protection required for her data.

3 Collaborative Anonymization: Requirements and Justification

A problem with centralized anonymization is that, if a respondent does not trust the data collector to properly use and/or anonymize her data, she may decide to provide false data (hence causing a response bias) or no data at all (hence causing a non-response bias). Local (also known as independent) anonymization is an alternative that is not free from problems either. As shown in Appendix B, permutation is essential to anonymization, but the permutation caused by a certain amount of masking depends not only on one's own record but on the values of the records of the other respondents. Hence, for a respondent anonymizing her own record in isolation it is hard to determine the amount of masking that yields a good trade-off between disclosure risk and information loss, *i.e.* that causes enough permutation but not more than enough permutation. A natural tendency is for each respondent to play it safe and overdo the masking just in case, which incurs more information loss than necessary.

To deal with the above shortcomings of centralized and local anonymization, we propose a new paradigm that we call collaborative data anonymization. Consider a set of respondents $R_1, \ldots, R_m$ whose data are to be collected. Each respondent is asked to report information about a set of attributes (some of them containing confidential/sensitive information). Since respondents place limited trust on the data collector, they may refuse to provide the collector with non-anonymized data. A more realistic goal is to generate, in a collaborative and distributed manner, an anonymized data set that satisfies the following two requirements: (i) it incurs no more information loss than the data set that would be obtained with the centralized paradigm for the same privacy level, and (ii) neither the respondents nor the data collector gain more knowledge about the confidential/sensitive attributes of a specific respondent than the knowledge contained in the final anonymized data set.

In general, the motivations for a respondent to contribute her data are not completely clear. A rational respondent will only contribute if the benefit she gets from participating compensates her privacy loss. It is not in our hands to determine what the motivations of the respondents are. However, since our collaborative approach achieves the same data utility as the centralized approach while improving the respondent's privacy vs the data collector, any respondent willing to participate under the centralized approach should be even more willing to participate under our collaborative scheme. More precisely, we can distinguish several types of respondents depending on their interests in the collected data and in their own privacy:

- A respondent without any interest in the collected data set is better off by declining to contribute.
- A respondent who is interested in the collected data and has no privacy concerns can directly supply her data and needs no anonymization (neither local, nor centralized nor collaborative).
- A respondent who is interested in the collected data but has privacy concerns will prefer the collaborative approach to the centralized and the local

approaches. Indeed, the collaborative approach outperforms the centralized approach in that the former offers privacy vs the data collector. Also, the collaborative approach outperforms the local approach in that it yields a collected anonymized data set with less information loss, that is, with higher utility.

Remark (co-utile anonymization). Note that the level of privacy protection obtained by a respondent affects the privacy protection that other respondents get. A basic approach for preserving the privacy of a specific respondent is based on hiding that respondent within a group of respondents. None of the respondents in such a group is interested making any of the respondents in the group re-identifiable, because that makes her own data more easily re-identifiable. For example, if one record in a k-anonymous group is re-identified, the probability of successful re-identification for the other group members increases from $1/k$ to $1/(k-1)$. This fact suggests that a respondent is interested not only in protecting her privacy, but also in helping other respondents in preserving theirs. This is the fundamental principle behind the notion of co-utility (see Appendix C): the best strategy to attain one's goal is to help others in attaining theirs. The fact that privacy protection turns out to be co-utile ensures that respondents will be willing to collaborate with each other to improve the protection of all the group.

4   Collaborative k-Anonymity

This section describes how to generate a k-anonymous data set in a distributed manner, such that none of the respondents releases more information than the one available on her in the final k-anonymous data set. To this end, some communication between the respondents is needed to determine the k-anonymous groups.

In general, there can be several combinations of attributes in a data set that together act like a quasi-identifier, that is, such that each combination of attributes can be used to re-identify respondents; for example, one might have a quasi-identifier *(Age, Gender, Birthplace)* and another quasi-identifer *(Instruction_level, City_of_residence, Nationality)*. Without loss of generality and for the sake of simplicity, we will assume there is a single quasi-identifier that contains all the attributes that can potentially be used in record re-identification. Note that this is the worst-case scenario. Let QI be the set of attributes in this quasi-identifier.

Quasi-identifier attributes are usually assumed to contain no confidential information, that is, the set of quasi-identifier attributes is assumed to be disjoint from the set of confidential/sensitive attributes. This assumption is reasonable, because it is equivalent to saying that the attacker's background information does not include sensitive information on any respondent (indeed, the attacker wants *to learn* sensitive information, so it is reasonable to assume that he does not yet know it). Certainly, there might be special cases in which the attacker knows and uses sensitive data for re-identification, but we will stick to the usual setting in which this does not happen.

Since the attributes in QI are non-confidential, respondents can share their values among themselves and with the data collector, so that all of them get the complete list of QI attribute values. Based on that list, the data collector or any respondent can generate the k-anonymous groups. We propose to delegate the generation of the k-anonymous groups to the data collector. There are two main reasons for this:

- *Utility.* The actual k-anonymous partition chosen may have an important impact over analyses that can be accurately performed on the k-anonymous data. The data collector is probably the one who knows best (even if often only partially) the intended use of the data and, thus, the one who can make the most appropriate partition in k-anonymous groups.
- *Performance.* Generating the k-anonymous groups is the most computationally intensive part of k-anonymity enforcement. Hence, by delegating this task to the data collector, respondents relieve themselves from this burden.

When respondents have some interest in using the anonymized data set, it is plausible to assume that any respondent will rationally collaborate to generate it. The level of protection that a respondent in a given k-anonymous group gets is dependent on the level of protection that the other respondents in the group get: as justified above in Sect. 3, k-anonymization is co-utile.

On the other side, the data collector may try to deviate from the algorithm. Because the generation of the k-anonymous partition has been delegated to the data collector, respondents must make sure before reporting confidential information that the partition computed and returned by the data collector satisfies the requirements of k-anonymity. That is, each respondent must check that her k-anonymous group comprises k or more respondents.

After verifying the partition returned by the data collector, the respondent uploads to the data collector the quasi-identifier attribute values of her k-anonymous group together with her confidential data. This communication must be done through an anonymous channel (*e.g.* Tor [2]) to prevent anyone (the data collector, an intruder or anyone else) from tracking the confidential data to any respondent.

The above described steps to collaboratively generate a k-anonymous data set are formalized in Protocol 1.

Protocol 1

1. Let $R_1, \ldots, R_m$ be the set of respondents. Let (qi_i, c_i) be the quasi-identifier and confidential attribute values of R_i, for $i = 1, \cdots, m$.
2. Each R_i uploads her qi_i to a central data store so that anyone can query for qi_i.
3. The data collector generates a k-anonymous partition $\{P_1, \ldots, P_p\}$ and uploads it to the central data store.
4. Each R_i checks that her k-anonymous group $P_{g(R_i)}$ contains k or more of the original quasi-identifiers.
   If that is not the case, R_i refuses to provide any confidential data and exits the protocol.

5. Each R_i sends $(P_{g(R_i)}, c_i)$ to the data collector through an anonymous channel.
6. With the confidential data collected, the data collector generates the k-anonymous data set.

Protocol 1 is compatible with any strategy to generate the k-anonymous partition. Possible strategies include:

- Methods reducing the detail of the quasi-identifier attributes. Options here are generalization and supression [11,12,14], or microaggregation [5].
- Methods breaking the connection between quasi-identifier attributes and confidential attributes. Among these we have Anatomy [19] (that splits the data into two tables, one containing the original quasi-identifier values and the other the original confidential attribute values, with both tables being connected through a group identifier attribute) and probabilistic k-anonymity [16] (that seeks to break the relation between quasi-identifiers and confidential attributes by means of a within-group permutation).

In fact, since the data collector and the respondents all know the exact values of the quasi-identifiers and the confidential attributes in each k-anonymous group, each of them can generate the k-anonymous data that suits her best.

In essence, the proposed protocol offers the same privacy protection as local anonymization (confidential data are only provided by the respondents in an anonymized form) while maintaining the data utility of centralized k-anonymization. At the respondents' side, there are only some minor additional communication and integrity checking costs.

We illustrate the steps of Protocol 1 for the respondents listed in the leftmost table of Fig. 1. In Step 2 each respondent uploads her quasi-identifiers. The uploaded data are shown in the center-left table of Fig. 1. At Step 3 the data collector analyzes the data uploaded in Step 2 and generates the partition in k-anonymous groups (for $k = 4$); that is, for each R_i, the data collector fixes the value of $g(R_i)$, the group assigned to R_i). This partition is shown in the center-right table of Fig. 1. For clarity, the records have been arranged in a way that the k-anonymous group P_1 contains the first $k = 4$ records and group P_2 contains the last $k = 4$ records; that is, $P_{g(R_1)} = P_{g(R_2)} = P_{g(R_3)} = P_{g(R_4)} = P_1$ and $P_{g(R_5)} = P_{g(R_6)} = P_{g(R_7)} = P_{g(R_8)} = P_2$. In Step 4 each respondent checks that her group contains k or more of the quasi-identifier values uploaded in Step 2. Since this condition holds for all respondents in the example of the figure, respondents proceed to Step 5. In Step 5 each respondent uploads, through an anonymous channel, the group identifier she has been assigned, $P_{g(R_i)}$, together with her value for the confidential/sensitive attribute. The result is shown in the rightmost table of Fig. 1. Here the layout of the rightmost table can be misleading: although we list in the i-th row the salary of R_i for $i = 1, \cdots 4$, any permutation of the four salaries could be listed (all four salaries in the P_1 group are indistinguishable). A similar comment holds for rows 5–8, in which we could list any permutation of the salaries in the P_2 group. At this point, the data collector (and the respondents) can generate the k-anonymous data set using the method they like best using that they see all tables in Fig. 1 except the leftmost one.

	QI		Sensitive		Step 2			Step 3			Step 5	
	Zip	Age	Salary		Zip	Age		Zip	Age			Salary
R_1	13053	28	35000		13053	28		13053	28		P_1	35000
R_2	13068	29	30000		13068	29	P_1	13068	29		P_1	30000
R_3	13068	21	20000		13068	21		13068	21		P_1	20000
R_4	13053	23	27000		13053	23		13053	23		P_1	27000
R_5	14853	50	40000		14853	50		14853	50		P_2	40000
R_6	14853	55	43000		14853	55	P_2	14853	55		P_2	43000
R_7	14850	47	48000		14850	47		14850	47		P_2	48000
R_8	14850	49	45000		14850	49		14850	49		P_2	45000

Fig. 1. Distributed collaborative k-anonymization. Step numbers refer to Protocol 1. Each row in the leftmost table is only seen by the corresponding respondent. The other three tables are entirely seen by all respondents and the data collector.

Distributed anonymization based on hiding in a group via manipulation of the quasi-identifiers has an important flaw. An attacker may try to simulate one or more respondents, in order to gain more insight into the k-anonymous groups. To thwart this kind of attack, we need to make sure that every respondent has a verified identity, possibly by having all respondents registered with some trusted authority. If that is not feasible, some mitigation measures can be put in place to make it more difficult for an attacker to adaptively fabricate quasi-identifier values similar to those of a target respondent in order to track her:

- One option is for the data store manager (maybe the data collector) to unlock the access to the quasi-identifiers list (of Step 2) only after every respondent has uploaded her quasi-identifiers. In this way, the attacker must generate his quasi-identifier values without knowing the quasi-identifier values of the other respondents. This option has the shortcoming that respondents need to trust the data store manager to perform the above access control.
- An alternative that does not require trust in any central entity is to have each respondent upload a commitment (in the cryptographic sense, [6]) to her quasi-identifiers before any actual quasi-identifier is uploaded. In this way, each respondent can check that none of the uploaded values was forged to target a specific respondent.

In the following section, we explore distributed anonymization based on masking the confidential attributes, rather than on hiding in a group via quasi-identifier manipulation.

5   Collaborative Masking of Confidential Data

Although k-anonymity is a popular privacy model, it has some important limitations. First of all, attribute disclosure is possible, even without re-identification, if the variability of the confidential attribute(s) within a k-anonymous group is small. Also, k-anonymity assumes that confidential attributes are not used in re-identification (*i.e.* that no confidential attribute is also a quasi-identifier), but this may not be the case if the attacker knows some confidential data. Moreover, we mentioned in the previous section that in our distributed generation of the k-anonymous data set, an attacker might simulate respondents to gain more insight into the k-anonymous groups. To deal with these issues, this section takes a different approach to generate the anonymized data set: instead of hiding within a group of respondents, each respondent masks her confidential data.

In this section, we relax the assumption that the set of quasi-identifier attributes and the set of confidential attributes are disjoint. The only assumption we make is that releasing the marginal distribution of a confidential attribute is not disclosive. What needs to be masked is the relation between a confidential attribute and any other attribute. Thus, we consider a data set with attributes $(A, C_1, \ldots, C_d)$ where C_j are confidential attributes for $j = 1, \cdots, d$ and A groups all non-confidential attributes.

Since the marginal distribution of confidential attributes is not disclosive, respondents can share the contents of each confidential attribute among themselves and with the data collector, so that all of them get the complete list of values for each confidential attribute. In this way, each respondent can evaluate the sensitivity of her value for each confidential attribute by taking into account the values of the other respondents for that attribute. From this sensitivity evaluation, the respondent can make a more informed decision regarding the amount of masking she needs to use.

Thus, we assume that each respondent R_i makes a decision about the amount of masking required for her confidential data and reports to the data collector the tuple $(a_i, c'_{1i} \ldots, c'_{di})$, where a_i is the original value of the non-confidential attributes and c'_{1i} the masked value of confidential attribute C_i. The fact that each respondent freely and informedly decides on the amount of masking required for her confidential data is a strong privacy guarantee (the respondent can enforce the level of permutation she wishes with respect to the original values). In fact, even if the data collector or any other entity recommend a specific amount of masking, respondents are free to ignore this recommendation. For a rational respondent, the selected level of masking is based on both privacy and utility considerations.

The reported masked data can be directly used to generate the masked data set. Better yet, by applying reverse mapping (see Appendix B for background on reverse mapping), the original marginal distribution of each confidential attribute can be recovered. This reverse mapping can be performed by the data collector and also by each respondent (because all respondents know the marginal distribution of the original attributes).

The previous discussion is formalized in Protocol 2.

Protocol 2

1. *Let $R_1, \ldots, R_m$ be the set of respondents. Let $(a_i, c_{i1} \ldots, c_{id})$ be the attribute values of R_i.*
2. *For each confidential attribute C_j, each respondent R_i uploads c_{ij} to a central data store through an anonymous channel.*

3. For each confidential attribute C_j, each respondent R_i analyzes all attribute values and decides on the amount of masking required for c_{ij}. Let c'_{ij} be the masked value.

4. Each respondent R_i uploads $(a_i, c'_{i1} \ldots, c'_{id})$ to the data store.

5. The data collector applies reverse mapping to the data uploaded in Step 4 in order to obtain the final anonymized data set. (The same can be done by each respondent.)

Although the reasons why Protocol 2 is safe have already been presented in the discussion prior to the algorithm formalization, a more systematic analysis is presented in the following proposition.

Proposition 1. *At the end of Protocol 2, nobody learns information about any respondent R_i that is more accurate than the masked data reported by R_i in Step 4.*

Proof. Apart from the release of the masked data in Step 4, the only step in which R_i releases data is Step 2. Since the data released in Step 2 are not anonymized, we need to make sure they cannot be linked back to R_i.

Because the uploads in Step 2 are performed through an anonymous channel, there is no way for an attacker to track the data transfers to any particular respondent. What is more, since each c_{ij} is separately uploaded through the anonymous channel, there is no way for the attacker to link to one another the values $c_{ij}, j = 1, \cdots, d$ corresponding to the same respondent R_i (if the attacker could link such values, he could reconstruct the original record of R_i).

Finally, since by assumption releasing the marginal distribution of each confidential attribute is not disclosive, there is no risk in uploading each c_{ij} in Step 2. The reason is that each c_{ij} carries less information than the marginal distribution of attribute C_j. (The release of a c_{ij} could be problematic if C_j contains confidential information and, at the same time, can be used in re-identification, but assuming that the marginals are not disclosive rules out this situation). □

We illustrate the steps of Algorithm 2 for the respondents listed in the leftmost table of of Fig. 2. We assume that Age and Salary are the confidential attributes. In Step 2 each respondent uploads to the central data store each of her values for the confidential attributes. Each respondent performs a separate upload through an anonymous channel for each of the confidential attributes. At the end of Step 2, the marginal distribution of the confidential attributes is available to the data collector and all respondents in the central data store, as illustrated in the center-left table of Fig. 2. In Step 3 each respondent can analyze the marginal distributions and decide on the amount of masking required for each confidential attribute. In this example Age is masked by adding a random value between −5 and 5, and Salary is masked by adding a random value between −5000 and 5000. Of course, each respondent could have applied a different masking. In Step 4 each respondent uploads the masked confidential attributes together with the rest of attributes (the non-confidential ones). This upload need not be done through an anonymous channel, because all confidential data are masked. The data set uploaded by respondents to the central data store at the end of Step 4 is shown in the center-right table of Fig. 2. In the final step, the data collector applies reverse mapping to each confidential attribute to recover the original marginal distributions, as illustrated in the rightmost table of Fig. 2.

		Sensitive		Step 2		Step 4			Step 5	
	Zip	Age	Salary	Age	Salary	Zip	Age	Salary	Zip	Age	Salary
R_1	13053	28	35000	28	35000	13053	29	37306	13053	29	40000
R_2	13068	29	30000	29	30000	13068	24	27765	13068	28	27000
R_3	13068	21	20000	21	20000	13068	18	18951	13068	21	20000
R_4	13053	23	27000	23	27000	13053	19	28151	13053	23	30000
R_5	14853	50	40000	50	40000	14853	51	36879	14853	50	35000
R_6	14853	55	43000	55	43000	14853	50	42631	14853	49	45000
R_7	14850	47	48000	47	48000	14850	52	45585	14850	55	48000
R_8	14850	49	45000	49	45000	14850	49	40390	14850	47	43000

Fig. 2. Distributed collaborative masking of the confidential attributes. Step numbers refer to Protocol 2. Each row in the leftmost table is only seen by the correspoding respondent. The other three tables are entirely seen by all respondents and the data collector.

6   Conclusions and Future Research

We have sketched two protocols for collaborative microdata anonymization. The first one assumes a clear separation between confidential attributes and quasi-identifiers, and seeks to attain k-anonymity. In the second one, no separation between quasi-identifiers and confidential attributes is assumed, and the goal is to sufficiently mask the confidential attributes. The main difference between both methods lies on the attributes that are masked to preserve privacy. On the one hand, collaborative k-anonymity masks the quasi-identifiers, thereby thwarting exact re-identification based on them. In this way, it should be preferred when we want to keep the values of the confidential attributes unmodified and limiting the probability of re-identification based on a preselected set of quasi-identifiers is viewed as sufficient protection. On the other hand, collaborative masking of confidential attributes only masks the values of the confidential attributes; hence, even if re-identification happens, the attacker is uncertain about the value of the confidential attributes. Thus, this protocol should be preferred when respondents are not comfortable with releasing fully accurate confidential data. For instance, this could be the case if the set of quasi-identifiers is not clear, a situation that occurs when intruders may know some confidential pieces of information (so that every attribute might be a quasi-identifier).

Compared to local anonymization, collaborative anonymization incurs less information loss and achieves the same privacy vs the data collector. Compared to centralized anonymization, collaborative anonymization requires less trust in the data collector and achieves the same data utility. Therefore, collaborative anonymization should be preferred by rational respondents to both local and centralized anonymization.

In a survey, the motivations for respondents to report data and report them truthfully to the data collector are in general unclear. As a rule, a rational respondent is willing to participate only if the benefit she obtains is greater than the potential harm due to privacy loss. If respondents are interested in the collected data set and they wish it to be as accurate as possible, then collaborative anonymization protocols are co-utile.

Future work will be devoted to develop collaborative anonymization protocols for a broader range of privacy models (beyond k-anonymity) and disclosure limitation techniques.

Acknowledgments and Disclaimer. The following funding sources are gratefully acknowledged: Templeton World Charity Foundation (grant TWCF0095/AB60 "CO-UTILITY"), Government of Catalonia (ICREA Acadèmia Prize to the second author and grant 2014 SGR 537), Spanish Government (project TIN2011-27076-C03-01 "CO-PRIVACY"), European Commission (projects FP7 "DwB", FP7 "Inter-Trust" and H2020 "CLARUS"). The second author leads the UNESCO Chair in Data Privacy. The views in this paper are the authors' own and do not necessarily reflect the views of the Templeton World Charity Foundation or UNESCO.

A k-Anonymity

k-Anonymity [14] is a privacy model that seeks to thwart re-identification of anonymized records. Central to k-anonymity is the notion of quasi-identifier attributes, also known as key attributes. Quasi-identifiers are attributes that, when considered separately, do not identify the respondent behind a record, but, which used in combination may allow an attacker to uniquely link that record to an external database containing identifiers (this database is the attacker's background knowledge). Such a unique linkage is called re-identification.

With the above setting in mind, k-anonymity can be defined as follows.

Definition 1 (k-Anonymity). *A protected data set is said to satisfy k-anonymity for $k > 1$ if, for each combination of values of quasi-identifier attributes, at least k records exist in the data set sharing that combination.*

If the quasi-identifiers considered by the data protector to enforce k-anonymity coincide with the quasi-identifiers that an attacker can use to link with his background knowledge, then k-anonymity reduces the probability of successful re-identification to $1/k$.

Of course, which attributes should be labeled as quasi-identifiers is debatable. At the very least, attributes that can be found in a public non-de-identified data sets (*e.g.* electoral rolls, phonebooks, etc.) must be taken as quasi-identifiers. However, this is not enough to prevent re-identification by attackers with additional knowledge.

B   Reverse Mapping

Reverse mapping [3, 13] is a post-masking technique that can be applied to any anonymized data set. The result is a *reverse-mapped* data set, constructed by taking each attribute of the anonymized data set at a time, and replacing the value of each record by the value in the original data set with equal rank.

Thus, reverse mapping requires knowing the marginal distribution of each of the attributes in the original data set. Hence, if the data collector wants to allow reverse mapping by parties other than himself, he must release those marginal distributions. And, for those distributions to be releasable, they must be assumed to be non-disclosive. The good news is that this is quite a reasonable assumption, as the distribution of an attribute essentially conveys statistical information (it is, in principle, unrelated to any specific individual). For the extreme cases in which a single value can be associated to a specific individual (*e.g.* the turnover of the largest company in a specific sector), prior masking of the marginal distribution would be needed (*e.g.* by top coding it).

The interesting point about the reverse mapping transformation is that it allows viewing any microdata anonymization method as being *functionally equivalent to permutation* (mapping the original data set to the reverse-mapped data set) *plus a small amount of noise* (mapping the reverse-mapped data set to the anonymized data set). The noise is necessarily small because it does not modify the ranks of the values: by construction, ranks in the reverse-mapped and the anonymized data set are the same. Therefore, *the essential anonymization principle turns out to be permutation.*

C   Co-utility

Consider a set of self-interested peers (having each a utility function, that is, a specific goal or a defined preference relation between a set of possible outcomes) that act strategically (each peer acts to seek an outcome that maximizes her utility, according to her knowledge of the environment).

Co-utility [4] models a kind of interaction between the peers in which it is in the best interest of each of them to help another peer in reaching her goal. The primary advantage of a co-utile system is that it does not require any external mechanism to enforce a particular outcome or coordinate the actions of the peers.

Co-utility can be formalized using game theory. To guarantee a specific interaction outcome without external enforcement, the outcome must be self-enforcing; in game-theoretic terms, it must be an equilibrium. An outcome is an equilibrium if no agent (peer) has incentives to change her strategy in that outcome; in other words, provided that all other agents keep their strategies unchanged, no agent can increase her utility by modifying her strategy.

If the utility of an outcome for an agent depended on the preferences of another agent, attaining an equilibrium would require each agent to report her preferences. On one side, this would increase the complexity of the system, because an agent may report untruthful preferences if she believes that doing so is going to yield a better outcome for her. On the other side, gathering all agents' preferences by a specific party or agent should be avoided if we want a truly distributed interaction. Following the above rationale, we define games amenable to co-utility as those in which the utility of each agent is independent of the preferences of other agents.

Definition 2 (Co-utility amenable game). *Let G be a sequential Bayesian game for n agents. We say that G is a co-utility-amenable game if the utility of any agent is independent of the types of the other agents, i.e., $\forall i, j$, with $i \neq j$ and $\forall t_j, t_j' \in T_j$, we have that $u_i(s_1, \ldots, s_j, \ldots, s_n, t_1, \ldots, t_j, \ldots, t_n) = u_i(s_1, \ldots, s_j, \ldots, s_n, t_1, \ldots, t_j', \ldots, t_n)$.*

Having defined a co-utility amenable game, we are ready to define when a protocol P that produces as output a strategy profile of the game is co-utile. An agent can be reluctant to play a strategy that is beneficial to herself if the strategy provides a much larger benefit to another agent. Because of that, different levels of co-utility can be distinguished, depending on whether agents maximize or just increase their utility by following the protocol. In *strict co-utility* each agent maximizes her utility and, thus, there is no reason for any agent to not to follow the protocol.

Definition 3 (Strict co-utility). *Let G be a co-utility amenable game for n agents. Let P be a self-enforcing protocol for G. We say P is a strictly co-utile protocol if $\forall i \in \{1, \ldots, n\}$, and $\forall s_1' \in S_1, \ldots, s_n' \in S_n$ and $\forall t_1 \in T_1, \ldots, t_n \in T_n$, we have that $u_i(s_1, \ldots, s_n, t_1, \ldots, t_n) \geq u_i(s_1', \ldots, s_n', t_1, \ldots, t_n)$, where the outcome of P is $(s_1, \ldots, s_n)$.*

Designing co-utile protocols is usually a matter of finding a group of peers with a sufficiently aligned set of preferences. As the focus of this paper is on data anonymization, we consider a set of privacy-conscious peers that are required to report some data, for instance, to answer a certain survey. Now, a rational privacy-conscious peer will report false data or report no data at all *unless she has some interest in the pooled responses by all peer to be as accurate as possible*. Hence, we will make the assumption that all peers are interested in obtaining an accurate data set.

Under the above assumption, one possible approach to designing a co-utile protocol can be based on each peer hiding within a group of peers when reporting her record. Note that hiding one's identity when reporting one's record helps other peers in the group to hide their own identities. Conversely, it is hard to hide in a group where none of the other members is anonymous.

References

1. Agrawal, S., Haritsa, J.R.: A framework for high-accuracy privacy-preserving mining. In: Proceedings of the 21st International Conference on Data Engineering (ICDE 2005), pp. 193–204. IEEE (2005)
2. Dingledine, R., Mathewson, N., Syverson, P.: Tor: The second-generation onion router. Naval Research Lab, Washington DC (2004)
3. Domingo-Ferrer, J., Muralidhar, K.: New directions in anonymization: permutation paradigm, verifiability by subjects and intruders, transparency to users. CoRR, abs/1501.04186 (2015)
4. Domingo-Ferrer, J., Soria-Comas, J., Ciobotaru, O.: Co-utility: self-enforcing protocols without coordination mechanisms. In: Proceeding of the 5th International Conference on Industrial Engineering and Operations Management (IEOM 2015), pp. 1–7. IEEE (2015)

5. Domingo-Ferrer, J., Torra, V.: Ordinal, continuous and heterogeneous k-anonymity through microaggregation. Data Min. Knowl. Discov. **11**(2), 195–212 (2005)
6. Goldreich, O.: Foundations of Cryptography. Basic Tools, vol. 1. Cambridge University Press, Cambridge (2001)
7. Hundepool, A., Domingo-Ferrer, J., Franconi, L., Giessing, S., Nordholt, E.S., Spicer, K., de Wolf, P.-P.: Statistical Disclosure Control. Wiley, Chichester (2012)
8. Jiang, W., Clifton, C.: Privacy-preserving distributed *k*-anonymity. In: Jajodia, S., Wijesekera, D. (eds.) Data and Applications Security 2005. LNCS, vol. 3654, pp. 166–177. Springer, Heidelberg (2005)
9. Jiang, W., Clifton, C.: A secure distributed framework for achieving k-anonymity. VLDB J. **15**(4), 316–333 (2006)
10. Jurczyk, P., Xiong, L.: Distributed anonymization: achieving privacy for both data subjects and data providers. In: Gudes, E., Vaidya, J. (eds.) Data and Applications Security XXIII. LNCS, vol. 5645, pp. 191–207. Springer, Heidelberg (2009)
11. LeFevre, K., DeWitt, D.J., Ramakrishnan, R.: Incognito: efficient full-domain k-anonymity. In: Proceedings of the 2005 ACM SIGMOD International Conference on Management of Data (SIGMOD 2005), pp. 49–60. ACM, New York (2005)
12. LeFevre, K., DeWitt, D.J., Ramakrishnan, R.: Mondrian multidimensional k-anonymity. In: Proceedings of the 22Nd International Conference on Data Engineering (ICDE 2006). IEEE Computer Society, Washington, DC (2006)
13. Muralidhar, K., Sarathy, R., Domingo-Ferrer, J.: Reverse mapping to preserve the marginal distributions of attributes in masked microdata. In: Domingo-Ferrer, J. (ed.) PSD 2014. LNCS, vol. 8744, pp. 105–116. Springer, Heidelberg (2014)
14. Samarati, P., Sweeney, L.: Generalizing data to provide anonymity when disclosing information. In: Proceedings of the 17th ACM SIGACT-SIGMOD-SIGART Symposium on Principles of Database Systems (PODS 1998), p. 188. ACM (1998)
15. Song, C., Ge, T.: Aroma: a new data protection method with differential privacy and accurate query answering. In: Proceedings of the 23rd ACM International Conference on Conference on Information and Knowledge Management (CIKM 2014), pp. 1569–1578. ACM, New York (2014)
16. Soria-Comas, J., Domingo-Ferrer, J.: Probabilistic k-anonymity through microaggregation and data swapping. In: Proceedings of the IEEE International Conference on Fuzzy Systems (FUZZ-IEEE 2012), pp. 1–8. IEEE (2012)
17. Wang, K., Fung, B.C.M., Dong, G.: Integrating private databases for data analysis. In: Kantor, P., Muresan, G., Roberts, F., Zeng, D.D., Wang, F.-Y., Chen, H., Merkle, R.C. (eds.) ISI 2005. LNCS, vol. 3495, pp. 171–182. Springer, Heidelberg (2005)
18. Warner, S.L.: Randomized response: a survey technique for eliminating evasive answer bias. J. Am. Stat. Assoc. **60**(309), 63–69 (1965)
19. Xiao, X., Tao, Y.: Anatomy: simple and effective privacy preservation. In: Proceedings of the 32nd International Conference on Very Large Data Bases (VLDB 2006), pp. 139–150 (2006)
20. Xiao, X., Tao, Y.: Personalized privacy preservation. In: Proceedings of the 2006 ACM SIGMOD International Conference on Management of Data (SIGMOD 2006), New York, NY, USA, pp. 229–240 (2006)

Generalization-Based k-Anonymization

Eva Armengol[1] and Vicenç Torra[2](✉)

[1] CSIC - Spanish Council for Scientific Research, IIIA - Artificial Intelligence
Research Institute, Campus UAB, 08193 Bellaterra, Catalonia, Spain
eva@iiia.csic.es
[2] University of Skövde, Skövde, Sweden
vtorra@his.se

Abstract. Microaggregation is an anonymization technique consisting
on partitioning the data into clusters no smaller than k elements and
then replacing the whole cluster by its prototypical representant. Most
of microaggregation techniques work on numerical attributes. However,
many data sets are described by heterogeneous types of data, i.e., numerical and categorical attributes. In this paper we propose a new microaggregation method for achieving a compliant k-anonymous masked file
for categorical microdata based on generalization. The goal is to build
a generalized description satisfied by at least k domain objects and to
replace these domain objects by the description. The way to construct
that generalization is similar that the one used in growing decision trees.
Records that cannot be generalized satisfactorily are discarded, therefore
some information is lost. In the experiments we performed we prove that
the new approach gives good results.

Keywords: k-anonymity · Generalization

1 Introduction

Data privacy is a key issue when data bases are published to disseminate information: it is important to protect the individuals and entities and, at the same
time, the data base has to be useful to extract representative information from
it. For this reason, the research on protection methods becomes of capital importance.

Masking methods form a family of methods that given a data base modify
it previous to its release so that published information is similar to the original
one but not equal. In this way, disclosure is more difficult but data is still useful
for analysis (i.e., information loss is low). To evaluate how good is a protection
method, there are two commonly used measures: information loss and disclosure
risk. As both measures are in contradiction, it is necessary to achieve a trade-off
between the two aspects. Different methods have been developed which differ on
how data is modified. They have different performance with respect to the level
of disclosure risk and information loss they achieve. See e.g. [6,10] for details on
data privacy and masking methods.

© Springer International Publishing Switzerland 2015
V. Torra and Y. Narakawa (Eds.): MDAI 2015, LNAI 9321, pp. 207–218, 2015.
DOI: 10.1007/978-3-319-23240-9_17

There are three general categories of protection methods: perturbative, non-perturbative and synthetic data generators. Perturbative methods perform some distortion of the original data by adding some error. Microaggregation, rank swapping and noise addition are examples of perturbative methods. Non-perturbative methods do not rely on distortion of the original data but on partial suppressions or reductions of detail. These methods include different algorithms based on generalization and suppression. Finally, synthetic data generators methods generate synthetic data that preserve some desired characteristics of the original data. In this paper we focus on *masking* methods and, more particularly, we introduce a non-perturbative method based on generalization.

Different definitions exist for assessing disclosure risk or for establishing when a data set can be released without compromising sensitive information. Identity and attribute disclosure are the two main types of disclosure. Record linkage has been used extensively for identity disclosure risk assessment [19,21]. See [1] for a study of the worst-case scenario using record linkage. Differential privacy [7] and k-anonymity, [17] are two definitions to establish a suitable level of privacy, the former focusing on attribute disclosure and the latter on identity disclosure. k-Anonimity is satisfied when for each records there are other $k - 1$ identical records, which avoids re-identification.

Files compliant with k-anonymity can be constructed by means of generalization, supression and microaggregation. See e.g. Mondrian and Incognito [12,13] as methods to achieve k-anonymity based on generalization. On the contrary, [5] is about the use of microaggregation for achieving k-anonymity.

In order that these methods lead to a file compliant with k-anonymity, they have to be applied to all the variables available to the intruder. When applied to subsets of these variables, k-anonymity is not ensured. In this type of situations disclosure risk assessment needs to be evaluated. Some microaggregation algorithms and Mondrian [15] have been used in this way.

Concerning the data included in the data bases, objects are represented by attributes. Different types of attributes have been considered in the literature on masking methods. For example, to name a few, there are methods for numerical, categorical (ordinal and nominal), time series, dates, text. In this work we focus on the case of categorical attributes.

Mondrian and Incognito [12,13], mentioned above, are examples of algorithms available for categorical data. [15] propose an algorithm to protect categorical attributes using clustering techniques. The original data are used to create clusters and then each cluster is protected independently. Li and Sankar [14] propose a protection method in two steps: first a linear programming formulation is applied in order to preserve the first-order marginal distribution, and then a simple Bayes-based swapping procedure assures the preservation of the joint distribution. Guo and Wu [8] investigate whether data mining or statistical analysis tasks can be conducted on randomized data when distortion parameters are not disclosed to data miners. There are also some approaches focusing on categorical attributes that use generalization. Thus, Wang et al. [20] explore the data generalization concept as a way to hide detailed information. Once the data is masked, standard data mining techniques can be applied without modification.

Some other authors, as Samarati and Sweeney [17] use ontologies of concepts that allow the generalization of the values of an attribute.

In this paper we propose a new method for achieving a compliant k-anonymous masked file for categorical microdata. The approach is based on generalization, building generalizations that accommodate k records and thus achieving k-anonymity. At the same time records that cannot be generalized satisfactorily are discarded. Compared with other methods in the literature, our approach is able to deal with missing values, and we do not need to start with an ontology of generalizations. In addition, our method is evaluated satisfactorily with respect to the performance of classifiers built from the protected data set. Note that classifiers are standard tools in machine learning to build models of the data.

The structure of the paper is as follows. In Sect. 2 we introduce the notation we use. In Sect. 3 the algorithm we propose for k-anonymization is explained in detail. In Sect. 4 we present the experiments we carried out on the Adult data set from the UCI Machine Learning Repository. The paper finishes with some conclusions.

2 Preliminaries

In this work we follow the notation and approach of common literature on microdata protection. Concerning the attributes, they can be divided into three classes:

- *Identifiers* are the attributes that unambiguously identify a single individual or entity (for instance, the passport number), so they are usually removed or encrypted.
- *Quasi-identifiers* attributes that are those that identify an individual with some degree of ambiguity, but a combination of quasi-identifiers provides an unambiguous identification of some records, so they have to be masked.
- *Confidential* attributes that are those containing sensitive information that could be useful for statitic analysis, so they are usually not modified.

In [17] authors prove that removing the identifiers is no enough to protect the identity of an individual. Therefore, the protection must be done on the quasi-identifiers.

3 An Algorithm for k-Anonymization

In this section we introduce an algorithm for k-anonymization. It is based on the generalization of a set of original records. Such generalization plays the role of a representative of a cluster in other microaggregation methods. The main goal is to search for a subset of records similar enough and set the attributes with different values to an *indifferent* value. The algorithm allows to choose the generalization degree, that is, how many attributes can have *indifferent* values. However, because we do not want to obtain generalizations very different of the

Function ANONYMIZE $(DB, K, C, long)$
;; DB := all records
;; C := set of confidential attributes
;; $long$:= minimum length for the new registers
Pt := set of patterns (initially $\emptyset$)
P_C := correct partition according to the confidential attributes
$analize\text{-}partition(DB, K, P_C, c1, long)$
end-function

Fig. 1. Initializations of the generalized-based k-anonymization. The main part of the process is to call the Analyze-partition function.

original records, in the experiments we work with anonymized records having only one or two *indifferent* values.

Figure 1 shows the Anonymize algorithm proposed to anonymize a data base. The input parameters are K, C and *long*, where K is the minimum number of records that have to be put together; C is the set of attributes that the algorithm considers as confidential, i.e., they have not been modified during the process of anonymization; and, *long* is the minimum length of the anonymized records, i.e., it controls number of *indifferent* values in order to prevent from too generalized records.

color	size
blue	small
green	big

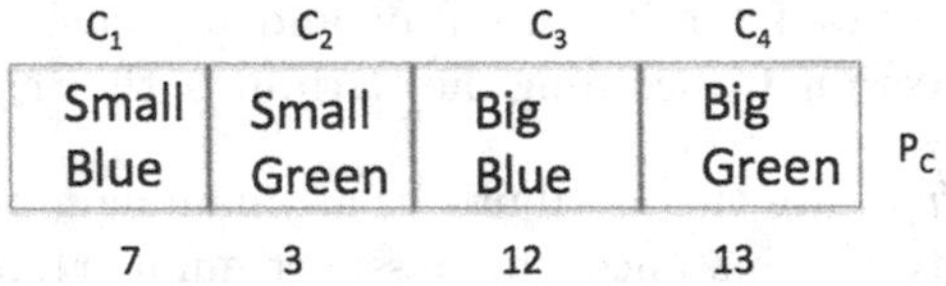

C_1	C_2	C_3	C_4	
Small Blue	Small Green	Big Blue	Big Green	P_C
7	3	12	13	

Fig. 2. Correct partition P_C induced taking into account two attributes: *color* and *size*. The correct partition has as many sets as the combination of values of both attributes. Numbers indicate how many records are contained in each partition set (Color figure online).

To illustrate the algorithm, let us suppose a data base composed of 35 records described by four attributes texture, material, color and size and that both color and size are confidential attributes. Figure 2 shows the values that these attributes can take. The first step is to build P_C, namely the *correct partition*, having as many sets as the number of possible combinations of values hold by the confidential attributes. In the example, P_C is composed of 4 sets since there are two confidential attributes taking two different values each. Let us suppose that the class C_1 has 7 records, the class C_2 has 3 records, the class C_3 has 12 records, and the class C_4 has 13 records. All the sets of P_C having less than K records are discarded. If we take $K = 5$, the records in the class C_2 are rejected. Notice that the rejection of these records can be seen as the rejection of outliers, since it means that there are few records with a given combination, so they could be easily re-identified. Then function Analyze-partition is called for each set of P_C. This function can be called several ways taking different *conditions*. This possibility will be explained in detail in Sect. 4.

```
Function Analize-partition (DB, K, Pᵢ, condition, long)
    loop
        for each set Cᵢ of Pᵢ do
            if Cardinality(Cᵢ) ≥ k then
                ident := analize-set(K, Cᵢ, condition, long)
            end-if
            if ident = ∅ then end process
                else DB := DB - ident
                     P'ᵢ := update(Pᵢ)
                     Analize-partition (DB, P'ᵢ, condition, long)
            end-if
        end-for
    end-loop
end-function
```

Fig. 3. The main goal of the Analize-partition function is to recursively call the function Analize-set for partition sets having more than K records.

Analyze-partition (Fig. 3) is a recursive function with three input parameters: a set of records DB (initially the whole data base), a partition P_i (initially P_C) and a *condition*. This function is a loop that analyzes each set C_i of P_i and, if the cardinality of that set is bigger than K, the function Analyze-set is called and returns the subset of records that are identified by some new created pattern. When no new records have been identified, the process ends; otherwise Analyze-partition is recursively called with an update of DB obtained by rejecting the newly identified records; and P'_n that is P_n without the identified records. The intuition behind this procedure is to induce partitions from the remaining records, i.e., those that have not still been used for generalization. Because at each step the number of records decreases, some sets of the partition can have a cardinality lower than K.

Analyze-set is also a recursive function (Fig. 4). In the first step, Analyze-set constructs a pattern AU using the anti-unification concept [2]. Such pattern is a record formed by all the attributes that have the same value in all the records in C_i. In this context, we call *description length* to the number of attributes describing AU. The remaining attributes are considered *indifferent*. If AU is either empty or has a length below *long* the input set has to be partitioned in subsets; otherwise there is a pattern long enough that satisfies more than K records. Notice that if AU is empty it means that the records in C_i have not attributes with common values. When the length of AU is lower than *long* it means that the pattern that could be extracted is too general, i.e., many attributes should have *indifferent* value. For instance, Fig. 5 shows the description of two records, namely *adult-8* and *adult-10*, described with 9 attributes (description-length = 9). Both objects have in common the value of only three of these attributes (i.e., the length of AU is 3), therefore this generalization is not useful since it has lost many information with respect to the complete description of a record.

Function Analize-set $(K, C_i, condition, long)$
    $AU :=$ anti-unification(C_i)
    if $A_c = \emptyset$ or description-length$(AU) < long$ then
        $P_A :=$ set of partitions on C_i induced by each quasi-confidential
        $P_{cond} :=$ subset of P_A selected according to $condition$
        for each $P_{a_i} \in P_{cond}$ do
            for each set C_j of P_{a_i} do
                if Cardinality$(C_j) \geq k$ then
                    $analize\text{-}set(K, C_j, condition, long)$
                end-if
        end-for
    else $Pt := Pt \cup \{AU\}$
        return C_i
    end-if
end-function

Fig. 4. The recursive function Analyze-set constructs the patterns that generalize more than K records.

When the elements in C_i have at least $long$ attributes with the same value, the function constructs a pattern (i.e., the anti-unification of the elements in C_i) and finishes by returning the set C_i.

Let us suppose now that in Analyze-set there is the situation such that AU is lower than $long$ and non-empty. The next step is, for each quasi-confidential attribute A_i describing the records, to induce a partition of the elements in C_i according to the values of A_i. The idea is to take a small number of records trying to find more commonalities among them, i.e., AU with an appropriate length. The way to do this is to take one of the remaining attributes and induce a partition according to the values that this attribute can take. Each partition has a certain number of sets. A partition with a high number of sets corresponds to a situation where each set has a small number of records but it is more likely than these records have many common attributes. Therefore, AU is specific and satisfied by a small number of records. Conversely partitions with small number of sets corresponds to a situation where each set has a high number of records

ADULT-8: (AGE B_25-29) (WORKCLASS PRIVATE) (EDUCATION BACHELORS)
(MARITAL-STATUS DIVORCED) (OCCUPATION PROF-SPECIALTY)
(RACE BLACK) (SEX FEMALE) (NATIVE-COUNTRY CUBA) (SALARY LESS_50)

ADULT-10: (AGE B_35-39) (WORKCLASS PRIVATE) (EDUCATION MASTERS)
(MARITAL-STATUS MARRIED-CIV-SPOUSE) (OCCUPATION EXEC-MANAGERIAL)
(RACE WHITE) (SEX FEMALE) (NATIVE-COUNTRY UNITED-STATES) (SALARY LESS_50)

AU: (WORKCLASS PRIVATE) (SEX FEMALE) (SALARY LESS_50))

Fig. 5. The records Adult-8 and Adult-8 are described by 9 attributes, however in only three of them they have the same value. The AU record shows these common attributes and their value.

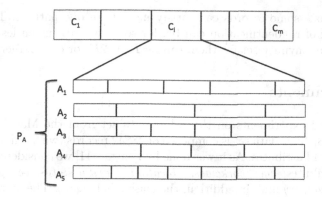

Fig. 6. From a set C_i of a partition, it can be induced a partition for each attribute A_j taking into account the values that such attribute holds in each one of the records of the set.

but it is more likely than these records have not many common attributes. This means that if we are able to find an AU of appropriate length (higher than *long*), it will generalize many objects. However, most of time this will not be the case and Analyze-set should be called again in order to reduce the number of records included in a set.

Let P_A be the set of all the induced partitions. This set could be ordered in several ways, so the input parameter *condition* indicates how the partitions in P_A have to be sorted. There are two simple sorts:

- to use $<$ as *condition*, meaning that partitions with the lowest number of sets are given first. In the example of Fig. 6 the attributes should be ordered in the following way: A_2, A_4, A_1, A_5, A_3 and A_6;
- to use $>$ as *condition*, meaning that partitions with the highest number of sets are given first. In the example of Fig. 6 the attributes should be ordered in the following way: A_6, A_3, A_1, A_5, A_2 and A_4.

Because several attributes can induce partitions with the same number of sets, let P_{cond} be the subset of P_A of the partitions with the same number of sets selected according to *condition*. In the example show in Fig. 6, $P_< = \{A_2, A_4\}$ and $P_> = \{A_6\}$.

Given a partition $P_{a_i} \in P_{cond}$, for each set C_j of P_{a_i} with cardinality higher than K, the function Analyze-set is recursively called. At the end of the process, Analyze-set has build a set of patterns each one representing the anonymization of at least K original records.

During the anonymization process, the sets with cardinality lower than K are discarded. This means that, at the end of the process, some records that do not satisfy any of the patterns can remain. For this reason, what we propose is to repeat the whole process on the subset of no identified records using a different *condition* (see Sect. 4).

Notice that, some records can satisfy more than one pattern. This implies that the risk of re-identification can be, for some records, much less than $1/K$ because the anonymity set for them can be K or $2K$, or even larger.

4   Experiments

We performed experiments on the Adult data set from the Machine Learning Repository [3]. This data set is composed of 48842 records (with unknown values) described by 14 atributes. As it was done by Iyengar [11] we considered only eight of these attributes: *age, workclass, education, marital status, occupation, race, sex, native country* and, in addition, the class label *salary*. The attribute *age* is numerical and we discretized it in intervals of 5 (i.e., [20–25), [25, 30) and so on). We also considered the labels *low-19* and *high-90* to include those records placed on both sides of the global age range. All the other attributes are considered categorical. As in [4] we considered the class *salary* as confidential, being all the other attributes quasi-confidential. Commonly, authors [4,11] discard around 3000 records due to unknown values. In our experiments we do not need to do so because the algorithm is able to deal with unknown values. The data set as it is downloaded from the Machine Learning Repository, is already split in a training set having 32561 records, and a test set having 16281 records.

In the experiments, we address the classification task. The goal is to classify people with salary up to 50 K and down to 50K. Therefore the correct partition has only two sets. To use the anonymization algorithm we need to fix the input parameters K and *long* and also to determine the *condition* under which the partitions are selected. We experimented with $K = 5, 10, 20$ and 30. Concerning *long*, we have set it to 2, that is to say, the maximum number of attributes that can have value *indifferent* in the patterns is 2. We also experimented with two combinations of *condition*:

- Combination $(<, >)$: first Analyze-partition is called with $<$. At the end of the process, the remaining records are used as input of Analyze-partition using $>$ as *condition*;
- Combination $(>, <)$: first Analyze-partition is called with $>$. At the end of the process, the remaining records are used as input of Analyze-partition using $<$ as *condition*.

Once the anonymization process is finished, what we observe first is that many records are discarded since they are not satisfied by any pattern. Thus, when $K = 5$ the number of discarded records goes from around 2500 to around 3300 depending on the combination of conditions. When K increases the number of discarded records also increases, since as long as the process advances, the sets of the partitions have low cardinality. This represents the lost of between a 7 % when $K = 5$ and to the 30 % when $K = 30$ of the records of the training set. Because we want to address the classification task, we have to evaluate how this lost affects the predictivity of the model we obtain after anonymization.

A simple way to test the equivalence of the original data base and the anonymized one for the classification task, is to induce a domain model (for instance, using a decision tree) for each data base and then evaluate the accuracy of the models on a test set. To construct the models of both the anonymized data base and the original one, we used the J48 inductive learning method, a clone of C4.5 [16] provided by Weka [9]. To test the accuracy of the models, we cannot use 10-fold cross-validation on the anonymized data base, but we have to use original records as test set. This is because we want to evaluate how well the model induced from the patterns represents the original records. Notice that a cross-validation on the anonymized data base will test the patterns (generalized records) instead of the original records. Thus, we carry out experiments with three test sets:

- *Experiment 1*: We randomly selected a subset of 11486 records, namely $S1$, from the original data base to act as test set.
- *Experiment 2*: The test set of 16281 records, namely *uci*, as it was downloaded from the UCI Repository.
- *Experiment 3*: The whole original data base (DB) is used as test set.

Therefore the process we carry out in the experiments has been the following:

1. To induce a decision tree from DB and evaluate the accuracy of that model on $S1$, DB itself, and *uci*;
2. To anonymize DB using the combination $(<, >)$, then induce a decision tree from the patterns that have been generated and, finally, evaluate the accuracy of that model on $S1$, DB, and *uci*;
3. To anonymize DB using the combination $(>, <)$, then induce a decision tree from the patterns that have been generated and, finally, evaluate the accuracy of that model on $S1$, DB, and *uci*.

Table 1 shows the accuracies obtained by each model with different values of K on each one of the test sets. Concerning the accuracy, we see that the combination $(>, <)$ is better than $(<, >)$ for all K except for $K = 30$. We have not any satisfactory explanation for this. However our intuition is that because the combination $(>, <)$ selects first the partitions having highest number of sets, the cardinality of these sets will be low. Probably, small sets are composed of more similar objects whose anti-unification will give a satisfactorily long pattern.

The combination $(<, >)$ tries to avoid overfitting by selecting the attributes inducing partitions with small number of sets. However, these sets have with high probability, records with very different descriptions.

When $K = 20$, both combinations have similar accuracy to the one given by the model obtained from DB. Notice that, despite for $K = 20$ there are more than 7000 records discarded, the accuracy is the best of all the combinations we tested. This is an unexpected result since it means that the anonymization is able to construct a satisfactory model of the domain with less records. Our explanation for this result is that the original database probably contains many outliers, i.e., records that have very particular combinations of values.

Table 1. Accuracy for $K = 5, 10, 20$ and 30 of the models with the original records (DB) and the anonymized ones $((<,>)$ and $(>,<))$ on three different test sets: $S1$ with around 16000 records randomly selected; the training set provided by the UCI Machine Learning Repository (uci); and, the whole data set (DB).

model	test	$K = 5$	$K = 10$	$K = 20$	$K = 30$
DB	$S1$	83.32	83.32	83.32	83.32
$(<,>)$	$S1$	73.19	72.51	80.38	80.58
$(>,<)$	$S1$	81.28	81.51	82.05	72.18
DB	uci	83.12	83.12	83.12	83.12
$(<,>)$	uci	73.23	72.38	80.34	80.82
$(>,<)$	uci	81.01	81.75	81.82	72.18
DB	DB	83.42	83.42	83.42	83.42
$(<,>)$	DB	73.01	72.12	80.27	80.71
$(>,<)$	DB	81.42	81.66	81.72	71.73

Table 2 shows the number of patterns generated by each combination and the number of discarded records for different values of K. The combination $(>,<)$ tends to discard lower number of records than the combination $(<,>)$. The combination $(<,>)$ seems to produce a high partitioning of the sets and, consequently, a higher number of discarded records than using $(>,<)$ since they achieve cardinality lower than K. Specially interesting is the combination $(>,<)$ with both $K = 20$ and $K = 30$ that generate lesser than 1000 patterns and, however, the induced model has an accuracy very near to the one obtained with the whole data set.

Table 2. Number of patterns $(patterns)$ and number of discarded records $(rest)$ for each one of the combinations.

model	$K = 5$		$K = 10$		$K = 20$		$K = 30$	
	patterns	rest	patterns	rest	patterns	rest	patterns	rest
$(<,>)$	2779	3299	1542	5231	821	7681	570	9671
$(>,<)$	10453	2578	4956	5042	2307	7396	1343	9613

5 Conclusions

In this paper we introduce a new method for k-anonymization of data bases where records are represented with attributes with categorical values. The approach is based on generalization, particularly we used the concept of anti-unification consisting on take only those attributes that take the same value in all the records of a given set. This approach is different from other generalization approaches in two main issues: first it is able to deal with unknown values; and, second it is not necessary to construct an ontology generalizing the values of the attributes. At the end of the process, we obtain an anonymized data base consisting on a set of patterns that have the same form than the original records but they are generalized. The generalization consists on setting some of the attributes to value unknown meaning that it is no important the value that they take. Because we only permit one or two unknown values in each pattern, we can assure that there is no overgeneralization, so we do not lost too much information.

There are some original records that cannot be generalized satisfactorily with any patterns, so they are discarded. We experimentally proved that the rejection of a (sometimes high) percentage of original records does not highly affect the predictive accuracy in classification tasks.

As future work we consider the application of this approach for microaggregation, in the sense that different subsets of attributes are generalized differently. This approach has already considered in [15] for other generalization algorithms with good results. In this microaggregation approach the analysis of disclosure risk is meaningful. In addition, as records may be generalized in several ways, the model described in [18] may be useful to study a worse case scenario.

Acknowledgments. This research is partially funded by the Spanish MICINN projects COGNITIO (TIN-2012-38450-C03-03), EdeTRI (TIN2012-39348-C02-01) and COPRIVACY (TIN2011-27076-C03-03), the grant 2009-SGR-1434 from the Generalitat de Catalunya, and the European Project DwB (Grant Agreement Number 262608).

References

1. Abril, D., Navarro-Arribas, G., Torra, V.: Supervised learning using a symmetric bilinear form for record linkage. Inf. Fusion **26**, 144–153 (2015)
2. Armengol, E., Plaza, E.: Bottom-up induction of feature terms. Mach. Learn. **41**, 259–294 (2000)
3. Bache, K., Lichman, M.: UCI machine learning repository (2013)
4. Bayardo, R.J., Agrawal, R.: Data privacy through optimal k-anonymization. In: Proceedings of the 21st International Conference on Data Engineering (ICDE 2005), pp. 217–228 (2005)
5. Domingo-Ferrer, J., Torra, V.: Ordinal, continuous and heterogeneous k-anonymity through microaggregation. Data Min. Knowl. Discov. **11**(2), 195–212 (2005)
6. Duncan, G.T., Elliot, M., Salazar, J.J.: Statistical Confidentiality. Springer, New York (2011)
7. Dwork, C.: Differential privacy. In: Bugliesi, M., Preneel, B., Sassone, V., Wegener, I. (eds.) ICALP 2006. LNCS, vol. 4052, pp. 1–12. Springer, Heidelberg (2006)
8. Guo, L., Wu, X.: Privacy preserving categorical data analysis with unknown distortion parameters. Trans. Data Priv. **2**(3), 185–205 (2009)
9. Hall, M., Frank, E., Holmes, G., Pfahringer, B., Reutemann, P., Witten, I.H.: The weka data mining software: an update. SIGKDD Explor. Newsl. **11**, 10–18 (2009)
10. Hundepool, A., Domingo-Ferrer, J., Franconi, L., Giessing, S., Nordholt, E.S., Spicer, K., de Wolf, P.-P.: Statistical Disclosure Control. Wiley, New York (2012)
11. Iyengar, V.S.: Transforming data to satisfy privacy constraints. In: Proceedings of the Eighth ACM SIGKDD International Conference on Knowledge Discovery and Data Mining, KDD 2002, pp. 279–288. ACM, New York (2002)

12. LeFevre, K., DeWitt, D.J., Ramakrishnan, R.: Multidimensional k-anonymity, Technical report 1521, University of Wisconsin (2005)
13. LeFevre, K., DeWitt, D.J., Ramakrishnan, R.: Incognito: efficient full-domain k-anonymity, SIGMOD 2005 (2005)
14. Li, X.-B., Sarkar, S.: Privacy protection in data mining: a perturbation approach for categorical data. Inf. Syst. Res. **17**(3), 254–270 (2004)
15. Marés, J., Torra, V.: Clustering-based categorical data protection. In: Domingo-Ferrer, J., Tinnirello, I. (eds.) PSD 2012. LNCS, vol. 7556, pp. 78–89. Springer, Heidelberg (2012)
16. Quinlan, J.R.: C4.5: Programs for Machine Learning. Morgan Kaufmann Publishers Inc., San Francisco (1993)
17. Samarati, P., Sweeney, L.: Protecting privacy when disclosing information: k-anonymity and its enforcement through generalization and suppression, SRI International Technical report (1998)
18. Tassa, T., Mazza, A., Gionis, A.: k-concealment: an alternative model of k-type anonymity. Trans. Data Priv. **5**(1), 189–222 (2012)
19. Torra, V., Stokes, K.: A formalization of record linkage and its application to data protection. Int. J. Uncertain. Fuzziness Knowl. Based Syst. **20**, 907–919 (2012)
20. Wang, K.: Bottom-up generalization: a data mining solution to privacy protection. Proc. ICDM **2004**, 249–256 (2004)
21. Winkler, W.E.: Re-identification methods for masked microdata. In: Domingo-Ferrer, J., Torra, V. (eds.) PSD 2004. LNCS, vol. 3050, pp. 216–230. Springer, Heidelberg (2004)

Logics

The Complexity of 3-Valued Łukasiewicz Rules

Miquel Bofill[1]([✉]), Felip Manyà[2], Amanda Vidal[2], and Mateu Villaret[1]

[1] Universitat de Girona, Girona, Spain
miquel.bofill@udg.edu
[2] Artificial Intelligence Research Institute (IIIA, CSIC), Bellaterra, Spain

Abstract. It is known that determining the satisfiability of n-valued Łukasiewicz rules is NP-complete for $n \geq 4$, as well as that it can be solved in time linear in the length of the formula in the Boolean case (when $n = 2$). However, the complexity for $n = 3$ is an open problem. In this paper we formally prove that the satisfiability problem for 3-valued Łukasiewicz rules is NP-complete. Moreover, we also prove that when the consequent of the rule has at most one element, the problem is polynomially solvable.

1 Introduction

The proof theory of many-valued logics has been deeply studied for a wide variety of logics [1, 8, 9]. Nevertheless, the development of satisfiability solvers has received less attention despite of the fact that, without competitive solvers, it is extremely difficult to apply many-valued logics to solve real-world problems.

Given the recent development of Satisfiability Modulo Theory-based (SMT-based) solvers for many-valued logics [2–4, 10, 11], there is the need to empirically evaluate and compare them with other existing approaches. Because of that we are interested in developing instance generators that produce instances of varying difficulty, as well as in analyzing the complexity of relevant fragments of many-valued logics. It is extremely difficult to advance in the development of fast satisfiability solvers without the availability of challenging benchmarks.

Before describing the related work and contributions of the present paper, let us recall that developing satisfiability solvers for Łukasiewicz logics is particularly interesting because $SAT^{\text{Ł}} \subsetneq SAT^{Bool}$, where $SAT^{\text{Ł}}$ is the set of formulas in Łukasiewicz logic that evaluate to 1 for some interpretation, and SAT^{Bool} is the set of satisfiable Boolean formulas [8]. This implies that some propositional formulas are satisfiable in Łukasiewicz logic whereas they are unsatisfiable in Boolean logic. However, in other relevant many-valued logics such as Gödel (G)

Research partially supported by the Generalitat de Catalunya grant AGAUR 2014-SGR-118, and the Ministerio de Economía y Competividad projects AT CONSOLIDER CSD2007-0022, INGENIO 2010, CO-PRIVACY TIN2011-27076-C03-03, EDETRI TIN2012-39348-C02-01 and HeLo TIN2012-33042. The second author was supported by Mobility Grant PRX14/00195 of the Ministerio de Educación, Cultura y Deporte.

© Springer International Publishing Switzerland 2015
V. Torra and Y. Narakawa (Eds.): MDAI 2015, LNAI 9321, pp. 221–229, 2015.
DOI: 10.1007/978-3-319-23240-9_18

and Product (Π) we have that $SAT^G = SAT^\Pi = SAT^{Bool}$ and, therefore, satisfiability testing in these logics can be proved directly with a Boolean SAT solver.

We have recently investigated, in [5], how the Conjunctive Normal Forms (CNFs) used by Boolean SAT solvers can be extended to Łukasiewicz logics. In a first attempt, we replaced the classical disjunction in Boolean CNFs with Łukasiewicz strong disjunction, and interpreted negation using Łukasiewicz negation. Interestingly, we proved that the satisfiability problem of these clausal forms has linear-time complexity,[1] regardless of the size of the clauses and the cardinality of the truth value set (assuming it is greater than two). This result is surprising because deciding the satisfiability of Boolean CNFs is NP-complete when there are clauses with at least three literals [7]. So, we identified a problem that is NP-complete in the Boolean case but has linear-time complexity in Łukasiewicz logic.

With the aim of producing computationally difficult instances, we defined a new class of clausal forms, called Łukasiewicz (Ł-)clausal forms, that are CNFs in which, besides replacing classical disjunction with Łukasiewicz strong disjunction, we allow negations above the literal level; i.e., clauses are strong disjunctions of terms, and terms are either literals or negated strong disjunctions of literals. We proved that, in this case, 3-SAT is NP-complete whereas 2-SAT has linear-time complexity.[2] Hence, we defined problems in Łukasiewicz logic that have the same complexity as their Boolean counterparts.

Independently of our work, Borgwardt et al. [6] investigated the complexity of finitely-valued Łukasiewicz rules (c.f. Sect. 2), and proved that the problem of deciding the satisfiability of such rules is NP-complete when the cardinality of the truth value set is at least four, but they left as an open problem the complexity of 3-valued Łukasiewicz rules. Analyzing the complexity of Łukasiewicz rules is appealing because this problem has linear-time complexity in the Boolean case whereas it is NP-complete for n-valued Łukasiewicz logics in which $n \geq 4$.

In this paper we prove that the satisfiability problem for 3-valued Łukasiewicz rules is NP-complete, solving this way an open problem. Moreover, we also prove that if the consequent of the rule has at most one element, the problem is polynomially solvable.

The paper is structured as follows. Section 2 defines basic concepts in Łukasiewicz logics, and n-valued Łukasiewicz rules. Section 3 shows that the satisfiability problem for 3-valued Łukasiewicz rules is NP-complete, but it is polynomially solvable in any finitely-valued Łukasiewicz logic if the consequent of the rule has at most one element. Section 4 concludes and points out future research directions.

---

[1] In the following, when we say linear-time complexity we mean that the complexity is linear in the size of the formula.

[2] When the number of literals per clause is fixed to k, the corresponding SAT problem is called k-SAT.

2   Preliminaries

This section formally defines the finitely-valued and infinitely-valued logics of Łukasiewicz, as well as the language of Łukasiewicz rules.

Definition 1. *A **propositional language** is a pair* $\mathbb{L} = \langle \Theta, \alpha \rangle$, *where* Θ *is a set of logical connectives and* $\alpha : \Theta \to \mathbb{N}$ *defines the arity of each connective. Connectives with arity 0 are called constants. A language* $\langle \Theta, \alpha \rangle$ *with a finite set of connectives* $\Theta = \{\theta_1, \ldots, \theta_r\}$ *is denoted by* $\langle \theta_1/\alpha(\theta_1), \ldots, \theta_r/\alpha(\theta_r) \rangle$.

Given a set of propositional variables $\mathcal{V}$, *the set* $L_\mathcal{V}$ *of* $\mathbb{L}$-*formulas over* $\mathcal{V}$ *is inductively defined as the smallest set with the following properties: (i)* $\mathcal{V} \subseteq L_\mathcal{V}$; *(ii) if* $\theta \in \Theta$ *and* $\alpha(\theta) = 0$, *then* $\theta \in L_\mathcal{V}$; *and (iii) if* $\phi_1, \ldots, \phi_m \in L_\mathcal{V}$, $\theta \in \Theta$ *and* $\alpha(\theta) = m$, *then* $\theta(\phi_1, \ldots, \phi_m) \in L_\mathcal{V}$.

Definition 2. *A **many-valued logic** $\mathcal{L}$ is a triplet* $\langle \mathbb{L}, N, A \rangle$ *where* $\mathbb{L} =< \Theta, \alpha >$ *is a propositional language,* N *is a truth value set, and* A *is an interpretation of the operation symbols that assigns to each* $\theta \in \Theta$ *a function* $A_\theta : N^{\alpha(\theta)} \to N$.

Many-valued logics are equipped with a non-empty subset D of N called the designated truth values, which are the truth values that are considered to affirm satisfiability.

Definition 3. *Let* $\mathcal{L}$ *be a many-valued logic. An **interpretation on** $\mathcal{L}$ is a function* $I : \mathcal{V} \to N$. I *is extended to arbitrary formulas* ϕ *in the usual way:*

1. *If* ϕ *is a logical constant, then* $I(\phi) = A_\phi$.
2. *If* $\phi = \theta(\phi_1, \ldots, \phi_r)$, *then* $I(\theta(\phi_1, \ldots, \phi_r)) = A_\theta(I(\phi_1), \ldots, I(\phi_r))$.

A formula ϕ *is **satisfiable** iff there is an interpretation such that* $I(\phi) \in D$.

Through this work, we focus on a particular family of many-valued logics: the Łukasiewicz logics. These were born from the generalization of a three valued logic proposed by J. Łukasiewicz in the early 20th century, and have been deeply studied both from theoretical and practical points of view. For a deeper study on these matters, see for instance [8].

The language of Łukasiewicz logic is given by

$$\mathbb{L}_{\text{Luk}} = \langle \bot/0, \top/0, \neg/1, \to /2, \wedge/2, \vee/2, \odot/2, \oplus/2 \rangle.$$

We refer to $\bot$ as bottom, to $\top$ as top, to $\neg$ as negation, to $\to$ as implication, to $\wedge$ as weak conjunction, to $\vee$ as weak disjunction, to $\odot$ as (strong) conjunction, and to $\oplus$ as (strong) disjunction.

Definition 4. *The **infinitely-valued Łukasiewicz logic**, denoted by* $[0, 1]_L$ *is the many-valued logic* $\langle \mathbb{L}_{Luk}, N, A \rangle$ *equipped with the set of designated values* $D = \{1\}$, *where* N *is the real unit interval* $[0, 1]$, *and the interpretation of the operation symbols* A *is given by:*

$$A_\perp = 0 \qquad\qquad A_\wedge(x,y) = \min\{x,y\}$$
$$A_\top = 1 \qquad\qquad A_\vee(x,y) = \max\{x,y\}$$
$$A_\neg(x) = 1 - x \qquad\qquad A_\odot(x,y) = \max\{0, x+y-1\}$$
$$A_\to(x,y) = \min\{1, 1-x+y\} \qquad\qquad A_\oplus(x,y) = \min\{1, x+y\}$$

The n-**valued Łukasiewicz logic**, denoted by $Ł_n$, is the logic defined from the infinitely-valued Łukasiewicz logic by restricting the universe of evaluation to the set $N_n = \{0, \frac{1}{n-1}, ..., \frac{n-1}{n-1}\}$. That is to say, $Ł_n = \langle Ł_{Luk}, N_n, A_Ł \rangle$ equipped with $D = \{1\}$. Note that the operations are well defined because N_n is a subalgebra of $[0,1]$ with the interpretation of the operation symbols A (for any operation A_* and any value/pair of values of N_n, the result of A_* over this/these values also belongs to N_n).

The function interpreting negation is called Łukasiewicz negation, the function interpreting strong conjunction is called Łukasiewicz t-norm, the function interpreting implication is called its residuum, and the function interpreting strong disjunction is called Łukasiewicz t-conorm.

We say that a logic $\mathcal{L}$ is a Łukasiewicz logic if it is either $[\mathbf{0,1}]_Ł$ or $Ł_n$ for some natural number n.

Given a Łukasiewicz logic $\mathcal{L}$, we denote by $SAT^{\mathcal{L}}$ the set of satisfiable formulas in $\mathcal{L}$; i.e.,

$$SAT^{\mathcal{L}} = \{\varphi : I(\varphi) = 1 \text{ for some interpretation } I \text{ on } \mathcal{L}\}.$$

The problem of deciding whether or not a formula belongs to the set $SAT^{\mathcal{L}}$ is called the $\mathcal{L}$-**satisfiability problem**.

Definition 5. *Given a finite truth value set N_n, an n-valued* **Łukasiewicz rule** *is an expression of one of the following two forms:*

- $x_1 \odot \cdots \odot x_k \to y_1 \odot \cdots \odot y_m \geq r$
- $x_1 \odot \cdots \odot x_k \to \perp \geq r'$

where $k \geq 0, m \geq 1, r, r' \in N_n$, and $x_1, \ldots, x_k, \ y_1, \ldots, y_m$ are propositional variables (if $k = 0$, $x_1 \odot ... \odot x_k$ stands for $\top$).

Definition 6. *An interpretation I satisfies a Łukasiewicz rule of the form*

$$x_1 \odot \cdots \odot x_k \to y_1 \odot \cdots \odot y_m \geq r$$

iff $I(x_1 \odot \cdots \odot x_k \to y_1 \odot \cdots \odot y_m) \geq r$, and a Łukasiewicz rule of the form

$$x_1 \odot \cdots \odot x_k \to \perp \geq r'$$

iff $1 - I(x_1 \odot \cdots \odot x_k) \geq r'$.

A set of n-valued Łukasiewicz rules is satisfiable iff there exists an interpretation that satisfies all the rules.

Remark 1. Łukasiewicz rules are called fuzzy Horn clauses by Borgwardt et al. [6], but we prefer not to refer to them as Horn clauses for the following reason: In Boolean propositional logic, a Horn clause is defined as a clause having at most one positive literal. Given a finite set of m Boolean Horn clauses of the form $x_1, \ldots, x_k \rightarrow y_i$, where $1 \leq i \leq m$ and all the clauses have the same antecedent, we have that such a set is equivalent to the clause $x_1, \ldots, x_k \rightarrow y_1, \ldots, y_m$, whose extension to Łukasiewicz logic corresponds to the first type of Łukasiewicz rules. However, in Łukasiewicz logic, a finite set of m rules of the form $x_1 \odot \cdots \odot x_k \rightarrow y_i \geq r$, where $1 \leq i \leq m$, is not equivalent to the rule $x_1 \odot \cdots \odot x_k \rightarrow y_1 \odot \cdots \odot y_m \geq r$.

In Sect. 3, we show that deciding the satisfiability of a set of 3-valued Łukasiewicz rules containing only rules of the form $x_1 \odot \cdots \odot x_k \rightarrow y_i \geq r$ or $x_1 \odot \cdots \odot x_l \rightarrow \bot \geq r'$ has linear-time complexity as in the Boolean case. However, deciding the satisfiability of the rules defined by Borgwardt et al. is polynomially solvable in the Boolean case, but is NP-complete in the Łukasiewicz case.

Remark 2. Observe that Łukasiewicz rules of the form $x_1 \odot \cdots \odot x_k \rightarrow y_1 \odot \cdots \odot y_m \geq r$ can be represented using strong disjunctions instead of strong conjuntions as $\neg x_1 \oplus \cdots \oplus \neg x_k \oplus \neg(\neg y_1 \oplus \cdots \oplus \neg y_m) \geq r$, and Łukasiewicz rules of the form $x_1 \odot \cdots \odot x_{k'} \rightarrow \bot \geq r'$ can be represented as $\neg x_1 \oplus \cdots \oplus \neg x_{k'} \geq r'$. So, Łukasiewicz rules are a fragment of the Łukasiewicz clausal forms defined in [5].

3 Complexity of the Satisfiability Problem of 3-Valued Łukasiewicz Rules

In this section we prove that the satisfiability problem of 3-valued Łukasiewicz rules is NP-complete, and give one subcase in which it can be solved in polynomial time.

Theorem 1. *The satisfiability problem of* 3-*valued Łukasiewicz rules is NP-complete.*

Proof. We will show that (i) this problem belongs to NP, and (ii) the Boolean 3-SAT problem is polynomially reducible to our problem.

The satisfiability problem of 3-valued Łukasiewicz rules clearly belongs to NP: given a set of rules, a nondeterministic algorithm can guess a satisfying interpretation and check that it satisfies the formula in polynomial time.

For what respects the second claim, let $\phi = \bigwedge_{i=1}^{n}(l_i^1 \vee l_i^2 \vee l_i^3)$ be a Boolean 3-SAT instance, where l_i^1, l_i^2, l_i^3 are literals over the set of Boolean variables $\{x_1, \ldots, x_m\}$. We construct the following set ϕ' of 3-valued Łukasiewicz rules from ϕ, over the set of three-valued variables $\{y_1, y_1', \ldots, y_m, y_m'\}$[3] as follows:

---

[3] Observe that only literals with positive polarity appear in Łukasiewicz rules, but the 3-SAT instance ϕ can contain occurrences of both positive and negative literals. Thus, we introduce the variable y_k' to simulate the literal $\neg x_k$, whereas y_k will simulate the literal x_k.

1. For every Boolean variable x_k, $1 \leq k \leq m$, we add to ϕ' the following two rules:
   $A_k)$ $y_k \odot y'_k \rightarrow \bot \geq \frac{1}{2}$
   $B_k)$ $\rightarrow y_k \odot y'_k \geq \frac{1}{2}$
   Observe that any interpretation I that satisfies the previous two rules has a very determined behaviour. From $B_k)$, it must hold that $I(y_k \odot y'_k) \geq \frac{1}{2}$, and thus, by definition of the Łukasiewicz conjunction operation, either both y_k and y'_k are interpreted to 1 or one of them is interpreted to $\frac{1}{2}$, while the other is interpreted to 1. Moreover, if I must satisfy also rule $A_k)$, it is not possible that y_k and y'_k are both interpreted to 1, so the only interpretations that meet all the requirements are those that send one of these variables to 1 and the other to $\frac{1}{2}$. In other words, exactly one of y_k and y'_k is evaluated to 1 in a satisfying interpretation, while the other is evaluated to $\frac{1}{2}$. The intuition behind this is that $I(y_k) = 1$ means that x_k is *true*, and $I(y'_k) = 1$ means that x_k is *false*.

2. Let ρ be the function that maps (Boolean) literals to three-valued variables given by:

$$\rho(l_i^j) = \begin{cases} y_k & \text{if } l_i^j = x_k \\ y'_k & \text{if } l_i^j = \neg x_k \end{cases}$$

Then, for every clause $\delta_i = l_i^1 \vee l_i^2 \vee l_i^3$ in ϕ, we add to ϕ' the rule

$C_{\delta_i})$ $\rho(\neg l_i^1) \odot \rho(\neg l_i^2) \odot \rho(\neg l_i^3) \rightarrow \bot \geq \frac{1}{2}$

Observe that this rule is only falsified if $\rho(\neg l_i^1) = \rho(\neg l_i^2) = \rho(\neg l_i^3) = 1$. This is equivalent to say that any interpretation that satisfies $C_{\delta_i})$ must send at least one $\rho(\neg l_i^1), \rho(\neg l_i^2)$ or $\rho(\neg l_i^3)$ to a value that is less than or equal to $\frac{1}{2}$ (and if this interpretation also satisfies the family or rules $\{A_k), B_k)\}_{1 \leq k \leq m}$, this value will be in fact equal to $\frac{1}{2}$). Notice that the clause $l_i^1 \vee l_i^2 \vee l_i^3$ is only falsified if a Boolean interpretation sets l_i^1, l_i^2, l_i^3 to *false* . Also notice that the clause $l_i^1 \vee l_i^2 \vee l_i^3$ is equivalent to the Boolean rule $\neg l_i^1 \wedge \neg l_i^2 \wedge \neg l_i^3 \rightarrow \bot$.

This reduction can clearly be performed in polynomial time, and the size of ϕ' is linear in the size of ϕ. We also prove that ϕ' is satisfiable if and only if ϕ is satisfiable.

First, let I' be an evaluation on $Ł_3$ such that I' satisfies ϕ'. By construction of ϕ', from rules $A_k)$ and $B_k)$ we have that either $I'(y_k) = 1$ and $I'(y'_k) = \frac{1}{2}$ or $I'(y'_k) = 1$ and $I'(y_k) = \frac{1}{2}$ (for each $1 \leq k \leq m$). Moreover, for each $\delta_i = l_i^1 \vee l_i^2 \vee l_i^3$ of ϕ, rule $C_{\delta_i})$ implies that I' sets to $\frac{1}{2}$ at least one of the literals $\rho(\neg l_i^1), \rho(\neg l_i^2), \rho(\neg l_i^3)$. Let then I be the Boolean interpretation defined by

$$I(x_k) = \begin{cases} \text{true} & \text{if } I'(y_k) = 1 \\ \text{false} & \text{if } I'(y'_k) = 1 \end{cases}$$

I satisfies at least one literal from each clause of ϕ. Indeed, consider without loss of generality that $I'(\rho(\neg l_i^1)) = \frac{1}{2}$.

- If $l_i^1 = x_j$, for some $1 \leq j \leq m$, then $\rho(\neg l_i^1) = y'_j$, and so, it is mandatory that $I'(y_j) = 1$, making $I(x_j) = I(l_i^1) = \text{true}$.

– If $l_i^1 = \neg x_j$ then $\rho(\neg l_i^1) = y_j$, and similarly, $I'(y_j') = 1$. This implies that $I(x_j) = false$, making $I(l_i^1) = true$.

Then, I is a Boolean evaluation satisfying ϕ.

For the other direction, assume that I is a Boolean interpretation satisfying ϕ. We construct a three-valued interpretation I' from I as follows:

$$I'(y_k) = \begin{cases} 1 & \text{if } I(x_k) = true \\ \frac{1}{2} & \text{if } I(x_k) = false \end{cases} \quad I'(y_k') = \begin{cases} \frac{1}{2} & \text{if } I(x_k) = true \\ 1 & \text{if } I(x_k) = false \end{cases}$$

I' is constructed in such a way that it naturally satisfies the family of rules $\{A_k), B_k)\}_{1 \leq k \leq m}$. On the other hand, for every clause $\delta_i = l_i^1 \vee l_i^2 \vee l_i^3$ in ϕ, since I satisfies ϕ, at least one of the literals l_i^1, l_i^2, l_i^3 is set to true. Assume without loss of generality that the satisfied literal is l_i^1. There are two possibilities:

– If $l_i^1 = x_s$ for some $1 \leq s \leq m$, (and so, $I(x_s) = true$), I' satisfies rule C_{δ_i}) because $I'(y_s') = \frac{1}{2}$ due to the fact that $I(x_s) = true$.
– If $l_i^1 = \neg x_s$ for some $1 \leq s \leq m$, (and so, $I(x_s) = false$), I' satisfies rule C_{δ_i}) because $I'(y_s) = \frac{1}{2}$ due to the fact that $I(x_s) = false$.

In all these cases, I' satisfies ϕ', which concludes the proof.

Remark 3. Observe that all the Łukasiewicz rules used in the reduction have either the consequent or the antecedent empty. So, the satisfiability problem of this fragment of 3-valued Łukasiewicz rules is NP-complete too. Also observe that this fragment corresponds to Łukasiewicz clausal forms only containing clauses of the form $(\neg x_1 \oplus \cdots \oplus \neg x_k) \geq r$ and of the form $(\neg x_1 \oplus \cdots \oplus \neg x_{k'}) \leq r'$.

Theorem 2. *The problem of deciding the satisfiability of a finite set ϕ of 3-valued Łukasiewicz rules containing only rules of the form $x_1 \odot \cdots \odot x_k \to y_i \geq r$ or $x_1 \odot \cdots \odot x_l \to \bot \geq r'$ is polynomially solvable.*

Proof. We can assume that there is no rule in which $r = 0$ or $r' = 0$ because such rules are tautologies and can be removed. Assume that r and r' are either $\frac{1}{2}$ or 1. If ϕ contains a rule of the form $x_j \to \bot \geq 1$, then x_j should be evaluated to 0 and, therefore, all the rules having x_j in the antecedent can be removed, and all the occurrences of x_j in the consequent of a rule can be substituted by $\bot$. This process is repeated until there are no more rules of the form $x_j \to \bot \geq 1$ or the empty rule is derived. If the empty rule is derived, it means that ϕ is unsatisfiable. Otherwise, we continue simplifying ϕ as follows: if ϕ contains a rule of the form $\to y_i \geq 1$, then y_i should be evaluated to 1 and, therefore, all the rules having y_i in the consequent can be removed, and all the occurrences of y_i in the antecedent of a rule can be removed too. This process is repeated until there are no more rules of the form $\to y_i \geq 1$ or the empty rule is derived. If the empty rule is derived, it means that ϕ is unsatisfiable.

Otherwise, ϕ is satisfied by the interpretation I that sets to $\frac{1}{2}$ all the variables. To see this, observe that the only remaining rules with exactly one literal are either of the form $x_j \to \bot \geq \frac{1}{2}$ or $\to y_i \geq \frac{1}{2}$ and, hence, they are satisfied by I.

Besides, any rule containing more than one literal is also satisfied by I: when there are at least two literals in the antecedent, the rule evaluates to 1 under I because $\frac{1}{2} \odot \frac{1}{2} = 0$; and when there is a literal in the antecedent and one literal in the consequent, the rule evaluates also to 1 under I because $\min\{1, 1 - \frac{1}{2} + \frac{1}{2}\} = 1$.

Since the number of rules with exactly one literal that can be derived is bounded by the total number of rules, deciding the satisfiability of ϕ can be performed in polynomial time.

Theorem 2 proves that the satisfiability problem of the generalization of Horn clauses to our setting is polynomial solvable. We claim that a linear-time algorithm could be implemented by adapting the Boolean linear-time unit propagation algorithm for Boolean CNFs described in [12].

4 Conclusions

We proved that the satisfiability problem of 3-valued Łukasiewicz rules is NP-complete, but is polynomially solvable when the rules have at most one literal in the consequent. Actually, to get intractable instances it is enough to require that the rules have an empty antecedent or an empty consequent. These results solve an open problem posed by Borgwardt et al. in [6], and the 3-valued Łukasiewicz rules become a challenging benchmark for Łukasiewicz satisfiability solvers.

As future work we propose to analyze the complexity of infinitely-valued Łukasiewicz rules, as well as identify —when testing the satisfiability of 3-valued Łukasiewicz rules with a fixed number of variables per rule, and generated uniformly at random— an easy-hard-easy pattern, and a phase transition phenomenon as the clause-to-variable ratio varies, similar to the ones identified for Łukasiewicz clausal forms in [5].

References

1. Aguzzoli, S., Gerla, B., Haniková, Z.: Complexity issues in basic logic. Soft Comput. **9**(12), 919–934 (2005)
2. Ansótegui, C., Bofill, M., Manyà, F., Villaret, M.: Building automated theorem provers for infinitely-valued logics with satisfiability modulo theory solvers. In: Proceedings of 42nd International Symposium on Multiple-Valued Logics (ISMVL), pp. 25–30. IEEE CS Press, Victoria (2012)
3. Ansótegui, C., Bofill, M., Manyà, F., Villaret, M.: Automated theorem provers for multiple-valued logics with satisfiability modulo theory solvers. Fuzzy Sets Syst. (2015). doi:10.1016/j.fss.2015.04.011
4. Ansótegui, C., Bofill, M., Manyà, F., Villaret, M.: SAT and SMT technology for many-valued logics. Multiple-Valued Logic Soft Comput. **24**(1–4), 151–172 (2015)
5. Bofill, M., Manyà, F., Vidal, A., Villaret, M.: Finding hard instances of satisfiability in Łukasiewicz logics. In: Proceedings of 45th International Symposium on Multiple-Valued Logics (ISMVL), page In press. IEEE CS Press, Waterloo (2015)
6. Borgwardt, S., Cerami, M., Peñaloza, R.: Many-valued Horn logic is hard. In: Proceedings of the First Workshop on Logics for Reasoning about Preferences, Uncertainty, and Vagueness, PRUV 2014, co-located with IJCAR 2014, Vienna, Austria, pp. 52–58 (2014)

7. Garey, M.R., Johnson, D.S.: Computers and Intractability: A Guide to the Theory of NP-completeness. Freeman, San Francisco (1979)
8. Hájek, P.: Metamathematics of Fuzzy Logic. Kluwer, Dordrecht (1998)
9. Metcalfe, G., Olivetti, N., Gabbay, D.M.: Proof Theory of Fuzzy Logics. Applied Logic Series, vol. 36. Springer, The Netherlands (2009)
10. Vidal, A.: NiBLoS: a nice BL-logics solver. Master's thesis, Universitat de Barcelona, Barcelona, Spain (2012)
11. Vidal, A., Bou, F., Godo, L.: An SMT-based solver for continuous t-norm based logics. In: Hüllermeier, E., Link, S., Fober, T., Seeger, B. (eds.) SUM 2012. LNCS, vol. 7520, pp. 633–640. Springer, Heidelberg (2012)
12. Zhang, H., Stickel, M.E.: An efficient algorithm for unit propagation. In: Proceedings of the Fourth International Symposium on Artificial Intelligence and Mathematics (AI-MATH'96), Fort Lauderdale, Florida pp. 166–169 (1996)

Information Theory for Subjective Logic

Tim Muller[1]([✉]), Dongxia Wang[1], and Audun Jøsang[2]

[1] Nanyang Technological University, Singapore, Singapore
t.j.c.muller@gmail.com
[2] University of Oslo, Oslo, Norway

Abstract. Uncertainty plays an important role in decision making. People try to avoid risks introduced by uncertainty. Probability theory can model these risks, and information theory can measure these risks. Another type of uncertainty is ambiguity; where people are not aware of the probabilities. People also attempt to avoid ambiguity. Subjective logic can model ambiguity-based uncertainty using opinions. We look at extensions of information theory to measure the uncertainty of opinions.

1 Introduction

Uncertainty exists in human decision making, where the results of our actions cannot be predicted, or we are not aware of the complete circumstances surrounding the decision. The unpredictability of the outcome of a test can be modelled using probability theory. For example, both calling heads/tails on the next coin flip (unpredictable) and guessing the side of a covered coin (unknown state), are naturally modelled as decisions with probabilities of a half. We name this kind of unpredictability as risk-based uncertainty.

Information theory [10] allows measuring how uncertain we are about the effects of a decision, in the form of entropy. People tend to be risk-averse [2], and they prefer to make decisions under low entropy. Concretely, most people prefer 1,000$ over an all-or-nothing coin flip for 2,000$. The former option has 0 bits entropy, and the latter has 1 bit entropy.

There is a type of uncertainty stronger than risk – ambiguity or Knightian uncertainty [5] – where the probabilities themselves are unknown. An example would be a coin with an unknown bias. Subjective logic is a formalism that addresses this type of uncertainty. People also tend to be ambiguity-avoiding; the Ellsberg paradox [2] (Sect. 2) shows that people may prefer a bigger known risk over a smaller unknown risk.

In this paper, we generalise information theory to cover subjective logic. As a consequence, entropy can be used to measure both types of uncertainty (rather than merely risk). Moreover, the information theory paradigm comes with a body of results, which may become useful for reasoning about ambiguity-based uncertainty. Cross entropy is an example of such a useful concept from information theory, as it allows measuring the difference between two settings with either type of uncertainty.

© Springer International Publishing Switzerland 2015
V. Torra and Y. Narakawa (Eds.): MDAI 2015, LNAI 9321, pp. 230–242, 2015.
DOI: 10.1007/978-3-319-23240-9_19

Table 1. Two sets of choices that comprise the Ellsberg paradox.

	Red	Black	Yellow
Option 1A	$100	$0	$0
Option 1B	$0	$100	$0

(a) Options for first choice

	Red	Black	Yellow
Option 2A	$100	$0	$100
Option 2B	$0	$100	$100

(b) Options for second choice

There are four types of extensions that we propose. The first two, pignistic entropy and aggregate uncertainty entropy, flatten ambiguity-based uncertainty to risk-based uncertainty. Pignistic entropy models a perfectly rational agent, interested in the expected risk, given an ambiguous situation; whereas aggregate uncertainty entropy models a paranoid agent, that assumes the worst-case reasonable risk. The final two, belief entropy and conceivability entropy, properly extend information theory to model ambiguity-based uncertainty. Both methods are based on extending surprisal. For belief entropy, surprisal is based on the beliefs of the agent; whereas for conceivability entropy, surprisal decreases with uncertainty. All four types coincide when there is no ambiguity-based uncertainty.

2   Ellsberg Paradox

The Ellsberg paradox [2] is a motivating example for uncertainty representation in subjective logic. The Ellsberg paradox shows that people make different decisions than rational risk-avoiding agents. We use the Ellsberg paradox as a running example throughout this paper.

Suppose you are shown an urn with 90 balls in it and you are told that 30 are red and that the remaining 60 balls are either black or yellow. One ball is selected at random and you are given the following choice: Option 1 A gives you $100 if a red ball was drawn and $0 of either a black or a yellow ball was drawn; option 1B gives you $100 if a black ball was drawn and $0 if a red or a yellow was drawn. Table 1a summarises the possible outcomes given the choices.

Experiments show that people strongly favour option 1 A over option 1B [2]. Assuming that people are rational, this implies that people believe that black balls are less probable than red.

Options 2 A and 2B are based on the exact same set-up. The amount of balls of each colour is equal to variant 1. Option 2 A pays $100 when either red or yellow is drawn, whereas 2B pays $100 when either black or yellow is drawn. Table 1b summaries these outcomes.

Experiments show that people strongly favour option 2B over option 2A [2]. Assuming that people are rational, this implies that people believe that the black balls are more probable than red. However, the set-up remains unchanged between variant 1 and 2. Thus, the choices made by the people cannot be explained as rational estimates of probabilities.

It is impossible to explain the choices using risk-based uncertainty, since the risks are perfectly symmetrical. The common explanation of the difference between variant 1 and 2 is ambiguity-based uncertainty. In variant 1, option 1a has no ambiguity-based uncertainty, as the odds are known to be one in three. In variant 2, option 2b is the ambiguity-free option, as the odds are known to be two in three. Options 1b and 2a have ambiguity-based uncertainty, as the real odds could be as low as 0 or $1/3$, respectively (if there are no black balls), or as high as $2/3$ or 1, respectively (if there are no yellow balls).

The Ellsberg paradox is the running example throughout this paper. We relate concepts from subjectively logic and the four types of entropy directly to the four choices of the Ellsberg paradox. A good entropy measure can describe the core difference between options 1 A or 2B and options 1B or 2A.

3   Opinion Representation in Subjective Logic

Random events have a set of possible outcomes. Each of these outcomes is assigned some probability. A user with incomplete knowledge, however, may not know these probabilities. Subjective logic introduces opinions to model users that estimate these probabilities.

The *domain* of an opinion is the set of outcomes of the underlying event. The elements of the domain are exclusive and exhaustive. The user realises that the underlying event can have only one outcome, and includes all possible outcomes in the domain.

A probability distribution assigns a (non-negative) probability to each of the outcomes. An opinion assigns a (non-negative) belief to each of the outcomes. Unlike the probability distribution, the sum of the beliefs may be less than one. The remainder is uncertainty.

An opinion of user A about an event with domain X is denoted ω_X^A. An opinion consists of a belief mass function $b_X^A : X \to [0, 1]$, such that $\sum_{x \in X} b_X^A(x) \le 1$, and a base rate function $a_X^A : X \to [0, 1]$, such that $\sum_{x \in X} a_X^A(x) = 1$. The uncertainty u_X^A is defined $1 - \sum_{x \in X} b_X^A(x)$. For the domain $X = \{x_1, \ldots, x_n\}$, we may denote an opinion as $\omega_X^A = (b_1, \ldots, b_n)$, to mean $b_X^A(x_1) = b_1, \ldots b_X^A(x_n) = b_n$.

The base rates denote the projected probabilities, in case of uncertainty. With a base rate, every opinion uniquely denotes a probability distribution, which we call the pignistic probabilities. The pignistic probability mass for $x \in X$ is computed $p_X^A(x) = b_X^A(x) + u_X^A \cdot a_X^A(x)$.

Barycentric coordinate systems can be used to visualise opinions. In a barycentric coordinate system the location of a point is specified as the centre of mass, or barycentre, of masses placed at its vertices [8]. A barycentric coordinate system with n axes is represented on a simplex with n vertices which has dimensionality $(n - 1)$. A triangle is a 2D simplex which has 3 vertices and is thus a barycentric system with 3 axes. A binomial opinion can be visualised as a point in a barycentric coordinate system of 3 axes represented by a 2D simplex which is in fact an equal sided triangle, as in Fig. 1. Here, the belief, disbelief and uncertainty axes go perpendicularly from each edge towards the respective opposite vertices denoted $x, \overline{x}$ and uncertainty. The base rate $a_X^A(x)$ is a point on the base line, and the projected probability $p_X^A(x)$ is determined by projecting the opinion point to the base

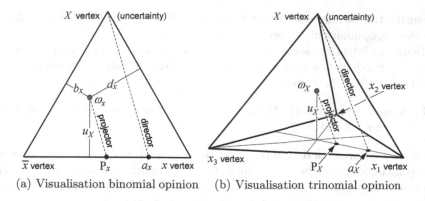

(a) Visualisation binomial opinion   (b) Visualisation trinomial opinion

Fig. 1. Barycentric visualisations of opinions.

line in parallel with the base rate director. The binomial opinion $\omega_x = (1/5, 2/5)$ with projected probability $p_X^A(x) = 1/2$ is shown as an example.

In case the opinion point is located at the left or right vertex of the triangle, i.e. with $b_X^A(\overline{x}) = 1$ or $b_X^A(x) = 1$ (and $u_X^A = 0$), then the opinion is equivalent to boolean TRUE or FALSE, in which case subjective logic becomes equivalent to binary logic. In case the opinion point is located on the baseline of the triangle, i.e. with $u_X^A = 0$, then the opinion is equivalent to a traditional probability, in which case subjective logic becomes equivalent to probabilistic logic.

In general, a multinomial opinion can be represented as a point inside a regular simplex. In particular, a ternary multinomial opinion can be represented inside a tetrahedron with a barycentric system of 4 axes, as shown in Fig. 1.

The tetrahedron is a 3D simplex. Assume the 3-domain $X = \{x_1, x_2, x_3\}$. Figure 1 shows a tetrahedron with the example multinomial opinion $\omega_X = (1/5, 1/5, 1/5)$, and base rate distribution $(6/8, 1/8, 1/8)$. The belief axes for x_1, x_2 and x_3 are omitted due to the difficulty of 3D visualisation.

Running Example 1. The Ellsberg paradox can be expressed elegantly in subjective logic. We can let the domain be {win, lose}. For choice 1A, the opinion is $(1/3, 2/3)$; for 1B $(0, 1/3)$; for 2A $(1/3, 0)$; and for 2B $(2/3, 1/3)$. The base rate $a(\text{win}) = 1/2$ is the most natural base rate – black balls are no more or less likely than yellow balls – but we generally consider arbitrary base rates for the Ellsberg paradox. The choices 1 A and 2B lead to opinions without uncertainty; their generalised entropy measure, therefore, equals the standard entropy measure. The choices 1B and 2 A have an amount of uncertainty; their various generalised entropy measures lead to different figures.

4   Information Theory

Subjective logic has an extensive set of operations that allow calculus with opinions. One particular operation represents constructing opinions based on recommendations. In [11], the authors show the use of (standard) information theory in measuring the usefulness of recommendations and the derived opinions. Information theory, opinions and uncertainty are intimately linked.

Before we introduce some extensions of information theory to cover subjective logic, we introduce the important standard notions. A more detailed discussion and treatment can be found, e.g., in [7].

Definition 1 (Surprisal). *The surprisal (or self-information) of an outcome x of a discrete random variable X is $I_X(x) = -\log(p_X(x))$.*

Surprisal measures the degree to which an outcome is surprising. The more surprising an outcome is, the more informative it is. In information theory, surprisal of an outcome is completely determined by the probability it happens. Usually, an outcome is more surprising if it is less likely to happen.

Definition 2 (Entropy). *The entropy of a discrete random variable X is the expected surprisal $H(X) = -\sum_x p_X(x)\log(p_X(x))$.*

Entropy measures the expected information carried with a random variable. In information theory, entropy of a random variable is decided by the uncertainty of its outcome in one test. A random variable has more entropy if all of its outcomes have more similar probabilities to happen.

Definition 3 (Cross Entropy). *The cross entropy of two discrete random variables X, Y is $H(X, Y) = -\sum_x p_X(x)\log(p_Y(x))$.*

The cross entropy measures the amount of surprisal obtained when you believe an event is distributed as Y, but in reality is distributed as X. This amount is not typically symmetric in Y and X. The cross entropy is minimised when Y is selected to be equal to X, in which case the believed distribution equals reality.

5   Pignistic Entropy

Subjective logic opinions model the subjective opinions of users. Users make decisions based on their opinions. We can imagine a user forced to make a decision, where he would decide one way if the probability is above a certain threshold, and the other way otherwise. The cut-off for the decision is called the *pignistic probability* of a belief in an opinion.

A user may have an opinion about a potentially unfair coin. The user believes that even unfair coins provide heads or tails at least 30 % of the time. Hence, his opinion ω is $(3/10, 3/10)$ (with uncertainty $4/10$). Since the user has no reason to prefer heads over tails (or vice versa), if he is forced to pick a probability distribution, then he assigns $1/2$ to both.

Pignistic entropy of an opinion ω_X^A characterised by belief mass function b_X^A and base rate a_X^A is based on the entropy of the associated pignistic probability distribution:

Definition 4 (Pignistic Entropy). *The pignistic entropy $H_p(\omega_X^A)$ is defined:*
$$-\sum_x p_X^A(x)\log(p_X^A(x)) = -\sum_x \left(b_X^A(x) + u_X^A \cdot a_X^A(x)\right)\log\left(b_X^A(x) + u_X^A \cdot a_X^A(x)\right).$$

The pignistic entropy is insensitive towards the change of uncertainty in an opinion:

Proposition 1. *Let ω_X^A and ω_X^B be two opinions, such that $u_X^A > u_X^B$ and for all x, $p_X^A(x) = p_X^B(x)$, then $H_p(\omega_X^A) = H_p(\omega_X^B)$.*

Proof. Proposition follows from the fact that H_p is completely determined by the pignistic probabilities, which are equal for ω_X^A and ω_X^B.

Running Example 2. The pignistic entropy models the way a rational agent would approach the Ellsberg paradox. As depicted in Table 1, if the base rate for black versus yellow is 50-50, then options 1 A and 1B have equal pignistic entropy, and options 2 A and 2B also have equal pignistic entropy. If the base rate is skewed towards black, then 1B and 2B are the superior choices, and 1 A and 2A if it is not. As expected, the pignistic entropy does not reflect the inherent desire to avoid ambiguity.

Entropy can not just be used to measure how much information there is, but also to compare the difference between two opinions. The cross entropy between ω_X^A and ω_X^B describes how well ω_X^B predicts ω_X^A:

Definition 5 (Pignistic Cross Entropy). *The pignistic cross entropy between ω_X^A and ω_X^B, $H_p(\omega_X^A, \omega_X^B)$ is defined:* $-\sum_x p_X^A(x) \log(p_X^B(x))$.

The pignistic cross entropy is insensitive towards the difference between the uncertainty of two opinions:

Proposition 2. *Let ω_X^A and $\omega_X^{A'}$ be two opinions, such that $u_X^A > u_X^{A'}$ and for all x, $p_X^A(x) = p_X^{A'}(x)$, and idem for ω_X^B and $\omega_X^{B'}$. Then $H_p(\omega_X^A, \omega_X^B) = H_p(\omega_X^{A'}, \omega_X^B) = H_p(\omega_X^A, \omega_X^{B'}) = H_p(\omega_X^{A'}, \omega_X^{B'})$.*

Proof. Proposition follows from the fact that H_p is completely determined by the pignistic probabilities, which are equal for ω_X^A and $\omega_X^{A'}$, and for ω_X^B and $\omega_X^{B'}$.

The pignistic cross entropy between two identical opinions is equal to the entropy of one of the opinions:

Proposition 3. $H_p(\omega_X, \omega_X) = H_p(\omega_X)$

To give an example of pignistic cross entropy, consider five opinions (with belief and disbelief): $\omega_X^A = (7/10, 3/10)$, $\omega_X^B = (3/10, 7/10)$, $\omega_X^C = (1/2, 1/10)$, $\omega_X^D = (1/10, 1/2)$, $\omega_X^E = (0, 0)$. We suppose the base rate is $1/2$. Their pignistic cross entropies are presented in Table 3. As ω_X^A (ω_X^B) and ω_X^C (ω_X^D) have the same pignistic probability distributions, their cross entropy is minimal. In this sense, the uncertainty component in ω_X^C (ω_X^D), which makes them different from ω_X^A (ω_X^B), is eliminated. Note that cross entropy between ω_X^E, which represents complete uncertainty, and any other opinions are the same. It implies that they are equally different when compared with the complete uncertain opinion. Such difference is actually smaller than that between two completely opposite opinions (e.g., ω_X^A and ω_X^B). Also note that the pignistic cross entropy measure is not symmetric.

Table 2. Pignistic entropy of the options in the Ellsberg paradox.

	Choice A	Choice B
Option 1	$-1/3 \log 1/3 - 2/3 \log 2/3$	$-2/3 \cdot a \log(2/3 \cdot a) - (1 - 2/3 \cdot a) \log(1 - 2/3 \cdot a)$
Option 2	$-(1/3 + 2/3 \cdot a) \log(1/3 + 2/3 \cdot a)$ $-2/3 \cdot (1 - a) \log(2/3 \cdot (1 - a))$	$-2/3 \log 2/3 - 1/3 \log 1/3$

Table 3. Pignistic cross entropy among five opinions.

	ω_X^A	ω_X^B	ω_X^C	ω_X^D	ω_X^E
ω_X^A	0.8813	1.3702	0.8813	1.3702	1.0000
ω_X^B	1.3702	0.8813	1.3702	0.8813	1.0000
ω_X^C	0.8813	1.3702	0.8813	1.3702	1.0000
ω_X^D	1.3702	0.8813	1.3702	0.8813	1.0000
ω_X^E	1.1258	1.1258	1.1258	1.1258	1.0000

The pignistic (cross) entropy ignores the uncertainty present in an opinion, and converts uncertainty to pignistic probability, before measuring the (cross) entropy. Pignistic entropy, therefore, accurately measures the entropy of the decisions of users with an opinion, but not the entropy of the opinion itself (nor the cross entropy between opinions). In the remainder of the paper, we want to study the entropy of the opinions including the uncertainty.

6   Aggregate Uncertainty Entropy

Dempster-Shafer theory [1] shares similarities with subjective logic. Dempster-Shafer theory also deals with beliefs and uncertainty. Extensions of information theory for Dempster-Shafer theory currently exist. The major variant is the aggregate uncertainty [6]. In this section, we discuss aggregate uncertainty, and translate it to subjective logic (Table 2).

A particular downside of pignistic entropy, is that an uncertainty plays no role in the amount of entropy. Intuitively, we should expect a more uncertain opinion not to have less entropy. The aggregate uncertainty entropy is the minimal extension of pignistic entropy that satisfies this requirement [6]:

Definition 6 (Aggregate Uncertainty Entropy). *Let F_X^A be the set of functions f with, for all x, $b_X^A(x) \leq f(x) \leq 1$ and $\sum_x f(x) = 1$. The aggregate uncertainty entropy $H_{au}(\omega_X^A)$ is defined:* $-\max_{f \in F_X^A} \sum_x f(x) \log(f(x))$.

The aggregate uncertainty entropy cannot decrease whenever uncertainty increases, even if the ratio of beliefs is affected:

Proposition 4. *Let ω_X^A and ω_X^B be two opinions, such that for all x, $b_X^A(x) > b_X^B(x)$, then $H_{au}(\omega_X^A) \leq H_{au}(\omega_X^B)$.*

Proof. As $b_X^A(x) > b_X^B(x)$, $F_X^A \subseteq F_X^B$, meaning maximal f in F_X^A is in F_X^B.

The functions f_X^A are all probability mass functions. Each probability mass function f_X^A has the property that it assigns a probability at least as great as the belief to each outcome. Thus, F_X^A is essentially the set of probability distributions that we believe may be the case. If we take the maximal element of the entropy using the different probabilities, then we satisfy the requirement that increasing uncertainty can never decrease entropy.

Running Example 3. The aggregate uncertainty entropy models the way a paranoid agent would approach the Ellsberg paradox. Specifically, the agent assumes that the Shannon entropy is maximised under constraints of his beliefs. For 1A and 2B, the beliefs fix the probabilities, but for 1B and 2A, the entropy is maximised by letting the probability of winning (and losing) be 1/2. As depicted in Table 1, the aggregate uncertainty entropy is independent of base rates (as it is based on Dempster-Shafer theory), and 1A and 2B score significantly better than 1B and 2A. This approach to the problem uses no notions of ambiguity, and has been suggested previously [3]. The problem with this view, is that the Ellsberg's experiment is purposely set-up to ensure the set-up remains unchanged between the two variants, whereas the maximal entropy cases of 1B and 2A are inconsistent.

Definition 7 (Aggregate Uncertainty Cross Entropy). *Let F_X^A and F_X^B as before, and f, g be* $\operatorname{argmax}_{f \in F_X^A} \sum_x f(x) \log(f(x))$, $\operatorname{argmax}_{g \in F_X^B} \sum_x g(x) \log(g(x))$ *The aggregate uncertainty cross entropy between ω_X^A and ω_X^B, $H_{au}(\omega_X^A, \omega_X^B)$ is defined:* $- \sum_x f(x) \log(g(x))$.

The aggregate uncertainty cross entropy between two identical opinions is equal to the entropy of one of the opinions:

Proposition 5. $H_{au}(\omega_X^A, \omega_X^A) = H_{au}(\omega_X^A)$

We compute aggregate uncertainty cross entropy between the opinions introduced in Table 3, and the results are presented in Table 4b. As $\operatorname{argmax}_{f \in F_X^A} \sum_x f(x) \log(f(x))$ are the same for ω_X^C, ω_X^D, ω_X^E, namely $f(x) = 0.5$ for all x,

Table 4. Ellsberg paradox and cross entropy for aggregate uncertainty entropy.

	Choice A	Choice B
Option 1	0.9183	1
Option 2	1	0.9183

(a) Ellsberg paradox.

	ω_X^A	ω_X^B	ω_X^C	ω_X^D	ω_X^E
ω_X^A	0.8813	1.3702	1.0000	1.0000	1.0000
ω_X^B	1.3702	0.8813	1.0000	1.0000	1.0000
ω_X^C	1.1258	1.1258	1.0000	1.0000	1.0000
ω_X^D	1.1258	1.1258	1.0000	1.0000	1.0000
ω_X^E	1.1258	1.1258	1.0000	1.0000	1.0000

(b) Cross entropy

any of five opinions has the same cross entropy with them. Also, due to symmetry between ω_X^A and ω_X^B, other three opinions have equal cross entropy with them. Using this cross entropy measure, opinions with partial uncertainty (ω_X^C, ω_X^D) seems to be the same as that with complete uncertainty (ω_X^E). Because they have the same distance with the other two deterministic opinions.

There are two major downsides to the aggregate uncertainty entropy. The first is theoretical, namely that the aggregate uncertainty is not a closed form expression. There is, however, research that addresses this specific problem to some degree [9]. The second downside is that the measure applies to Dempster-Shafer theory, which has a subtly different interpretation of uncertainty (specifically, that the probability mass must be over the belief mass). In the next two sections, we study how subjective logic's interpretation of uncertainty impacts the definition of entropy.

7   Ambiguity Entropy

Rather than using the entropy based on risk as a proxy for ambiguity-based uncertainty entropy, we can directly encode beliefs and ambiguity-based uncertainty into surprisal. An interesting question is whether uncertainty leads to surprisal. Two opposing interpretations are that total uncertainty means that everything is maximally surprising, or that nothing is surprising at all. We demonstrate that which interpretation is appropriate depends on the context.

Before introducing the two types of ambiguity entropy, we introduce an overarching definition of surprisal: $-\log(b_X^A(x) + c \cdot u_X^A)$. The definition contains a parameter $c \in [0, 1]$, which determines the amount of surprisal from uncertainty. The special cases for c are when $c = 0$ (or $c \approx 0$) and when $c = 1$. If the uncertainty is zero, then all choices of c collapse into one, which equals the standard definition of surprisal. If the uncertainty is non-zero, then possible interpretations of surprisal start to diverge. In the next two sections, we formally analyse the two edge cases, belief entropy and conceivability entropy.

In [4], Klir explores a similar idea, where belief entropy parallels confusion ambiguity and conceivability entropy parallels dissonance ambiguity. The fundamental difference is that [4] considers Dempster-Shafer theory, and therefore cannot use the base-rate that subjective logic has. As a consequence, his notions cannot use projected probabilities. His notions are further removed from classical notions in information theory, as he cannot use the expected surprisal.

7.1   Belief Entropy

A user has an opinion with beliefs. If the belief in an outcome is low, then the user thinks it is unlikely that the outcome will happen. To encode this, we can define surprisal based on the belief mass, by letting the belief in x be: $-\log(b_X^A(x))$. This equates to $-\log(b_X^A(x) + c \cdot u_X^A)$, when $c = 0$.

We take the natural definition of entropy as the expected surprisal. Entropy of an opinion should measure the expected surprisal of beliefs.

Definition 8 (Belief Entropy). *The belief entropy $H_b(\omega_X^A)$ is defined as:* $-\sum_x p_X^A(x)\log(b_X^A(x))$.

The belief entropy has several nice properties. The first property is that, with the pignistic probabilities remaining constant, the entropy strictly increases when the uncertainty increases:

Proposition 6. *Let ω_X^A and ω_X^B be two opinions, such that $u_X^A > u_X^B$ and for all x, $p_X^A(x) = p_X^B(x)$, then $H_b(\omega_X^A) > H_b(\omega_X^B)$.*

The second property is that, unlike aggregate uncertainty entropy, the entropy of a completely uncertain opinion is strictly larger than the entropy of any pignistic entropy:

Proposition 7. *Let ω_X^A be complete uncertainty; $u_X^A = 1$. Then $H_b(\omega_X^A) > H_p(\omega_X^B)$, for all ω_X^B.*

As the entropy of uncertainty strictly exceeds the entropy of any other opinion in subjective logic, uncertainty contains less information than any other opinion.

Running Example 4. The belief entropy directly models the beliefs on the agent, where ambiguity reduces the sum of the beliefs. Table 1 shows the belief entropies associated with each of the choices. It is interesting to note that 1B and 2 A are both assigned infinite belief entropy. The reason is that the participant has no reason to assume it is even possible to win, so the surprisal upon winning is the global maximum of surprisal; positive infinity.

Table 5. Ellsberg paradox and cross entropy for belief entropy.

	Choice A	Choice B
Option 1	0.9183	∞
Option 2	∞	0.9183

(a) Ellsberg paradox.

	ω_X^A	ω_X^B	ω_X^C	ω_X^D	ω_X^E
ω_X^A	0.8813	1.3702	1.6966	2.6253	∞
ω_X^B	1.3702	0.8813	2.6253	1.6966	∞
ω_X^C	0.8813	1.3702	1.6966	2.6253	∞
ω_X^D	1.3702	0.8813	2.6253	1.6966	∞
ω_X^E	1.1258	1.1258	2.1610	2.1610	∞

(b) Cross entropy

Note that in Table 5a, choices 1B and 2 A have infinite entropy. The reason is that they contain the terms $-2/3a\log(0)$ and $2/3(1 - a)\log(0)$, which equate to infinity except when $a = 0$ or $a = 1$, respectively. Intuitively, the cause is that we have zero belief in winning or losing, respectively, although both winning and losing have a non-zero probability of occurring (except for extreme base rates). The interesting aspect of the extreme base rates is that they would remove the ambiguity uncertainty altogether. Since we would know the outcome under uncertainty, there is no ambiguity-based uncertainty to measure.

That choices 1B and 2 A have infinite entropy may be desirable for one reason: the entropy exceeds that of any opinions without zero-belief events. However, the downside is that a one-in-a-million event with zero belief and a certain event with zero belief yield the exact same entropy: infinite bits. Consider the general definition of ambiguity surprisal, $-\log(b_X^A(x) + c \cdot u_X^A)$, where c converges to 0 (and the expected surprisal to belief entropy), then the entropy converges to infinity slowly. We can consider $c = \epsilon$, for some small $\epsilon > 0$, then the entropy remains finite. The addition of ϵ hardly affects the entropy of opinions without zero-beliefs. For example, $(0.5, 0.1)$, with base rate $1/2$, has $-0.7 \log(0.5 + \epsilon) - 0.3 \log(0.1 + \epsilon) \approx -0.7 \log(0.5) - 0.3 \log(0.1) \approx 1.6966$ bits entropy. For opinions with zero-beliefs, we get a more fine-grained measure of entropy. For example, $(0.5, 0)$ and $(0, 0)$, both with base rate $1/2$, have $-0.75 \log(0.5 + \epsilon) - 0.25 \log(\epsilon) \approx -0.25 \log(\epsilon)$ and $-0.5 \log(\epsilon) - 0.5 \log(\epsilon) = -\log(\epsilon)$ bits of entropy, and the latter is four times more bits entropy. Thus, the belief entropy can be made more fine-grained without loss of generality.

The belief entropy can be extended to belief cross entropy:

Definition 9 (Belief Cross Entropy). *The belief cross entropy between* ω_X^A *and* ω_X^B, $H_b(\omega_X^A, \omega_X^B)$ *is defined:* $-\sum_x p_X^A(x) \log(b_X^B(x))$.

We compute the belief cross entropy between the opinions introduced in Table 3, and the results are presented in Table 5b. Some equalities in the table can be easily derived, based on the Definition 9. Note that cross entropy between any opinions and ω_X^E, which means complete uncertainty, is infinity. This is not reasonable, as explained below.

The belief cross entropy measures the information distance from one opinion to the other. Intuitively, when an uncertain opinion conflicts with another opinion, this may not surprise us, whereas two conflicting and certain opinions would be a surprise. Unfortunately, this is not the intuition captured by the definition of belief cross entropy. Belief cross entropy measures the information gap between two opinions, and uncertainty introduced large quantities of entropy, allowing for bigger information gaps. In the next section, we introduce a measure of entropy that models the intuition of entropy that is suitable for cross entropy.

7.2   Conceivability Entropy

In belief entropy, the belief in an outcome determines the surprisal. However, we can imagine that users are not surprised when they are uncertain. To encode this, we can define surprisal based on the belief mass plus the uncertainty, by letting the belief in x be: $-\log(b_X^A(x))$. This equates to $-\log(b_X^A(x) + c \cdot u_X^A)$, when $c = 1$. Note that $b_X^A(x) + u_X^A = 1 - \sum_{x' \neq x} b_X^A(x)$, so conceivability can be seen as the converse of belief.

The entropy can be derived from the surprisal:

Definition 10 (Conceivability Entropy). *The conceivability entropy* $H_c(\omega_X^A)$ *is defined:* $-\sum_x p_X^A(x) \log(b_X^A(x) + u_X^A)$.

When the opinion is complete uncertainty, surprisal is zero, as all outcomes are fully conceivable. For this reason, viewing surprisal as the opposite of information does not make sense here (unlike the other notions of entropy, such as belief entropy and Shannon entropy),

Table 6. Ellsberg paradox and cross entropy for conceivability entropy.

	Choice A	Choice B
Option 1	0.9183	$2/3a \log(2/3)$
Option 2	$2/3(1 - a) \log(2/3)$	0.9183

(a) Ellsberg paradox.

	ω_X^A	ω_X^B	ω_X^C	ω_X^D	ω_X^E
ω_X^A	0.8813	1.3702	0.4064	0.7456	0
ω_X^B	1.3702	0.8813	0.7456	0.4064	0
ω_X^C	0.8813	1.3702	0.4064	0.7456	0
ω_X^D	1.3702	0.8813	0.7456	0.4064	0
ω_X^E	1.1258	1.1258	0.5760	0.5760	0

(b) Cross entropy

However, the conceivability entropy notion is suitable for cross entropy:

Definition 11 (Conceivability Cross Entropy). *The conceivability cross entropy between* ω_X^A *and* ω_X^B, $H_c(\omega_X^A, \omega_X^B)$ *is defined:* $-\sum_x p_X^A(x) \log(b_X^B(x) + u_X^B)$.

Conceivability cross entropy is a more useful measure of distance between opinions than belief cross entropy. More uncertain opinions tend to have a shorter distance. The reason why conceivability cross entropy is a better measure for distance, is that we want to measure whether it is "conceivable" that an opinion describes another opinion. We can see the concrete numbers in Table 6b. The cross entropy of opinions with similar pignistic probabilities is lower, but the amount of uncertainty correlates more strongly. The distance from any opinion to complete uncertainty is 0.

8   Conclusion

To understand decision making, not only must we analyse uncertainty introduced by risk, but also the uncertainty about risk (ambiguity). Standard notions of Shannon entropy in information theory can measure the former, but not the latter. We extend information theory to capture subjective logic – a framework to deal with uncertainty about ambiguity – in four ways.

Two of the extensions of information theory remove ambiguity before measuring entropy. The first extension, pignistic entropy, models rational agents. The second extension, aggregate uncertainty entropy, models paranoid agents.

However, the interesting extensions model ambiguity, rather than remove it. The final two extensions, belief entropy and conceivability entropy, are two sides of the same coin. Belief entropy is suitable for measuring entropy of both risk and ambiguity. Conceivability entropy is more suited for measuring cross entropy.

All extensions are related using Ellsberg paradox as a running example, and the different entropies provide insights into the paradox. Moreover, the different entropies can be generalised to cross entropy – a measure of the quality of an opinion, given a valid opinion. Cross entropy can be used for analysing the quality of opinions in systems that use subjective logic.

References

1. Dempster, A.P.: Upper and lower probabilities induced by a multivalued mapping. Ann. Math. Statist. **38**(2), 325–339 (1967)
2. Ellsberg, D.: Risk, ambiguity, and the savage axioms. Q. J. Ecomonics **75**, 643–669 (1961)
3. Jøsang, A.: A logic for uncertain probabilities. Int. J. Uncertainty, Fuzziness Knowl.-Based Syst. **9**(3), 279–311 (2001)
4. Klir, G.J.: Where do we stand on measures of uncertainty, ambiguity, fuzziness, and the like? Fuzzy Sets and Syst. **24**(2), 141–160 (1987)
5. Knight, F.H.: Risk, Uncertainty, and Profit. Library of Economics and Liberty, New York (1921)
6. Maeda, Y., Ichihashi, H.: An uncertainty measure with monotonicity under the random set inclusion. Int. J. Gen. Syst. **21**(4), 379–392 (1993)
7. McEliece, R.J.: Theory of Information and Coding, 2nd edn. Cambridge University Press, New York (2001)
8. Möbius, A.F.: Der barycentrische Calcul. Johann Ambrosius Barth, Leipzig (1827). Re-Published by Georg Olms Verlag. Hildesheim, New York 1976
9. Pouly, M.: Generalized information theory based on the theory of hints. In: Liu, W. (ed.) ECSQARU 2011. LNCS, vol. 6717, pp. 299–313. Springer, Heidelberg (2011)
10. Shannon, C.: A mathematical theory of communication. Bell Syst. Tech. J. **27** (1948)
11. Wang, D., Muller, T., Irissappane, A.A., Zhang, J., Liu, Y.: Using information theory to improve the robustness of trust systems. In: Proceedings of the Fourteenth International Conference on Autonomous Agents and Multiagent Systems (2015)

Author Index

Printed in the United States
By Bookmasters